Reading Essentials
Course 3

Send all inquiries to:
McGraw-Hill Education
8787 Orion Place
Columbus, OH 43240-4027

ISBN: 978-0-07-889390-2
MHID: 0-07-889390-9

Printed in the United States of America.

1 2 3 4 5 6 7 8 9 10 REL 15 14 13 12 11 10

Table of Contents

To the Student

In today's world, a knowledge of science is important for thinking critically, solving problems, and making decisions. But understanding science can sometimes be a challenge.

Reading Essentials takes the stress out of reading, learning, and understanding science. This book covers important concepts in science, offers ideas for how to learn the information, and helps you review what you have learned.

In each lesson you will find:

Before You Read

- **What do you think?** asks you to agree or disagree with statements about topics that are discussed in the lesson.

Read to Learn

This section describes important science concepts with words and graphics. In the margins you can find a variety of study tips and ideas for organizing and learning information.

- **Study Coach and Mark the Text** offer tips for finding the main ideas in the text.
- **Foldables® Study Organizers** help you divide the information into smaller, easier-to-remember concepts.
- **Reading Check** questions ask about concepts in the lesson.
- **Think It Over** elements help you consider the material in-depth, giving you an opportunity to use your critical thinking skills.
- **Visual Check** questions relate specifically to the art and graphics used in the text. You will find questions that get you actively involved in illustrating the information you have just learned.
- **Math Skills** reinforces the connection between math and science.
- **Academic Vocabulary** defines important words that help you build a strong vocabulary.
- **Word Origin** explains the English background of a word.
- **Key Concept Check** features ask the Key Concept questions from the beginning of the lesson.
- **Interpreting Tables** includes questions or activities that help you interact with the information presented.

After You Read

This final section reviews key terms and asks questions about what you have learned.

- The **Mini Glossary** assists you with science vocabulary.
- Review questions focus on the key concepts of the lesson.
- **What do you think now?** gives you an opportunity to revisit the *What do you think?* statements to see if you changed your mind after reading the lesson.

See for yourself—***Reading Essentials*** makes science enjoyable and easy to understand.

Scientific Problem Solving

Scientific Inquiry

Understanding Science

A clear night sky is one of the most beautiful sights on Earth. The stars seem to shine like a handful of diamonds scattered on black velvet. Why do the stars seem to shine more brightly some nights than others?

Did you know that when you ask questions, such as the one above, you are practicing science? **Science** *is the investigation and exploration of natural events and of the new information that results from those investigations.* Like a scientist, you can help shape the future by accumulating knowledge, developing new technologies, and sharing ideas with others.

Throughout history, people of many different backgrounds, interests, and talents have made scientific contributions. Sometimes they overcame a limited education and excelled in science. One example is Marie Curie. She was a scientist who won two Nobel prizes in the early 1900s for her work with radioactivity. As a young student, Marie was not allowed to study at the University of Warsaw in Poland because she was a woman. Despite this obstacle, she made significant contributions to science. X-rays, radioactive cancer treatments, and nuclear-power generation are some of the technologies that Marie Curie and her associates made possible through their pioneering work.

Branches of Science

Scientific study is organized into several branches, or parts. The three branches that you will study in middle school are physical science, Earth science, and life science. Each branch focuses on a different part of the natural world. ✔

Physical Science Physics and chemistry is the study of matter and energy. Physical scientists ask questions such as

- What happens to energy during chemical reactions?
- How does gravity affect roller coasters?
- What makes up protons, neutrons, and electrons?

Key Concepts
- What are some steps used during scientific inquiry?
- What are the results of scientific inquiry?
- What is critical thinking?

◣ Study Coach

Building Vocabulary Work with another student to write a question about each vocabulary term in this lesson. Answer the questions and compare your answers. Reread the text to clarify the meaning of the terms.

✔ **Reading Check**

1. Name three branches of science.

_____

_____

_____

Earth Science Earth scientists study the many processes that occur on Earth and deep within Earth. Earth scientists ask questions such as

- What are the properties of minerals?
- How is energy transferred on Earth?
- How do volcanoes form?

Life Science Life scientists study all organisms and the many processes that occur in them. Life scientists ask questions such as

- How do plant cells and animal cells differ?
- How do animals survive in the desert?
- How do organisms in a community interact?

What is scientific inquiry?

When scientists conduct investigations, they often want to answer questions about the natural world. To do this, they use scientific inquiry—a series of skills used to answer questions. You might have heard these steps called "the scientific method." However, there is no one scientific method. In fact, scientists can use the skills for conducting an investigation in any order. One possible sequence is shown in the figure below and at the bottom of the next page. Like a scientist, you also perform scientific investigations every day. You also will do investigations throughout this course. ✅

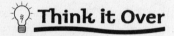

Think it Over

2. Apply Write your own life science question based on something you have observed in nature.

_____

_____

_____

_____

✅ **Reading Check**

3. Define What is scientific inquiry?

_____

_____

_____

One Process of Scientific Inquiry

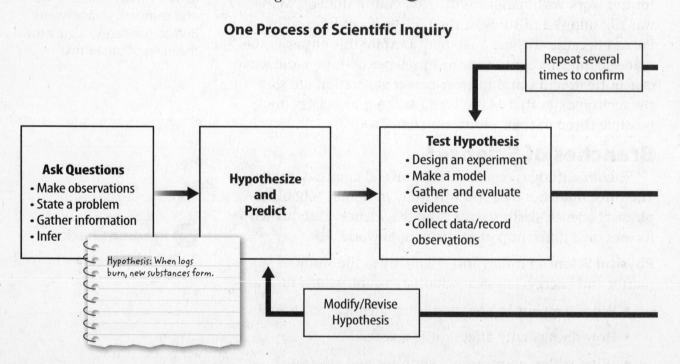

Ask Questions

Imagine warming yourself near a campfire. As you throw wood onto the fire, you see that the fire releases smoke and light. These are **observations**—*the results of using one or more of your senses to gather information and taking note of what occurs.* Observations often lead to questions. You ask yourself, "When logs burn, what happens to the wood? Do the logs disappear? Do they change in some way?"

When observing the fire, you might recall from an earlier science course that matter can change form, but it cannot be created or destroyed. Therefore, you infer that the logs do not just disappear. They must undergo some type of change. An **inference** *is a logical explanation of an observation that is drawn from prior knowledge or experience.*

Hypothesize and Predict

After making observations and inferences, you decide to investigate further. Like a scientist, you might develop a hypothesis. *A **hypothesis** is a possible explanation for an observation that can be tested by scientific investigations.* Your hypothesis about what happens might be: When logs burn, new substances form because matter cannot be destroyed. When scientists state a hypothesis, they often use it to make predictions to help test their hypothesis. *A **prediction** is a statement of what will happen next in a sequence of events.* Scientists make predictions based on what information they think they will find when testing their hypothesis. Based on your hypothesis, you might predict that if the logs burn, then the substances that make up the logs change into other substances.

Copyright © Glencoe/McGraw-Hill, a division of The McGraw-Hill Companies, Inc.

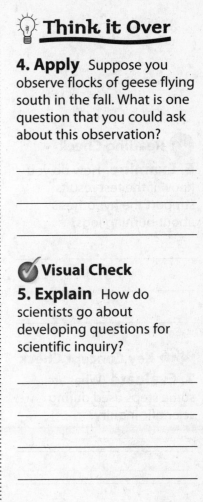

Think it Over

4. Apply Suppose you observe flocks of geese flying south in the fall. What is one question that you could ask about this observation?

_____

_____

_____

Visual Check

5. Explain How do scientists go about developing questions for scientific inquiry?

_____

_____

_____

_____

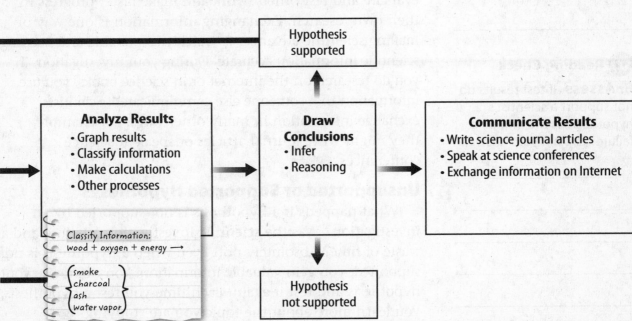

Hypothesis supported

Analyze Results
- Graph results
- Classify information
- Make calculations
- Other processes

Draw Conclusions
- Infer
- Reasoning

Communicate Results
- Write science journal articles
- Speak at science conferences
- Exchange information on Internet

Classify Information:
wood + oxygen + energy ⟶

{ smoke
charcoal
ash
water vapor }

Hypothesis not supported

Test Hypothesis and Analyze Results

How could you test your hypothesis? When you test a hypothesis, you often test your predictions. If a prediction is confirmed, then it supports your hypothesis. If your prediction is not confirmed, you might modify your hypothesis and retest it. To test your predictions and hypothesis, you could design an experiment to find out what substances make up wood. Then you could determine what makes up the ash, smoke, and other products that formed after the burning process. You also could research this topic and possibly find answers on reliable science Web sites or in science books. ✓

After doing an experiment or research, you need to analyze your results and findings. You might make additional inferences after reviewing your data. If you find that new substances form when wood burns, your hypothesis is supported. If new products do not form, your hypothesis is not supported.

Draw Conclusions

After analyzing your results, you can begin to draw conclusions about your investigation. A conclusion is a summary of the information gained from testing a hypothesis. Like a scientist does, you should test and retest your hypothesis several times to make sure the results are consistent.

Communicate Results

Sharing the results of a scientific inquiry is an important part of science. By exchanging information, scientists can evaluate and test others' work and make faster progress in their own research. Exchanging information is one way of making scientific advances as quickly as possible and keeping scientific information accurate. During your investigation, if you do research on the Internet or in science books, you use information that someone else communicated. Scientists exchange information in many other ways. For example, they might write journal articles or speak at science conferences. 🔑

Unsupported or Supported Hypotheses

What happens if a hypothesis is not supported by an investigation? Was the scientific investigation a failure and a waste of time? Absolutely not! Even when a hypothesis is not supported, you gain valuable information. You can revise your hypothesis and test it again. Each time you test a hypothesis, you learn more about the topic you are studying. ✓

✓ Reading Check

6. Examine How do you know if the test results support the hypothesis about burning logs?

_____

_____

_____

🔑 Key Concept Check

7. Evaluate What are some steps used during scientific inquiry?

_____

_____

_____

✓ Reading Check

8. Assess If test results do not support a scientist's hypothesis, is the inquiry a failure? Explain your answer.

_____

_____

_____

_____

_____

Scientific Theory

When many tests over many years support a hypothesis (or a group of closely related hypotheses), a scientific theory can develop. *A* **scientific theory** *is an explanation of observations or events that is based on knowledge gained from many observations and investigations.* A scientific theory does not develop from just one hypothesis but from many hypotheses connected by a common idea. For example, the kinetic molecular theory described below explains the behavior and energy of particles that make up a gas.

Scientific Law

A scientific law is different from a societal law, which is an agreement on a set of behaviors. *A* **scientific law** *is a rule that describes a repeatable pattern in nature.* A scientific law does not explain why or how the pattern happens. It only states that it will happen. For example, when you drop a ball, it always will fall toward Earth. This is a repeated pattern that relates to the law of universal gravitation. The figure below explains how a theory and a law differ.

✔️ **Visual Check**

9. Contrast How is a scientific law different from a scientific theory?

_____

_____

_____

_____

Scientific Law v. Scientific Theory

Kinetic Molecular Theory

The kinetic molecular theory explains how particles that make up a gas move in constant, random motions. A particle moves in a straight line until it collides with another particle or with the wall of its container.

The kinetic molecular theory also assumes that the collisions of particles in a gas are elastic collisions. An elastic collision is a collision in which no kinetic energy is lost. Therefore, kinetic energy among gas particles is conserved.

Law of Conservation of Energy

The law of conservation of energy states that in any chemical reaction or physical change, energy is neither created nor destroyed. The total energy of particles before and after collisions is the same.

However, this scientific law, like all scientific laws, does not explain *why* energy is conserved. It simply states that energy is conserved.

Scientific Law v. Scientific Theory

Both are based on repeated observations and can be rejected or modified.

A scientific law states that an event *will* occur. For example, energy will be conserved when particles collide. It does not explain why an event will occur or how it will occur. Scientific laws work under specific conditions in nature. A law stands true until an observation is made that does not follow the law.

A scientific theory is an explanation of *why* or *how* an event occurred. For example, collisions of particles of a gas are elastic collisions. Therefore, no kinetic energy is lost. A theory can be rejected or modified if someone observes an event that disproves the theory. A theory will never become a law.

Results of Scientific Inquiry

Why do you and others ask questions and investigate the natural world? Just as scientific questions vary, so do the results of science. Most often, the purpose of a scientific investigation is to develop new materials and technology, discover new objects, or find answers to questions, as described below. 🔑

New Materials and Technology Every year, corporations and governments spend millions of dollars on research and the design of new materials and technologies. **Technology** *is the practical use of scientific knowledge, especially for industrial or commercial use.*

For example, scientists hypothesize and predict how new materials will make bicycles and cycling gear lighter, more durable, and more aerodynamic. Using wind tunnels, scientists test these new materials to see whether they improve the cyclist's performance. If the cyclist's performance improves, their hypotheses are supported. If the performance does not improve or if it doesn't improve enough, scientists will revise their hypotheses and conduct more tests. ✓

New Objects or Events Scientific investigations also lead to newly discovered objects or events. For example, NASA's *Hubble Space Telescope* captured images of two colliding galaxies. The galaxies have been nicknamed the mice because of their long tails. The tails are composed of gases and young, hot blue stars. If computer models are correct, these galaxies will combine in the future and form one large galaxy.

Answers to Questions Often scientists launch investigations to answer *who, what, when, where,* or *how* questions. For example, research chemists investigate new substances, such as substances found in mushrooms and bacteria. New drug treatments for cancer, HIV, and other diseases might be found using new substances.

Other scientists look for clues about what causes diseases. Some scientists investigate whether a disease can be passed from person to person, or they try to find out when the disease first appeared.

🔑 Key Concept Check

10. Identify What are the results of scientific inquiry?

_____

_____

_____

_____

✓ Reading Check

11. Relate Why would the development of a more durable material for a bicycle be considered technology?

_____

_____

_____

FOLDABLES®

Create a two-tab book to discuss the importance of evaluating scientific information.

Why is it important to...

| ...be scientifically literate? | ...use critical thinking? |

Evaluating Scientific Information

Do you ever read advertisements, articles, or books that claim to contain scientifically proven information? Are you able to determine if the information is true and scientific instead of pseudoscientific (information that is incorrectly represented as scientific)? Whether you are reading printed media or watching commercials on TV, it is important that you are skeptical, identify facts and opinions, and think critically about the information. **Critical thinking** *is comparing what you already know with the information you are given in order to decide whether you agree with it.*

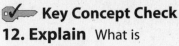 **Key Concept Check**

12. Explain What is critical thinking?

_____

_____

_____

✅ **Visual Check**

13. Appraise Highlight claims in the advertisement that make you skeptical.

Evaluating Information

Skepticism
Have you heard the saying, if it sounds too good to be true, it probably is? To be skeptical is to doubt the truthfulness of something. A scientifically literate person can read information and know that it misrepresents the facts. Science often is self-correcting because someone usually challenges inaccurate information and tests scientific results for accuracy.

Critical Thinking
Use critical thinking skills to compare what you know with the new information given to you. If the information does not sound reliable, either research and find more information about the topic or dismiss the information as unreliable.

Be A Rock Star!
Do you dream of being a rock star?

Sing, dance, and play guitar like a rock star with the new Rocker-rific Spotlight. A new scientific process developed by Rising Star Laboratories allows you to overcome your lack of musical talent and enables you to perform like a real rock star.

This amazing new light actually changes your voice quality and enhances your brain chemistry so that you can sing, dance, and play a guitar like a professional rock star. Now, there is no need to practice or pay for expensive lessons. The Rocker-rific Spotlight does the work for you.

Dr. Sammy Truelove says, "Never before has lack of talent stopped someone from achieving his or her dreams of being a rock star. This scientific breakthrough transforms people with absolutely no talent into amazing rock stars in just minutes. Of the many patients that I have tested with this product, no one has failed to achieve his or her dreams."

Disclaimer: This product was tested on laboratory rats and might not work for everyone.

Identifying Facts and Misleading Information
Misleading information often is worded to sound like scientific facts. A scientifically literate person can recognize fake claims and quickly determine when information is false.

Identify Opinions
An opinion is a personal view, feeling, or claim about a topic. Opinions cannot be proven true or false. And, an opinion might contain inaccurate information.

Science cannot answer all questions.

It might seem that scientific inquiry is the best way to answer all questions. But there are some questions that science cannot answer. Questions that deal with beliefs, values, personal opinions, and feelings cannot be answered scientifically. This is because it is impossible to collect scientific data on these topics. ✓

Science cannot answer questions such as

- Which video game is the most fun to play?
- Are people kind to others most of the time?
- Is there such a thing as good luck?

Safety in Science

Scientists know that using safe procedures is important in any scientific investigation. When you begin scientific inquiry, you should always wear protective equipment. You also should learn the meaning of safety symbols, listen to your teacher's instructions, and learn to recognize potential hazards.

Mini Glossary

critical thinking: comparing what you already know with the information you are given in order to decide whether you agree with it

hypothesis: a possible explanation for an observation that can be tested by scientific investigations

inference: a logical explanation of an observation that is drawn from prior knowledge or experience

observation: the results of using one or more of your senses to gather information and taking note of what occurs

prediction: a statement of what will happen next in a sequence of events

science: the investigation and exploration of natural events and of the new information that results from those investigations

scientific law: a rule that describes a repeatable pattern in nature

scientific theory: an explanation of observations or events that is based on knowledge gained from many observations and investigations

technology: the practical use of scientific knowledge, especially for industrial or commercial use

1. Review the terms and their definitions in the Mini Glossary. Write a sentence that explains how a hypothesis and an observation are related.

_____

_____

2. Write the letter of each statement in the correct location in the diagram to compare and contrast scientific law and scientific theory.

   a. based on repeated observations

   b. explains why an event occurred

   c. can be rejected or modified

   d. states that an event will occur

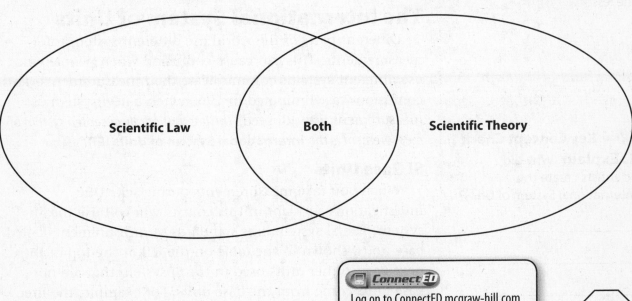

Scientific Law   Both   Scientific Theory

Log on to ConnectED.mcgraw-hill.com and access your textbook to find this lesson's resources.

END OF LESSON

Scientific Problem Solving

Measurement and Scientific Tools

Key Concepts 🔑

- Why did scientists create the International System of Units (SI)?
- Why is scientific notation a useful tool for scientists?
- How can tools, such as graduated cylinders and triple-beam balances, assist physical scientists?

Mark the Text ▶

Make an Outline As you read, highlight the main idea under each heading. Then use a different color to highlight a detail or an example that might help you understand the main idea. Use your highlighted text to make an outline with which to study the lesson.

🔑 **Key Concept Check**

1. Explain Why did scientists create the International System of Units?

_____

_____

_____

·············· **Read to Learn** ··············

Description and Explanation

Suppose you work for a company that tests how cars perform during crashes. You might use various scientific tools to measure the acceleration of cars as they crash into other objects. The measurements you record are descriptions of the results of the crash tests. *A* **description** *is a spoken or written summary of observations.* A description of an event details what you observed.

Later, your supervisor asks you to write a report that interprets the measurements you took during the crash tests. *An* **explanation** *is an interpretation of observations.* An explanation explains why or how the event occurred. As you write your explanation, you make inferences about why the crashes damaged the vehicles in specific ways.

Notice that there is a difference between a description and an explanation. When you describe something, you report your observations. When you explain something, you interpret your observations.

The International System of Units

Different parts of the world use different systems of measurements. This can cause confusion when people who use different systems communicate their measurements. This confusion was eliminated in 1960 when a new system of measurement was adopted. *The internationally accepted system of measurement is the* **International System of Units (SI).** 🔑

SI Base Units

When you take measurements during scientific investigations and labs in this course, you will use the SI system. The SI system uses standards of measurement, called base units, shown in the table on the left at the top of the next page. Other units used in the SI system that are not base units come from the base units. For example, the liter, used to measure volume, was derived from the base unit for length.

SI Base Units	
Quantity Measured	**Unit (symbol)**
Length	meter (m)
Mass	kilogram (kg)
Time	second (s)
Electric current	ampere (A)
Temperature	Kelvin (K)
Substance amount	mole (mol)
Light intensity	candela (cd)

Prefixes	
Prefix	**Meaning**
Mega- (M)	1,000,000 or (10^6)
Kilo- (k)	1,000 or (10^3)
Hecto- (h)	100 or (10^2)
Deka- (da)	10 or (10^1)
Deci- (d)	0.1 or $\left(\frac{1}{10}\right)$ or (10^{-1})
Centi- (c)	0.01 or $\left(\frac{1}{100}\right)$ or (10^{-2})
Milli- (m)	0.001 or $\left(\frac{1}{1,000}\right)$ or (10^{-3})
Micro- (μ)	0.000001 or $\left(\frac{1}{1,000,000}\right)$ or (10^{-6})

SI Unit Prefixes

Older systems of measurement usually had no common factor that related one unit to another. The SI system eliminated this problem.

The SI system is based on multiples of ten. You can convert any SI unit to another unit by multiplying by a power of ten. Prefixes represent factors of ten, as shown in the table on the right above. For example, the prefix *milli-* means 0.001 or 10^{-3}. So, a milliliter is 0.001 L, or 1/1,000 L. Another way to say this is: 1 L is 1,000 times greater than 1 mL.

Converting Among SI Units

It is easy to convert from one SI unit to another. You either multiply or divide by a factor of ten. You also can use proportion calculations to make conversions. For example, suppose a rock has a mass of 17.5 grams. To convert that measurement to kilograms, follow these steps:

Step 1 Determine the correct relationship between grams and kilograms. There are 1,000 g in 1 kg.

$$\frac{1 \text{ kg}}{1,000 \text{ g}}$$

$$\frac{x}{17.5 \text{ g}} = \frac{1 \text{ kg}}{1,000 \text{ g}}$$

$$x = \frac{(17.5 \text{ g})(1 \text{ kg})}{1,000 \text{ g}}; \quad x = 0.0175 \text{ kg}$$

Step 2 Check your units. The unit *grams* divides in the equation, so the answer is 0.0175 kg.

Interpreting Tables

2. Recognize If you were measuring the amount of electric current in a wire, what SI base unit would you use?

_____

_____

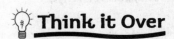 **Think it Over**

3. Apply Write the name of the unit that is 1/100 of a meter.

_____

_____

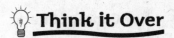 **Think it Over**

4. Solve Convert 305 meters to kilometers.

_____

_____

_____

Interpreting Tables

5. Recognize In Trial 1, which student's measurement is most accurate? Why?

_____

_____

_____

Key Concept Check

6. Explain Why is scientific notation a useful tool for scientists?

_____

_____

_____

Measurement and Uncertainty

In science, the terms *precision* and *accuracy* have different meanings. Precision describes how similar or close repeated measurements are to each other. Accuracy describes how close a measurement is to an accepted value.

The table below illustrates the difference between precision and accuracy. Three students were asked to find the density of sodium chloride (NaCl). In three trials, each student measured the volume and the mass of NaCl. Then, they calculated the density for each trial and calculated the mean, or average. Student A's measurements are the most precise because they are closest to each other. Student C's measurements are the most accurate because they are closest to the scientifically accepted value. Student B's measurements are neither precise nor accurate. They are not close to each other or to the accepted value.

Student Density and Error Data (Accepted value: Density of sodium chloride, 21.65 g/cm³)		
	Student A	**Student B**	**Student C**
	Density	**Density**	**Density**
Trial 1	21.20 g/cm³	20.92 g/cm³	21.90 g/cm³
Trial 2	21.30 g/cm³	22.27 g/cm³	21.85 g/cm³
Trial 3	21.25 g/cm³	21.10 g/cm³	21.46 g/cm³
Average	21.25 g/cm³	21.43 g/cm³	21.73 g/cm³

Tools and Accuracy

No measuring tool provides a perfect measurement. All measurements have some degree of uncertainty. Some tools or instruments produce more accurate measurements. For example, suppose a glass cylinder is graduated, or marked off, in 1-mL increments. Suppose a glass beaker is graduated in 50-mL increments. Liquid measurements taken with the graduated cylinder will have greater accuracy.

Scientific Notation

Suppose you are writing a report that includes Earth's distance from the Sun—149,600,000 km—and the density of the Sun's lower atmosphere—0.000000028 g/cm³. These numbers take up too much space in your report, so you use **scientific notation**—*a method of writing or displaying very small or very large numbers in a short form*. To write the numbers above in scientific notation, use the steps on the next page.

How to Write in Scientific Notation

Step 1 Write the original number.

> A. 149,600,000
>
> B. 0.000000028

Step 2 Move the decimal point to the right or the left to make the number larger than 1 but smaller than 10. Count the number of decimal places moved and note the direction.

> A. 1.49600000 = 8 places to the left
>
> B. 00000002.8 = 8 places to the right

Step 3 Rewrite the number and delete all extra zeros to the right or to the left of the decimal point.

> A. 1.496
>
> B. 2.8

Step 4 Write a multiplication symbol and the number *10* with an exponent. The exponent should equal the number of places that you moved the decimal point in step 2. If the decimal point moved to the left, the exponent is positive. If it moved to the right, the exponent is negative.

> A. 1.496×10^8
>
> B. 2.8×10^{-8}

Percent Error

The densities recorded in the table on the previous page are experimental values because they were calculated during an experiment. Each of these values has some error because the scientifically accepted value for NaCl density is 21.65 g/cm³. Percent error can help you determine the size of your experimental error. **Percent error** *is the expression of error as a percentage of the accepted value.*

Percent Error Equation

$$\text{percent error} = \frac{|\text{experimental value} - \text{accepted value}|}{\text{accepted value}} \times 100\%$$

Scientific Tools

As you conduct scientific investigations, you will use tools to measure quantities. Some of the tools commonly used in science are described below.

Science Journal

Use a science journal to record observations, write questions and hypotheses, collect data, and analyze the results of scientific inquiry. All scientists record the information they learn while conducting investigations.

Math Skills ÷

A student in the laboratory measures the boiling point of water at 97.5°C. If the accepted value for the boiling point of water is 100.0°C, what is the percent error?

a. This is what you know:
experimental value = 97.5°C
accepted value = 100.0°C

b. You need to find:
percent error

c. Use this formula:

percent error =

$$\frac{\left|\begin{array}{c}\text{experimental} \\ \text{value}\end{array} - \begin{array}{c}\text{accepted} \\ \text{value}\end{array}\right|}{\text{accepted value}}$$

$\times 100\%$

d. Substitute the known values into the equation and perform the calculations.

percent error =

$$\frac{|97.5° - 100.0°|}{100.0°} \times 100\%$$

$= 2.50\%$

7. Solve for Percent Error Calculate the percent error if the experimental value of the density of gold is 18.7 g/cm³ and the accepted value is 19.3 g/cm³.

Balances

A balance is used to measure the masses of an object. Units often used for mass are kilograms (kg), grams (g), and milligrams (mg). Two common types of balances are the electronic balance and the triple-beam balance. In order to get accurate measurements when using a balance, it is important to calibrate the balance often.

Glassware

Laboratory glassware is used to hold or measure the volume of liquids. Flasks, beakers, test tubes, and graduated cylinders are just some of the different types of glassware available. Volume usually is measured in liters (L) and milliliters (mL).

Thermometers

A thermometer is used to measure the temperature of substances. Although Kelvin is the SI unit of measurement for temperature, in the science classroom, you often measure temperature in degrees Celsius (°C). Never stir a substance with a thermometer because it might break. If a thermometer does break, tell your teacher immediately. Do not touch the broken glass or the liquid inside the thermometer.

Calculators

A handheld calculator is a scientific tool that you might use in math class. But you also can use it in the lab and in the field (real situation outside the lab) to make quick calculations using your data.

Computers

For today's students, it is difficult to think of a time when scientists—or anyone—did not use computers in their work. Scientists can collect, compile, and analyze data more quickly using computers. Scientists use computers to prepare research reports and to share their data and ideas with investigators worldwide.

Hardware refers to the physical components of a computer, such as the monitor and the mouse. Computer software refers to the programs that run on computers, such as word processing, spreadsheet, and presentation programs.

You can attach electronic probes to computers and handheld calculators to record measurements. There are probes for collecting different kinds of information, such as temperature and the speed of objects.

Reading Check

8. Identify What is a balance used to measure?

_____

_____

Reading Check

9. Choose What instrument would you use to measure the temperature of a liquid?

_____

_____

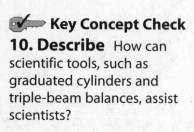
Key Concept Check

10. Describe How can scientific tools, such as graduated cylinders and triple-beam balances, assist scientists?

_____

_____

_____

Copyright © Glencoe/McGraw-Hill, a division of The McGraw-Hill Companies, Inc.

Additional Tools Used by Physical Scientists

You can use pH paper to quickly estimate the acidity of a liquid substance. The paper changes color when it comes into contact with an acid or a base.

A hot plate is a small heating device that can be placed on a table or a desk. Hot plates are used to heat substances in the laboratory.

Scientists use stopwatches to measure the time it takes for an event to occur. The SI unit for time is seconds (s). However, for longer events, the units *minutes (min)* and *hours (h)* can be used.

You use a spring scale to measure the weight or the amount of force applied to an object. The SI unit for weight is the newton (N).

✓ Reading Check

11. Choose If you were measuring the amount of force required to move a boulder, what SI unit would you use?

_____

Mini Glossary

description: a spoken or written summary of observations

explanation: an interpretation of observations

International System of Units (SI): the internationally accepted system of measurement

percent error: the expression of error as a percentage of the accepted value

scientific notation: a method of writing or displaying very small or very large numbers in a short form

1. Review the terms and their definitions in the Mini Glossary. Write a sentence that explains the difference between a description and an explanation.

_____

_____

2. Identify the quantity and SI unit symbols that each scientific tool measures.

Scientific Tool	Used to Measure . . .	SI Units and Symbols
triple-beam balance		
graduated cylinder		
thermometer	temperature of substances	Kelvin (K)
stopwatch		
spring scale		

3. Describe one main point that you highlighted as you read and a detail or an example that helped you understand the main point.

_____

Connect ED

Log on to ConnectED.mcgraw-hill.com and access your textbook to find this lesson's resources.

END OF LESSON

Scientific Problem Solving

Case Study

Copyright © Glencoe/McGraw-Hill, a division of The McGraw-Hill Companies, Inc.

·············· **Read to Learn** ··············

The Minneapolis Bridge Failure

On August 1, 2007, the center section of the Interstate-35W (I-35W) bridge in Minneapolis, Minnesota, suddenly collapsed. A major portion of the bridge fell more than 30 m into the Mississippi River. There were more than 100 cars and trucks on the bridge at the time, including a school bus carrying over 50 students. Several people were killed and many more were injured.

The failure of this 8-lane, 581-m long interstate bridge came as a surprise to almost everyone. Drivers do not expect a bridge to drop out from underneath them. The design and engineering processes that bridges undergo are supposed to ensure that bridge failures do not happen.

Controlled Experiments

After the 2007 bridge collapse, investigators had to determine why the bridge failed. To do this, they used a process of scientific inquiry similar to the one you read about in Lesson 1. The investigators designed controlled experiments to help them answer questions and test their hypotheses. A controlled experiment is a scientific investigation that tests how one factor affects another. You might conduct controlled experiments to help discover answers to questions, to test hypotheses, or to collect data. ✓

Identifying Variables and Constants

When conducting an experiment, you must identify factors that can affect the experiment's outcome. *A* **variable** *is any factor that can have more than one value.*

Types of Variables In controlled experiments, there are two kinds of variables. *The* **independent variable** *is the factor that you want to test. It is changed by the investigator to observe how it affects a dependent variable. The* **dependent variable** *is the factor you observe or measure during an experiment.*

Key Concepts

- Why are evaluation and testing important in the design process?
- How is scientific inquiry used in a real-life scientific investigation?

◄ Study Coach

Preview Headings Before you read the lesson, preview all the headings. Make a chart and write a question for each heading beginning with *What* or *How*. As you read, write the answers to your questions.

✓ **Reading Check**
1. Identify What does a controlled experiment test?

_____

_____

_____

Factors that Do Not Vary *Constants are the factors in an experiment that do not change.* You can change the independent variable to observe how it affects the dependent variable. Without constants, two independent variables could change at the same time, and you would not know which variable affected the dependent variable.

Experimental Groups

A controlled experiment usually has at least two groups. *The* **experimental group** *is used to study how a change in the independent variable changes the dependent variable. The* **control group** *contains the same factors as the experimental group, but the independent variable is not changed.* Without a control, it is impossible to know if your observations result from the variable you are testing or some other factor. ✓

This case study will explore how the investigators used scientific inquiry to determine why the bridge collapsed. The tables like the one on the next page provide additional information and show what a scientist might write in a science journal.

Simple Beam Bridges

Before you read about the bridge-collapse investigation, think about the structure of bridges. The simplest type of bridge is a beam bridge. This type of bridge has one horizontal beam across two supports. A beam bridge might be constructed across a small creek. A disadvantage of beam bridges is that they tend to sag in the middle if they are too long.

Truss Bridges

A truss bridge often spans long distances. This type of bridge is supported only at its two ends, but a series of interconnected triangles, or trusses, strengthens it. The I-35W bridge was a truss bridge designed in the early 1960s. Straight beams connected to triangular and vertical supports held up the deck. These supports held deck of the bridge, or the roadway. The beams in the bridge's deck and the supports came together at structures known as gusset plates. ✓

The steel gusset plates joined the triangular and vertical trusses to the overhead roadway beams. These beams ran along the deck of the bridge. This area, where the truss structure connects to the roadway at a gusset plate, is called a node. The gusset plates at each node are critical pieces that hold the bridge together. ✓

✓ Reading Check

2. State What is the purpose of a control group in an experiment?

_____

_____

_____

✓ Reading Check

3. Explain Why are trusses often used for bridges that span long distances?

_____

_____

_____

✓ Reading Check

4. Describe What are the gusset plates of a bridge?

_____

_____

Bridge Failure Observations

After the bridge collapsed, the local sheriff's department handled the initial recovery of the collapsed bridge. Finding, freeing, and identifying victims was a higher priority than preserving evidence about why the bridge collapsed. Emergency rescue workers also damaged the collapsed structure. The unintentional damage to the bridge made evaluating why the collapse occurred more difficult.

However, investigators eventually recovered the entire structure. The investigators labeled each part with the location where it was found. They also noted the date when they removed each piece. Investigators then moved the pieces to a nearby park. There, they placed the pieces in their relative original positions. Examining the reassembled structure, investigators found physical evidence they needed to determine where the breaks in each section occurred. ✓

The investigators had an additional source of information about the collapse. A motion-activated security camera recorded the bridge collapse. The video showed about 10 seconds of the collapse. This revealed the sequence of events that destroyed the bridge. Investigators used this video to help pinpoint where the collapse began. This video gave investigators additional clues as to how and why the bridge failed. ✓

✓ **Reading Check**

5. Consider How did investigators go about studying the bridge?

_____

_____

_____

_____

_____

Observe and Gather Information
Scientists often observe and gather information about an object or an event before proposing a hypothesis. This information is recorded or filed before the investigation.
Observations: • Recovered parts of the collapsed bridge • A video showing the sequence of events as the bridge fails and falls into the river

✓ **Reading Check**

6. Identify What important information did investigators gain from the surveillance video?

_____

_____

_____

Asking Questions

Asking questions and seeking answers to those questions is a way that scientists formulate hypotheses. One or more factors could have caused the bridge to fail. Was the original bridge design faulty? Were bridge maintenance and repair poor or lacking? Was there too much weight on the bridge at the time of the collapse? Each of these questions was studied to determine why the bridge collapsed. Did one or a combination of these factors cause the bridge to fail?

Interpreting Tables

7. Point Out How do scientists use information about an object or event?

_____

_____

_____

Gathering Information and Data

Investigators reviewed the modifications made to the bridge since it opened in 1967. In 1977, engineers noticed that salt used to deice the bridge during winter weather was causing the reinforcement rods in the roadway to weaken. To protect the rods, engineers applied a thicker layer of concrete to the surface of the bridge roadway.

Analysis after the collapse revealed that this extra concrete increased the dead load on the bridge by about 13.4 percent. A load can be a force applied to the structure from the structure itself (dead load) or from temporary loads such as traffic, wind gusts, or earthquakes (live load). ✓

Investigators recorded this qualitative and quantitative data. **Qualitative data** *uses words to describe what is observed.* **Quantitative data** *uses numbers to describe what is observed.* ✓

More modifications were made to the bridge in 1998. The bridge that was built in the 1960s did not meet current safety standards. Analysis showed that the changes made to the bridge during this renovation further increased the dead load on the bridge by about 6.1 percent.

An Early Hypothesis

At the time of the bridge collapse in 2007, the bridge was undergoing additional renovations. Piles of sand and gravel, a water tanker filled with over 11,000 L of water, a cement tanker, a concrete mixer, and other equipment, supplies, and workers were assembled on the bridge. In addition to these renovation materials, normal vehicle traffic was on the bridge. Did these renovations, materials, and traffic overload the bridge, causing the center section to collapse? Only a thorough analysis could answer this question.

Collect and Process Data
When gathering information or collecting data, scientists might perform an experiment, create a model, gather and evaluate evidence, or make calculations.
Qualitative Data: A thicker layer of concrete was added to the bridge to reinforce rods.
Quantitative Data: • The concrete increased the load on the bridge by 13.4 percent. • The modifications in 1998 increased the load on the bridge by 6.1 percent. • At the time of the collapse in 2007, the load on the bridge increased by another 20 percent.

Hypothesis

A hypothesis is a possible explanation for an observation that can be tested by scientific investigations.

Hypothesis: The bridge failed because it was overloaded.

Computer Modeling

Engineers used computer models to analyze the structure and loads on the bridge. Using computer-modeling software, investigators entered data from the Minnesota bridge into a computer. The computer performed numerous mathematical calculations. After thorough modeling and analysis, it was determined that the bridge was not overloaded.

Revising the Hypothesis

Evaluations conducted in 1999 and 2003 provided additional clues as to why the bridge might have failed. As part of the study, investigators took numerous pictures of the bridge structure. The photos revealed bowing of the gusset plates at the eleventh node from the south end of the bridge. Investigators labeled this node *U10*.

Gusset plates are designed to be stronger than the structural parts they connect. It is possible that the bowing of the plates indicated a problem with the gusset plate design. Previous inspectors and engineers missed this warning sign. ✓

The accident investigators found that some recovered gusset plates were fractured, while others were not damaged. If the bridge had been properly constructed, none of the plates should have failed. But inspection showed that some of the plates failed very early in the collapse.

After evaluating the evidence, the accident investigators formulated the hypothesis that the gusset plates failed, which lead to the bridge collapse. Now investigators had to test this hypothesis.

Hypothesis

1. ~~The bridge failed because it was overloaded.~~

2. The gusset plates failed, which lead to the bridge collapse.

🔆 Think it Over

11. Name one advantage of using computer modeling software.

_____

_____

_____

✓ Reading Check

12. Describe What evidence caused investigators to be concerned about the U10 node?

_____

_____

_____

Interpreting Tables

13. Identify Highlight the investigators' new hypothesis.

_____

_____

_____

Testing the Hypothesis

The investigators knew the load limits of the bridge. To calculate the load on the bridge when it collapsed, they estimated the combined weight of the bridge and the traffic on the bridge. The investigators divided the load on the bridge when it collapsed by the load limits of the bridge to find the demand-to-capacity ratio. The demand-to-capacity ratio provides a measure of a structure's safety.

Interpreting Tables

14. Recognize What is the demand-to-capacity ratio?

_____

_____

_____

_____

Test the Hypothesis
• Compare the load on the bridge when it collapsed with the load limits of the bridge at each of the main gusset plates. • Determine the demand-to-capacity ratios for the main gusset plates. • Calculate the appropriate thicknesses of the U10 gusset plates.
Independent Variables: actual load on bridge and load bridge was designed to handle.
Dependent Variable: demand-to-capacity ratio

Analyzing Results

As investigators calculated the demand-to-capacity ratios for each of the main gusset plates, they found that the ratios were particularly high for the U10 node. The U10 plate failed earliest in the bridge collapse. The table below shows the demand-to-capacity ratios for a few of the gusset plates at some nodes. A value greater than 1 means the structure is unsafe. Notice how high the ratios are for the U10 gusset plate compared to the other plates.

Interpreting Tables

15. Explain Why does this data suggest that the U10 gusset plate was unsafe?

_____

_____

_____

_____

Node-Gusset Plate Analysis

Gusset Plate	Thickness (cm)	*Demand-to-Capacity Ratios for the Upper-Node Gusset Plates					
		Horizontal loads			Vertical loads		
U8	3.5	0.05	0.03	0.07	0.31	0.46	0.20
U10	**1.3**	**1.81**	**1.54**	**1.83**	**1.70**	**1.46**	**1.69**
U12	2.5	0.11	0.11	0.10	0.71	0.37	1.15
*A value greater than 1 indicates the plates are unsafe.							

🔑 **Key Concept Check**

16. Analyze Why are evaluation and testing important in the design process?

_____

_____

_____

Further calculations showed that the U10 plates were not thick enough to support the loads they were supposed to handle. They were about half the required thickness. 🔑

Analyzing Results
The U10 gusset plates should have been twice as thick as they were to support the bridge.

Drawing Conclusions

Over the years, modifications to the I-35W bridge added more load to the bridge. On the day of the accident, traffic and the concentration of construction vehicles and materials added still more load. Investigators concluded that if the U10 gusset plates were properly designed, they would have supported the added load.

When investigators examined the original records for the bridge, they were unable to find any detailed gusset plate specifications. They could not determine whether undersized plates were used because of a mistaken calculation or some other error in the design process. The only thing that they could conclude with certainty was that undersized gusset plates could not reliably hold up the bridge.

The Federal Highway Administration and the National Transportation Safety Board published the results of their investigations. These published reports now provide scientists and engineers with valuable information they can use in future bridge designs. These reports are good examples of why it is important for scientists and engineers to publish their results and to share information. 🔑

Conclusions

The bridge failed because the gusset plates were not properly designed and they could not carry the load that they were supposed to carry.

🔑 **Key Concept Check**
17. Identify Give three examples of the scientific inquiry process that was used in this investigation.

_____

_____

_____

_____

_____

Mini Glossary

constant: factors in an experiment that a scientist does not change

control group: the part of an experiment that contains the same factors as the experimental group, but the independent variable is not changed

dependent variable: the factor a scientist observes or measures during an experiment

experimental group: used to study how a change in the independent variable changes the dependent variable

independent variable: the factor in an experiment that a scientist wants to test that is changed to see how it affects a dependent variable

qualitative data: observations in an experiment described in words

quantitative data: observations in an experiment described in numbers

variable: any factor that can have more than one value

1. Review the terms and their definitions in the Mini Glossary. Write a sentence that gives an example of qualitative data gathered by investigators of the bridge collapse.

_____

_____

2. Identify the steps in the investigation of the bridge collapse by writing the following phrases in the correct sequence in the flow chart.

analyzing results   **testing a hypothesis**   **asking questions**

gathering information   **drawing conclusions**   **making a hypothesis**

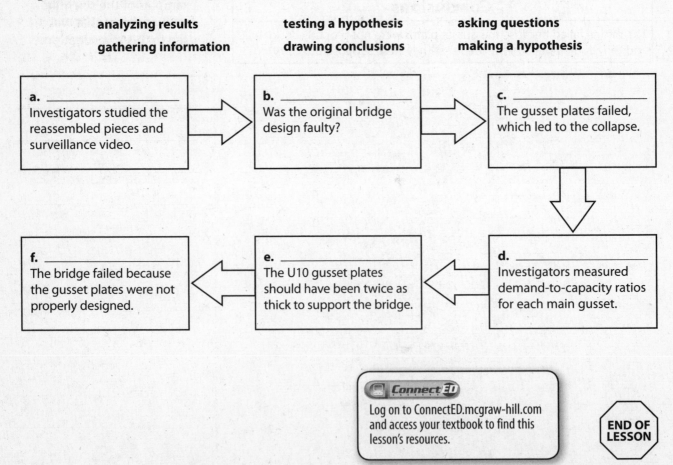

a. _____
Investigators studied the reassembled pieces and surveillance video.

b. _____
Was the original bridge design faulty?

c. _____
The gusset plates failed, which led to the collapse.

f. _____
The bridge failed because the gusset plates were not properly designed.

e. _____
The U10 gusset plates should have been twice as thick to support the bridge.

d. _____
Investigators measured demand-to-capacity ratios for each main gusset.

Connect ED
Log on to ConnectED.mcgraw-hill.com and access your textbook to find this lesson's resources.

END OF LESSON

Describing Motion

Position and Motion

·············· **Before You Read** ··············

What do you think? Read the two statements below and decide whether you agree or disagree with them. Place an A in the Before column if you agree with the statement or a D if you disagree. After you've read this lesson, reread the statements to see if you have changed your mind.

Before	Statement	After
	1. Displacement is the distance an object moves along a path.	
	2. The description of an object's position depends on the reference point.	

··············· **Read to Learn** ···············

Describing Position

How would you describe where you are right now? You might say that you are sitting one meter to the left of your friend. You might explain that you are at home, which is two houses north of your school.

What do these descriptions have in common? Each states your location relative to a certain point. This point is called the reference point. *A* **reference point** *is the starting point you choose to describe the location, or position, of an object*. The reference point in the first example is your friend. In the second example, it is your school. These descriptions <u>specify</u> your location relative to a certain point. You compared your location to reference points, your friend and your school.

Each description of your location also includes your distance and direction from the reference point. Describing your location in this way defines your position. In the first example, the distance is one meter. The direction is to the left, and the reference point is your friend. In the second example, the distance is two houses. The direction is north, and the reference point is your school. *A* **position** *is an object's distance in a certain direction from a reference point*. A complete description of your position includes a distance, a direction, and a reference point. ✓

Key Concepts

- How does the description of an object's position depend on a reference point?
- How can you describe the position of an object in two dimensions?
- What is the difference between distance and displacement?

◂ **Study Coach**

Create an Outline As you read, make an outline to summarize the information in the lesson. Use the main headings in the lesson as the main headings in your outline. Use your outline to review the lesson.

ACADEMIC VOCABULARY
specify
(verb) to indicate or identify

✓ **Reading Check**

1. Name two ways you could describe your position right now.

_____

_____

_____

_____

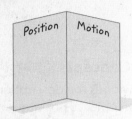
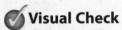 **Visual Check**

2. Interpret There are two reference points in the drawing. The arrows show the distances and directions from the different reference points. How do you know which reference point is farther from the table?

_____

_____

 Key Concept Check

3. Summarize How does the description of an object's position depend on a reference point?

_____

_____

_____

 **Visual Check**

4. Locate State the position of the museum relative to the library.

_____

_____

Using a Reference Point to Describe Position

Suppose you are planning a family picnic at the park shown below. How would you describe the position of the picnic table you reserved? First, choose a reference point that is easy to find. In this park, a statue is a good choice. Next, describe the direction of the table from the reference point. The direction is toward the slide. Finally, suppose the table is about 10 m from the statue. You would say that the position of the table is about 10 m from the statue, toward the slide.

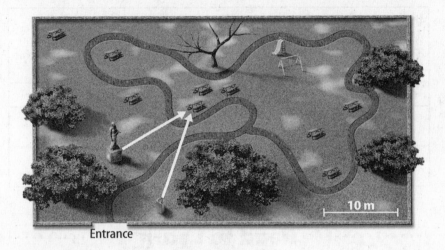

Entrance

Changing the Reference Point

Look again at the drawing of the park. Suppose you choose the drinking fountain as the reference point instead of the statue. Now the direction of the table is toward the dead tree. The distance from the drinking fountain to the table is about 12 m. You can tell your family that the table is about 12 m from the drinking fountain toward the dead tree. The table did not change position. Your description of its position and the reference point are different.

The Reference Direction

Sometimes direction is described using the words *positive* or *negative*. The reference direction is the positive (+) direction. The opposite direction is the negative (−) direction. Suppose east is the reference direction in the diagram below. The museum's entrance is 80 m east of a bus stop. The library is 40 m west of the bus stop. You could say that the museum is +80 m from the bus stop and the library is −40 m from the bus stop. Using the words *positive* or *negative* to describe direction can be useful for explaining changes in an object's position.

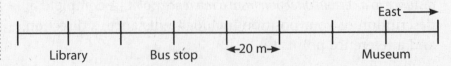

Library | Bus stop | ←20 m→ | Museum

East →

Describing Position in Two Dimensions

Sometimes you need to use more than one reference direction to describe an object's position. When you describe position using two directions, you are using two dimensions.

Reference Directions in Two Dimensions

The figure below is a map of a city. It shows positions in two dimensions. To describe a position on the map, you might choose north and east or south and west as the reference directions.

Sometimes north, south, east, and west are not the most useful reference directions. Imagine that you are looking at a skyscraper. You might describe a certain window as "up" and "to the left."

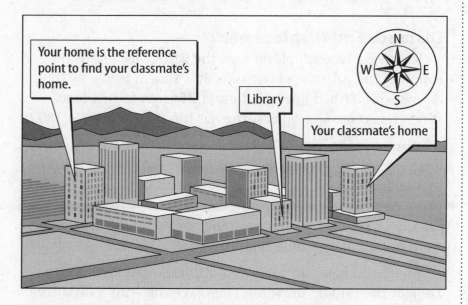

Your home is the reference point to find your classmate's home.

Library

Your classmate's home

Locating a Position in Two Dimensions

Suppose you want to locate your classmate's home on the map above. To find a position in two dimensions, first choose a reference point. You could choose your home as a reference point. Next, give specific reference directions. In this case, you would use south and east. Then, determine the distance along each reference direction. On the map, your classmate's home is one block south and four blocks east of your home.

Describing Changes in Position

Sometimes you need to describe how an object's position changes. Suppose a boat is floating on a lake. How do you know whether the boat has moved throughout the day? You know this when its position changes relative to, or compared to, something else. **Motion** *is the process of changing position.*

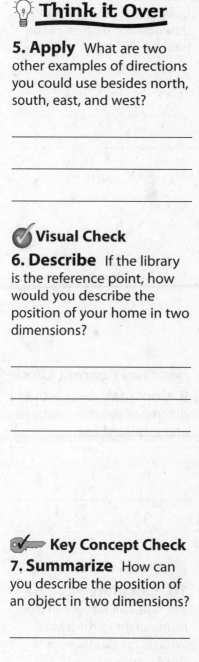

Think it Over

5. Apply What are two other examples of directions you could use besides north, south, east, and west?

_____

_____

Visual Check

6. Describe If the library is the reference point, how would you describe the position of your home in two dimensions?

_____

_____

_____

Key Concept Check

7. Summarize How can you describe the position of an object in two dimensions?

_____

_____

_____

Motion Relative to a Reference Point

In the figure below, is the man in motion? Use the fishing pole as the reference point. The positions of the man and the pole do not change relative to each other. This means the man does not move relative to the pole. When the buoy is the reference point, the man's distance from the buoy changes. The man is in motion relative to the buoy.

Visual Check

8. Describe Using the buoy as a reference point, describe the motion of the boat.

_____

_____

_____

Distance and Displacement

Suppose a baseball player runs the bases. Distance is the length of the path the player runs. It is shown below by the arrows with dashed lines. He runs 90 ft to get to first base. When he gets to second base, he has run 90 ft + 90 ft = 180 ft.

Displacement *is the difference between the initial (first) position and the final position of an object*. Displacement is shown below by the arrows with solid lines. The initial position is home plate. At first base, the player's distance and displacement are the same. He has run a distance of 90 ft and he is 90 ft from home plate, where he started.

At second base, distance and displacement are different. Look at the middle drawing. The player has run a distance of 180 ft, but his displacement is 127 ft from home plate.

In the drawing on the far right below, the player has circled the bases. He has run a distance of 360 ft (90 ft × 4), but his displacement is 0 ft. His starting position and ending position are the same—home plate.

Distance depends on the path taken. Only the starting and ending positions matter in displacement. Notice that distance and displacement are equal only if the motion is in one direction.

Key Concept Check

9. Contrast What is the difference between distance and displacement?

_____

_____

_____

Visual Check

10. Calculate Suppose the baseball player ran from home plate to third base. What is his distance and displacement?

Distance: _____

Displacement: _____

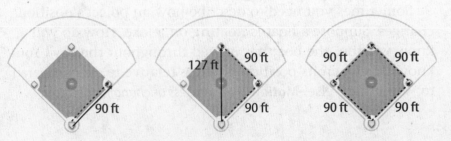

Mini Glossary

displacement: the difference between the initial (first) position and the final position of an object

motion: the process of changing position

position: an object's distance in a certain direction from a reference point

reference point: the starting point you choose to describe the location, or position, of an object

1. Review the terms and their definitions in the Mini Glossary. Write a sentence identifying three things you need to include in a description of any object's position. Then choose an object in the room. Write another sentence describing the position of the object.

_____

_____

_____

_____

2. In the diagram below, the theater is your reference point. Your reference direction is east. The mall is located +2 blocks from the theater. The park is located −3 blocks from the theater. In the diagram, circle the locations of the mall and park and label them.

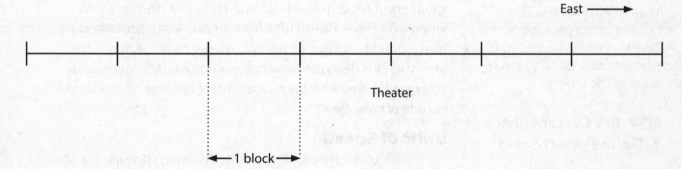

East ⟶

Theater

⟵ 1 block ⟶

3. How did your outline of this lesson help you understand position and motion?

_____

_____

What do you think NOW?

Reread the statements at the beginning of the lesson. Fill in the After column with an A if you agree with the statement or a D if you disagree. Did you change your mind?

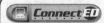 Connect ED

Log on to ConnectED.mcgraw-hill.com and access your textbook to find this lesson's resources.

END OF LESSON

Describing Motion

Speed and Velocity

What do you think? Read the two statements below and decide whether you agree or disagree with them. Place an A in the Before column if you agree with the statement or a D if you disagree. After you've read this lesson, reread the statements to see if you have changed your mind.

Before	Statement	After
	3. Constant speed is the same thing as average speed.	
	4. Velocity is another name for speed.	

Key Concepts
- What is speed?
- How can you use a distance-time graph to calculate average speed?
- What are ways velocity can change?

Study Coach

Discuss What You Read
Work with a partner. Read a paragraph to yourselves. Then discuss what you learned in the paragraph. Continue until you and your partner understand the main ideas of the lesson.

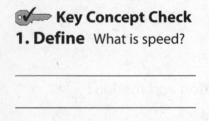 **Key Concept Check**
1. Define What is speed?

_____

_____

_____

················· **Read to Learn** ··············

What is speed?

How fast do you walk when you are hungry and you smell good food in the kitchen? How fast do you move when you have a difficult chore to do? Sometimes you move quickly. Other times you might move slowly. One way you can describe how fast you move is to determine your speed. **Speed** *is the measure of the distance an object travels per unit of time.*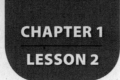

Units of Speed

To calculate speed, divide the distance traveled by the time it takes to travel that distance. The units of speed are units of distance divided by units of time. The SI unit for speed is meters per second (m/s).

There are other units of speed. Different units of distance and time can be used to express the speed an object is moving. The speed of an airplane might be expressed in kilometers per hour (km/h). The speed of a car on a highway might be expressed in miles per hour (mph).

The typical speed of an airplane can be expressed as 245 m/s, or 882 km/h, or 548 mph. Similarly, the typical speed of a person walking can be expressed as 1.3 m/s, or 4.7 km/h, or 2.9 mph.

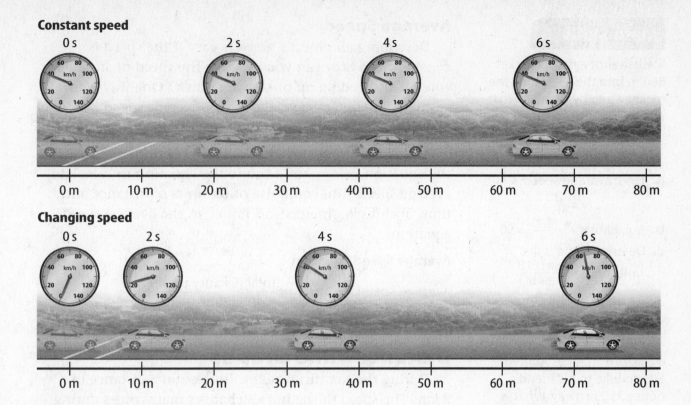

Constant speed

0 s · 2 s · 4 s · 6 s

0 m 10 m 20 m 30 m 40 m 50 m 60 m 70 m 80 m

Changing speed

0 s · 2 s · 4 s · 6 s

0 m 10 m 20 m 30 m 40 m 50 m 60 m 70 m 80 m

Constant Speed

What happens to a car's speed as the driver moves from one place to another? If the car is traveling on a highway, it might move at a steady speed. As a car moves away from a stop sign, the car's speed increases. The driver slows the car down to pull into a parking space.

Think specifically about a time when a car's speed does not change. For example, the car in the top figure is moving at the same speed. In 1 s, the car moves 11 m. In 2 s, it moves 22 m. Each second, the car moves 11 m. Because the car moves the same distance each second, its speed is not changing. The speed of the car is constant. **Constant speed** *is the rate of change of position in which the same distance is traveled each second.* The car is moving at a constant speed of 11 m/s.

Changing Speed

The bottom half of the figure shows a car moving away from a stop sign with increasing speed. Between 0 s and 2 s, the car at the bottom travels about 10 m. However, between 4 s and 6 s, the car travels more than 30 m. The car travels a different distance each second. Its speed is changing.

If an object's speed is not constant, you might want to know its speed at a certain moment. **Instantaneous speed** *is speed at a specific instant in time.* You can see a car's instantaneous speed on its speedometer.

Copyright © Glencoe/McGraw-Hill, a division of The McGraw-Hill Companies, Inc.

✔ **Visual Check**

2. Identify Look at the speedometer of the bottom car. What is its instantaneous speed at 6 s?

_____

✔ **Reading Check**

3. Calculate How would the distance the car travels each second change if it were slowing down?

_____

_____

Melissa shot a model rocket 360 m into the air. It took the rocket 4 s to fly that far. What was its average speed?

distance: $d = 360$ m

time: $t = 4$ s

a. Find: average speed: v

$$v = \frac{d}{t}$$

b. Substitute: $v = \frac{360}{4} = 90$

c. Determine the units:

units of $v = \frac{\text{units of } d}{\text{units of } t}$

$= \frac{m}{s} = $ m/s

The rocket's average speed $= 90$ m/s

4. Solve It takes Ahmed 50 s to bike to his friend's house 250 m away. What is his average speed?

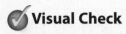 **Visual Check**

5. Apply How far did Horse A travel in 80 s?

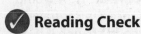 **Reading Check**

6. State How is constant speed shown on a distance-time graph?

Average Speed

Describing an object's speed is easy if the speed is constant. But how can you describe the speed of an object when it is speeding up or slowing down? One way is to calculate the object's average speed. **Average speed** *is the total distance traveled divided by the total time taken to travel that distance.*

The SI unit for speed, meters per second (m/s), is used in the equation below. You can use the equation to calculate average speed. You could use other units of distance and time, such as kilometers and hours, in the average speed equation.

Average Speed Equation

average speed (in m/s) $= \dfrac{\text{total distance (in m)}}{\text{total time (in s)}}$

$$v = \frac{d}{t}$$

Distance-Time Graphs

During the Kentucky Derby, horses run a distance of 2 km. The speed of the horses changes many times during the race. You can create a graph that charts these changes in speed. But sometimes it is helpful to graph an object's motion if its speed does not change.

The graph below describes what the motion of horse A and horse B might be if their speeds did not change. Notice that distance measurements are made every 20 seconds. Follow the height of the line from the left side of the graph to the right side. You can see how the distance that each horse ran changed over time.

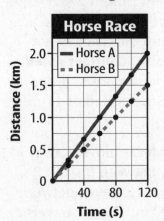

Graphs like the one on the left show how one measurement compares to another. In the study of motion, the two measurements that are compared are time and distance. Graphs that show comparisons between time and distance are called distance-time graphs. Notice that the change in the distance the horses ran around the track is the same each second on the graph. This means the horses were moving at a constant speed. Constant speed is shown as a straight line on distance-time graphs. ✓

Comparing Speeds on a Distance-Time Graph

You can use distance-time graphs to compare the motion of two objects. The distance-time graph on the previous page compares the motion of two horses that ran the Kentucky Derby.

The motion of horse A is shown by the solid line. The motion of horse B is shown by the dashed line. When horse A reached the finish line at 2 km from the starting point, horse B was 1.5 km from the starting point of the race.

Recall that average speed is distance traveled divided by time. Horse A traveled a greater distance than horse B in the same amount of time. This means that horse A had greater average speed.

Compare how steep the lines are on the graph. The measure of steepness is the slope. The line for horse A is steeper than the line for horse B. Steeper lines on distance-time graphs indicate faster speeds.

Using a Distance-Time Graph to Calculate Speed

You can use distance-time graphs to calculate the average speed of an object. The graph below represents the motion of a trail horse traveling at a constant speed. The steps on the graph explain how to calculate average speed from a distance-time graph.

Key Concept Check

7. Explain How can you use a distance-time graph to calculate average speed?

_____

_____

Visual Check

8. Compare How does the horse's average speed from 60 s to 120 s compare to its average speed from 120 s to 180 s?

_____

_____

_____

Complete steps 1–4. Then divide the difference in distance by the difference in time: 300 m/60 s = 5 m/s. This value is the average speed of the horse from 60 s to 120 s.

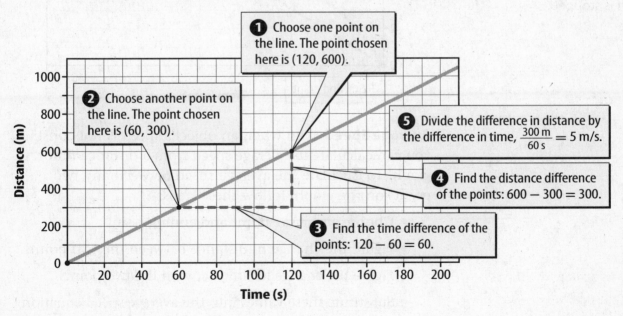

❶ Choose one point on the line. The point chosen here is (120, 600).

❷ Choose another point on the line. The point chosen here is (60, 300).

❺ Divide the difference in distance by the difference in time, $\frac{300\,m}{60\,s}$ = 5 m/s.

❹ Find the distance difference of the points: 600 − 300 = 300.

❸ Find the time difference of the points: 120 − 60 = 60.

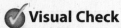

Distance-Time Graph and Changing Speed

A straight line on a distance-time graph means that the object moved at a constant speed. The graph for the motion of a train is shown below. The graph is different because the speed of the train is not constant. The train speeded up, slowed down, and stopped. When the changes in speed are graphed, the train's motion is shown as a curved line.

Slowing Down Notice the shape of the solid line on the graph. The line rises steeply at first. Then it levels off between 2 and 3 minutes. This means that the train is slowing down.

Stopping Between 3 and 5 minutes, the graph line is horizontal. The train's distance from the starting point remains at 4 km. This horizontal line means that there is no motion. The train stopped between minutes 3 and 5.

Speeding Up Between about 5 minutes and 10 minutes, the slope of the line increases. The upward curve means that the train was speeding up.

Visual Check

9. Identify Highlight the area on the graph where the train is stopped.

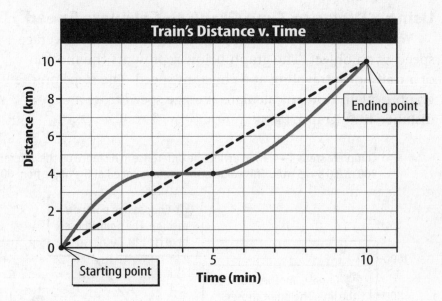

Average Speed Even when an object's speed is not constant, you can calculate its average speed from a distance-time graph. To calculate the average speed between any two points on the graph, follow these steps:

- Choose a starting point and ending point.

- Figure the change in distance between the two points.

- Figure the change in time between the two points.

- Substitute these values into the average speed equation.

The slope of the dashed line shown above represents the train's average speed between 0 minutes and 10 minutes.

Velocity

Describing the speed of a moving object does not completely describe its motion. Motion is more than just speed. It also includes direction. **Velocity** *is the speed and the direction of a moving object.*

Representing Velocity

In Lesson 1, an arrow represented the displacement of an object from a reference point. An arrow can also be used to represent the velocity of an object. The length of the arrow indicates the speed. A greater speed is shown by a longer arrow. The arrow points in the direction of the object's motion.

Imagine two people walking in the same direction at the same speed. They would have equal velocities. Arrows representing the people would point in the same direction and be the same length. But if the people were walking in opposite directions at the same speed, they would have different velocities. Arrows representing them would be the same length but be pointing in opposite directions.

Changes in Velocity

Look at the bouncing ball in the figure below. The arrows show the velocity of the ball. From one position to the next, the arrows change direction and length. This means that the velocity is changing. Velocity changes when the speed changes, the direction changes, or both the speed and the direction change. The velocity of the ball changes continually because both the speed and the direction of the ball change as the ball bounces.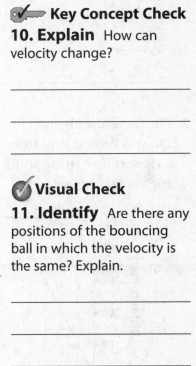

Velocity

Copyright © Glencoe/McGraw-Hill, a division of The McGraw-Hill Companies, Inc.

Key Concept Check

10. Explain How can velocity change?

_____

_____

_____

Visual Check

11. Identify Are there any positions of the bouncing ball in which the velocity is the same? Explain.

_____

_____

_____

_____

After You Read

Mini Glossary

average speed: the total distance traveled divided by the total time taken to travel that distance

constant speed: the rate of change of position in which the same distance is traveled each second

instantaneous speed: the speed at a specific instant in time

speed: the measure of the distance an object travels per unit of time

velocity: the speed and the direction of a moving object

1. Review the terms and their definitions in the Mini Glossary. Write a one-sentence example of two objects traveling at the same speed but with different velocities.

_____

_____

2. On the graph below, find point (20, 200). Label it *A*. Next find point (40, 400). Label it *B*. Draw a dotted line from point *A* along the 200 km line to the 40 s line. Next, draw a dotted line from point *B* along the 40 s line to the 200 km line. Use this graph to solve for average speed.

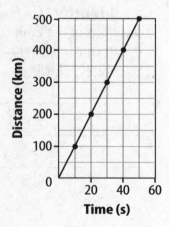

a. Difference in distance between points *A* and *B*: _____

b. Difference in time between points *A* and *B*: _____

c. Average speed = _____ ÷ _____

d. Average speed = _____ km/s

3. Which idea that you and your partner discussed was hardest to understand? Write a question and the answer on the lines below to show that you understand the idea.

_____

_____

What do you think NOW?

Reread the statements at the beginning of the lesson. Fill in the After column with an A if you agree with the statement or a D if you disagree. Did you change your mind?

Log on to ConnectED.mcgraw-hill.com and access your textbook to find this lesson's resources.

END OF LESSON

Describing Motion

Acceleration

·············· **Before You Read** ··············

What do you think? Read the two statements below and decide whether you agree or disagree with them. Place an A in the Before column if you agree with the statement or a D if you disagree. After you've read this lesson, reread the statements to see if you have changed your mind.

Before	Statement	After
	5. You can calculate average acceleration by dividing the change in velocity by the change in distance.	
	6. An object accelerates when either its speed or its direction changes.	

·············· **Read to Learn** ··············

Acceleration—Changes in Velocity

Recall that velocity changes if either speed or direction changes. When a car moves faster, its velocity increases. When a car slows down, its velocity decreases. If the car turns, its velocity changes because its direction changes.

When the car's velocity changes, the car is accelerating. **Acceleration** *is a measure of the change in velocity during a period of time.* An object accelerates when its velocity changes as a result of increasing speed, decreasing speed, or a change in direction. ✓

Representing Acceleration

Like velocity, acceleration has a direction. Acceleration can be represented by an arrow. The figures on the next page show different ways an object can accelerate.

The length of the solid arrows shows the amount of acceleration. The direction of an acceleration arrow depends on whether velocity is increasing or decreasing.

Changing Speed

Two ways an object can accelerate is that it can speed up or slow down. Velocity and acceleration are shown in the top two figures on the next page.

Key Concepts

- What are three ways an object can accelerate?
- What does a speed-time graph indicate about an object's motion?

▸ Mark the Text

Identify the Main Ideas
Highlight the main idea in each paragraph. Underline the details that support the main idea.

✓ **Reading Check**
1. Define What is acceleration?

_____

_____

FOLDABLES

Use a two-tab book to summarize information about the changes in velocity that can occur when an object is accelerating.

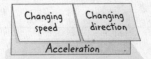

Ways an Object Can Accelerate

Speeding Up

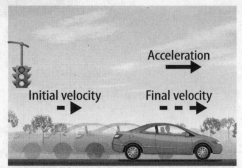

Slowing Down

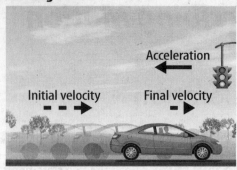

Changing Direction

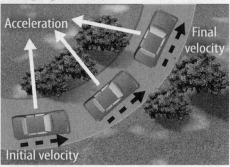

✓ **Visual Check**

2. State If the car in the Speeding Up figure moved faster, how would the acceleration arrow change?

_____

_____

_____

Speeding Up In the speeding up figure, the arrow for the car's initial velocity is short. This means the car is moving slowly at first. The car's speed increases, so the final velocity arrow is longer. As velocity increases, the car accelerates. Notice that the arrows for velocity and acceleration point in the same direction.

Slowing Down In the slowing down figure, the car moves fast at first. The arrow for its initial velocity is long. The arrow for the final velocity is shorter. This means that the car is slowing down. The arrows for velocity and acceleration point in opposite directions. When velocity decreases, acceleration and velocity are in opposite directions, so the acceleration arrow and velocity arrow point in the opposite directions.

Changing Direction

The car in the bottom figure has a constant speed, so the velocity arrow is the same length at each point in the turn. But the car's velocity changes because its direction changes. Because velocity changes, the car is accelerating. Notice that the acceleration arrows point toward the inside of the curve. Compare each velocity arrow to the next one in the picture. Notice that the change is always toward the inside of the curve. To show this direction of change, the acceleration arrows point toward the inside of the curve. ✓

✓ **Key Concept Check**

3. Summarize What are three ways an object can accelerate?

_____

_____

_____

Calculating Acceleration

The equation below shows how to calculate acceleration. Acceleration is a change in velocity during a time interval divided by the time interval during which the velocity changes.

The equation uses speed, not velocity. If acceleration is positive, the acceleration is in the direction of the motion. If acceleration is negative, acceleration is in the opposite direction of the motion.

Acceleration is usually measured in SI units. The SI units are meters per second per second, which is meters per second squared (m/s^2).

Acceleration Equation

acceleration (in m/s^2) =

$$\frac{\text{final speed (in m/s)} - \text{initial speed (in m/s)}}{\text{total time (in s)}}$$

$$a = \frac{v_f - v_i}{t}$$

Speed-Time Graphs

In the last lesson, you learned how to use a distance-time graph to show an object's speed. You can also use a speed-time graph to show how speed changes over time. Just like a distance-time graph, time is plotted on the horizontal axis, which is the x-axis. Speed is plotted on the vertical axis, which is the y-axis.

Object at Rest

An object at rest is not moving. Its speed is always zero. As a result, the speed-time graph for an object at rest is a horizontal line at $y = 0$, as shown in the graph below.

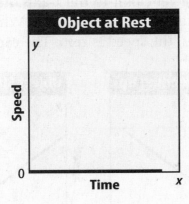

Object at Rest

Speed

Time

Math Skills

A bicyclist started from rest along a straight path. After 2.0 s, his speed was 2.0 m/s. After 5.0 s, his speed was 8.0 m/s. What was his acceleration during the time 2.0 s to 5.0 s?

a. This is what you know:

initial speed: $v_i = 2.0$ m/s

final speed: $v_f = 8.0$ m/s

total time: $t = 5.0\ s - 2.0\ s$
$= 3.0\ s$

b. You need to find acceleration: a

c. Use this formula:

$$a = \frac{v_f - v_i}{t}$$

d. Substitute the values for v_i, v_f, and t into the formula, subtract, then divide:

$$a = \frac{8.0 - 2.0}{3.0}$$

$$= \frac{6.0}{3.0} = 2.0$$

e. Determine the units:

units of $a = \dfrac{\text{units of } v}{\text{units of } t}$

$= \frac{m/s}{s} = m/s^2$

The acceleration of the bicyclist = 2.0 m/s^2.

4. Solve Aidan drops a rock from a cliff. After 4.0 s, the rock is moving at 39.2 m/s. What is the acceleration of the rock?

_____

_____

 **Visual Check**

5. Compare Why is the line horizontal?

_____

_____

_____

Constant Speed

Suppose a farm machine is moving through a field at a constant speed. At every point in time, its speed is the same. The graph below shows an object moving at a constant speed. Because the speed does not change, the line on the graph is horizontal. The object's speed is represented by the distance from the *x*-axis. If the line is farther from the *x*-axis, it is moving faster. If the line is closer to the *x*-axis, it is moving slower.

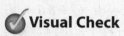
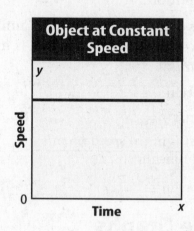

Speeding Up

Suppose the speed of a plane increases at a steady rate as it takes off. The graph of its speed might look like the speed-time graph on the left below. The line begins close to the *x*-axis when speed is low. The line slants up and to the right as speed increases.

Slowing Down

Suppose a car slows down at a steady rate and stops. The graph might look like the graph on the right below. At first, the car is moving at a high speed. As its speed decreases, the points on the line get closer to the *x*-axis. As a result, the graph line slants down and to the right. When the line touches the *x*-axis, the speed is zero. The car is stopped.

Key Concept Check

7. Identify What does a speed-time graph show about the motion of an object?

_____

_____

_____

Visual Check

8. Illustrate Circle the point on each graph at which the object is stopped.

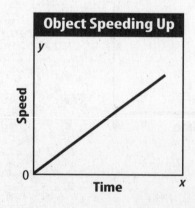

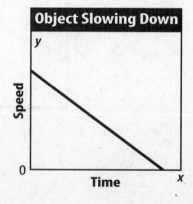

Limits of Speed-Time Graphs

You have read that distance-time graphs show the speed of an object. But they do not describe the direction in which an object is moving. In the same way, a speed-time graph shows only the relationship between speed and time. Distance-time graphs and speed-time graphs do not show what happens when direction changes. For example, a slalom skier moves from side to side through the slalom gates as he works his way down the slope. The skier's velocity changes partly as a result of his change of direction. Neither graph would indicate this change of direction. ✅

Summarizing Motion

Think about the many ways you can now describe a walk down a hallway at your school. You can describe your position by your direction and distance from a reference point. You can compare your distance and your displacement. This will help you find your average speed. You know that you have an instantaneous speed. You can also tell when you are walking at a constant speed. You can describe your velocity by your speed and direction. Whenever your velocity is changing, you know you are accelerating.

Copyright © Glencoe/McGraw-Hill, a division of The McGraw-Hill Companies, Inc.

✅ **Reading Check**

9. Name one weakness of both a distance-time graph and a speed-time graph.

_____

_____

_____

Mini Glossary

acceleration: a measure of the change in velocity during a period of time

1. Review the term and its definition in the Mini Glossary. Write two sentences that describe changes in acceleration on a roller-coaster ride.

_____

_____

_____

_____

2. Fill in the blanks with *A, B,* or *C* from the speed-time graph.

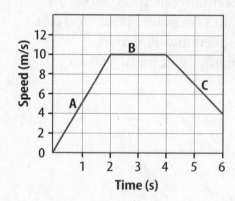

a. Line _____ represents an object slowing down.

b. Line _____ represents an object moving at a constant speed.

c. Line _____ represents an object speeding up.

What do you think NOW?

Reread the statements at the beginning of the lesson. Fill in the After column with an A if you agree with the statement or a D if you disagree. Did you change your mind?

 **Connect ED**

Log on to ConnectED.mcgraw-hill.com and access your textbook to find this lesson's resources.

END OF LESSON

The Laws of Motion

Gravity and Friction

·············· **Before You Read** ··············

What do you think? Read the two statements below and decide whether you agree or disagree with them. Place an A in the Before column if you agree with the statement or a D if you disagree. After you've read this lesson, reread the statements to see if you have changed your mind.

Before	Statement	After
	1. You pull on objects around you with the force of gravity.	
	2. Friction can act between two unmoving, touching surfaces.	

·············· **Read to Learn** ··············

Types of Forces

Think about all the things you pushed or pulled today. You might have pushed toothpaste out of a tube. Maybe you pulled out a chair to sit down. *A push or a pull on an object is called a* **force.** An object or a person can apply a force to another object or person. Some forces are applied only when objects touch. Other forces are applied even when objects do not touch.

Contact Forces

You have probably seen a musician strike the keys of a piano or an athlete hit a ball with a bat. In these cases, the person or object applies a force to an object that it touches. *A* **contact force** *is a push or a pull on one object by another that is touching it.*

Contact forces can be weak. When you press the keys on a computer keyboard, the contact force is weak. Contact forces can also be strong. The plates, or rock slabs, that make up Earth's crust can apply strong contact forces against each other. Over long periods of time, these forces can be strong enough to form mountain ranges if one plate pushes another plate upward.

Key Concepts

- What are some contact forces and some noncontact forces?
- What is the law of universal gravitation?
- How does friction affect the motion of two objects sliding past each other?

Mark the Text

Building Vocabulary As you read this lesson, underline each key term about the laws of motion. Then highlight information about each term to help you review the lesson later.

💡 **Think it Over**

1. Describe Give an example of a force.

_____

_____

_____

Noncontact Forces

When you drop an object, it falls toward the floor. This downward force occurs even though nothing is touching the object. *A force that one object can apply to another object without touching it is a* **noncontact force.** Gravity, which pulls on objects, is a noncontact force. The magnetic force, which attracts certain metals to magnets, is also a noncontact force. The electric force is another noncontact force. 🔑

Strength and Direction of Forces

Forces have both strength and direction. If you push your textbook away from you, it probably slides across the desk. What happens if you push down on your book? It probably does not move. You can use the same strength of force in both cases. Different things happen each time because the direction of the force is different.

Arrows can be used to show forces. The length of an arrow shows the strength of the force. A longer arrow indicates a stronger force than a shorter arrow. The direction that an arrow points shows the direction in which the force was applied.

Force is measured in newtons (N). When you lift a stick of butter, you apply about 1 N of force. When you lift a 2-L bottle of water, you apply about 20 N of force. If you use arrows to show these forces, the water's arrow would be 20 times longer than the butter's arrow.

What is gravity?

Objects fall to the ground because Earth exerts an attractive force on them. You also exert an attractive force on objects. **Gravity** *is an attractive force that exists between all objects that have mass.* **Mass** *is the amount of matter in an object.* Mass is often measured in kilograms (kg).

The Law of Universal Gravitation

In the late 1600s, an English scientist and mathematician, Sir Isaac Newton, developed the law of universal gravitation. The law of universal gravitation states that all objects are attracted to each other by a gravitational force. The strength of the force depends on the mass of each object and the distance between them. 🔑

Copyright © Glencoe/McGraw-Hill, a division of The McGraw-Hill Companies, Inc.

Key Concept Check

2. Identify What are some contact forces and some noncontact forces?

_____

_____

_____

FOLDABLES

Use a two-tab book to organize your notes on gravity and friction.

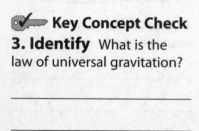

Gravity

Friction

Key Concept Check

3. Identify What is the law of universal gravitation?

_____

_____

_____

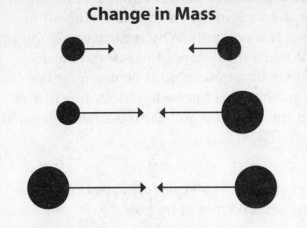

Change in Mass	Change in Distance
Gravitational force increases if the mass of at least one of the objects increases.	The gravitational force between objects decreases as the objects move apart.

Gravitational Force and Mass The way in which the mass of objects affects gravity is shown above on the left. When the mass of one or both objects increases, the gravitational force between them also increases. Look at each pair of objects in the figure on the left above. Notice that the force arrows for each pair of objects are the same length, even when one object has less mass. Each object exerts the same attraction on the other object.

Gravitational Force and Distance The effect of distance on gravity is shown above on the right. The attraction between objects decreases as the distance between the objects increases. For example, if your mass is 45 kg, the gravitational force between you and Earth is about 440 N. On the Moon, about 384,000 km away, the gravitational force between you and Earth would only be about 0.12 N. ✓

Weight—A Gravitational Force

Earth has more mass than any object near you. As a result, the gravitational force that Earth exerts on you is greater than the force exerted by any other object.

Weight *is the gravitational force exerted on an object.* Near Earth's surface, an object's weight is the gravitational force exerted on the object by Earth. Because weight is a force, it is measured in newtons.

The Relationship Between Weight and Mass An object's weight is proportional to its mass. For example, if one object has twice the mass of another object, it also has twice the weight. ✓

Visual Check

4. Interpret How does the gravitational force between objects change if one object increases in mass but the other does not increase?

_____

_____

_____

Reading Check

5. Explain What effect does distance have on gravity?

_____

_____

_____

Reading Check

6. Describe What is the relationship between mass and weight?

_____

_____

_____

☑ Reading Check

7. Explain Why is the gravitational force that a friend exerts on you less than the gravitational force exerted on you by Earth?

_____

_____

_____

Weight and Mass High Above Earth Astronauts in orbit around Earth are not weightless. Their weight is about 90 percent of what it is on Earth. Why is there no <u>significant</u> change in weight when the distance increases so much? Earth is so large that an astronaut must be much farther away before the gravitational force will change much. The distance between the astronaut and Earth is small compared to the size of Earth. ☑

Friction

Friction *is a force that resists the motion of two surfaces that are touching.* There are several types of friction.

Static Friction

Static friction prevents surfaces from sliding past each other. The box on the left below does not move because the strength of the static friction is equal to the force a person is applying to the box. Up to a limit, the strength of static friction changes to match the applied force. If the person increases the applied force, the static friction will also increase. The box still will not move.

When static friction reaches its limit between the surfaces, the box will move. The box on the right below moves because the two people are pushing with greater force than the static friction between the box and the floor.

Sliding Friction

Sliding friction opposes the motion of surfaces sliding past each other. As long as the box is sliding, the sliding friction does not change. If the pushing force increases, the box will slide faster. If the two people stop pushing, sliding friction will cause the box to slow down and stop.

☑ Visual Check

8. Visualize Do frictional forces act in the same direction or in the opposite direction to the applied force?

_____

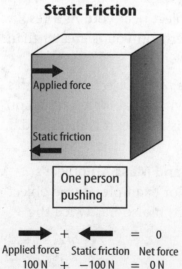

Static Friction

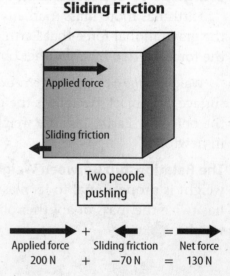

Sliding Friction

Fluid Friction

Fluid friction is friction between a surface and a fluid. A fluid is any material that flows. For example, water and air are fluids. Fluid friction between air and a surface is air resistance.

The amount of fluid friction on an object depends on the amount of surface area that faces the oncoming fluid. The greater the facing surface area, the greater the fluid friction is on the object. Imagine that you drop a crumpled paper and a flat paper. The crumpled paper will fall faster than the flat paper because the crumpled paper has less surface area facing the oncoming air. The crumpled paper has less air resistance than the flat paper.

What causes friction?

If you rub your hands together when they are soapy, they will slide past each other easily. If you rub your hands together when they are dry, you will feel more friction. Your dry hands will not slide past each other as easily as when they are soapy.

What causes friction between surfaces? Microscopic dips and bumps cover all surfaces. The dips and bumps on one surface catch on the dips and bumps on the other surface when they slide past each other. This microscopic roughness slows sliding and is a source of friction.

In addition, small particles—atoms and molecules—make up all surfaces. These particles contain weak electrical charges. When a positive charge on one surface slides by a negative charge on the other surface, there is attraction between the particles. This attraction slows sliding and is another source of friction between the surfaces. ✓

Reducing Friction

Soap acts as a lubricant. When you rub soapy hands together, the soapy water slightly separates the surfaces of your hands. This creates less contact between the microscopic dips and bumps and between the electrical charges of your hands. Friction decreases with less contact.

Like soap on your hands, motor oil also acts as a lubricant. It reduces friction between the moving parts of a car's engine. With less friction, surfaces can slide past each other more easily. Reducing an object's surface area also reduces the fluid friction between the object and the fluid.

Think it Over

9. Specify How can fluid friction be decreased?

_____

_____

_____

Key Concept Check

10. Describe How does friction affect the motion of two objects sliding past each other?

_____

_____

Reading Check

11. Identify What are two causes of friction?

_____

_____

_____

Mini Glossary

contact force: a push or a pull on one object by another that is touching it

force: a push or a pull on an object

friction: a force that resists the motion of two surfaces that are touching

gravity: an attractive force that exists between all objects that have mass

mass: the amount of matter in an object

noncontact force: a force that one object can apply to another object without touching it

weight: the gravitational force exerted on an object

1. Review the terms and their definitions in the Mini Glossary. Write a sentence comparing contact forces and noncontact forces.

_____

_____

_____

2. Label each of the diagrams below with the type of friction that is represented.

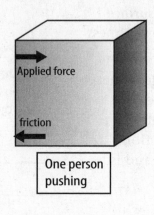

Applied force

friction

One person pushing

_____

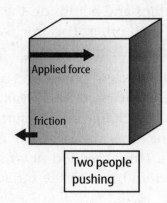

Applied force

friction

Two people pushing

_____

3. Give an example of something you could do to reduce friction.

_____

_____

What do you think NOW?

Reread the statements at the beginning of the lesson. Fill in the After column with an A if you agree with the statement or a D if you disagree. Did you change your mind?

ConnectED

Log on to ConnectED.mcgraw-hill.com and access your textbook to find this lesson's resources.

END OF LESSON

The Laws of Motion

Newton's First Law

··············· Before You Read ··············

What do you think? Read the two statements below and decide whether you agree or disagree with them. Place an A in the Before column if you agree with the statement or a D if you disagree. After you've read this lesson, reread the statements to see if you have changed your mind.

Before	Statement	After
	3. Forces acting on an object cannot be added.	
	4. A moving object will stop if no forces act on it.	

················ Read to Learn ················
Identifying Forces

Imagine a bird of prey that lives near a lake. It dives through the air at a high speed toward a fish swimming in the water. It moves its legs forward to grab the fish with its talons. It then uses its wings to climb high into the air. The bird then slows its speed and lands softly on the edge of the nest, near the young birds waiting for it.

Forces helped the bird change speed and direction. Recall that a force is a push or a pull. Some forces are contact forces, such as air resistance. Other forces are noncontact forces, such as gravity. When an object moves, it often has several different forces acting on it at the same time. To understand the motion of an object, you need to identify the forces acting on it. In this lesson you will read how forces change the motion of objects.

Combining Forces—The Net Force

Imagine that you are trying to move a piece of heavy furniture. If you push on it by yourself, you will have to push hard to get it to move. But if you ask a friend to push with you, you do not have to push as hard. When two or more forces act on an object, the forces combine. *The combination of all the forces acting on an object is the* **net force.** The way in which forces combine depends on the direction of the forces applied to the object.

Key Concepts

• What is Newton's first law of motion?

• How is motion related to balanced and unbalanced forces?

• What effect does inertia have on the motion of an object?

Mark the Text

Underline Main Ideas As you read, underline the main ideas under each heading. After you finish reading, review the main ideas that you have underlined.

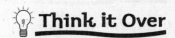 **Think it Over**

1. Analyze When a soccer ball is in flight, what forces are acting on it?

_____

_____

Combining Forces in the Same Direction

When the forces acting on an object are in the same direction, the net force is the sum of the individual forces. The direction of the net force is the same direction as the forces you add together.

Because forces have direction, you have to specify a reference direction when you combine forces. For example, if two people are pushing on a dresser from the left side, you would probably choose "to the right" as the reference direction. Both forces would then be positive. One person pushes the dresser with a force of 200 N to the right. The other person pushes with a force of 100 N to the right. To calculate the net force, add the two forces together, as shown below. The net force is 300 N to the right.

$$\longrightarrow + \longrightarrow = \longrightarrow \text{net force}$$
$$200\text{ N} \quad + \quad 100\text{ N} \quad = \quad 300\text{ N}$$

The force exerted on the dresser is the same as if one person pushed on the dresser with a force of 300 N to the right. ✓

Combining Forces in Opposite Directions

When forces act in opposite directions on an object, the net force is still the sum of the forces. Imagine that two people push on a dresser in opposite directions. One pushes with a force of 200 N to the right. The other person pushes with a force of 100 N to the left. You choose "to the right" as the reference direction. A force in that direction, then, is positive. A force in the opposite direction is negative. The net force is the sum of the positive and negative forces, as shown below. The net force is 100 N to the right.

$$\longrightarrow + \longleftarrow = \longrightarrow \text{net force}$$
$$200\text{ N} \quad + \quad -100\text{ N} \quad = \quad 100\text{ N}$$

Balanced and Unbalanced Forces

If two people push in opposite directions, but with the same amount of force, the net force on the object is zero. The effect is the same as if there were no forces at all acting on the object. *Forces acting on an object that combine and form a net force of zero are* **balanced forces.** As shown in the figure on the opposite page, balanced forces do not change the motion of an object. *Forces acting on an object that combine and form a net force that is not zero are* **unbalanced forces.**

REVIEW VOCABULARY

reference direction
a direction that you choose from a starting point to describe an object's position

✓ Reading Check

2. Describe How do you calculate the net force on an object if two forces are acting on it in the same direction?

_____

_____

💡 Think it Over

3. Interpret What does a negative number in the net-force equation mean?

_____

_____

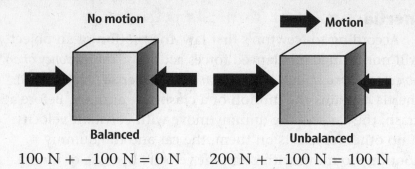

No motion   Motion

Balanced   Unbalanced

$100\ N + {-100}\ N = 0\ N$ $200\ N + {-100}\ N = 100\ N$

Newton's First Law of Motion

Sir Isaac Newton studied how forces affect the motion of objects. He developed three rules that are known as Newton's laws of motion. *According to* **Newton's first law of motion,** *if the net force on an object is zero, the motion of the object does not change.* As a result, balanced forces and unbalanced forces have different results when they act on an object. 🔑

Balanced Forces and Motion

According to Newton's first law of motion, balanced forces cause no change in the motion of an object. This is true when an object is at rest or in motion. A dresser is at rest before anyone pushes on it. It remains at rest when balanced forces are applied.

Both gravity and air resistance act on the motion of a parachutist. Gravity pulls the parachutist toward the ground. Air resistance against the parachute slows the fall. When the air resistance and gravity on the parachutist are balanced, the parachutist moves downward with a constant velocity known as terminal velocity. Recall that velocity is the speed and the direction of motion. Terminal velocity is the constant velocity reached when air resistance equals the force of gravity on a falling object.

Unbalanced Forces and Motion

Newton's first law of motion applies only to balanced forces acting on an object. When unbalanced forces act on an object, the object's velocity changes. If an object is at rest, unbalanced forces cause it to start moving. If an object is already moving, unbalanced forces cause its velocity to change. 🔑

✔️ **Visual Check**
4. Evaluate Why doesn't the box on the left move?

_____

_____

🔑 **Key Concept Check**
5. Define What is Newton's first law of motion?

_____

_____

_____

🔑 **Key Concept Check**
6. Explain How is motion related to balanced and unbalanced forces?

_____

_____

_____

Inertia

According to Newton's first law, the motion of an object will not change if balanced forces act on it. *The tendency of an object to resist a change in its motion is called* **inertia** (ihn UR shuh). Inertia explains the motion of a crash-test dummy. Before a crash, the car and the dummy move with constant velocity. If no other force acts on them, the car and the dummy continue to move with constant velocity because of inertia. The car crashing into a wall results in an unbalanced force on the car, and the car stops. The dummy continues moving forward because of its inertia.

Why do objects stop moving?

Think about how friction and inertia together affect an object's movement. A book sitting on a table, for example, stays in place because of inertia. When you push the book, the force you apply to the book is greater than static friction between the book and the table. The book moves in the direction of the greater force. If you stop pushing, friction stops the book.

If there were no friction between the book and the table, inertia would keep the book moving. According to Newton's first law, the book would continue to move at the same speed in the same direction as your push.

On Earth, friction can be reduced but not totally removed. For an object to start moving, a force greater than static friction must be applied to it. To keep the object in motion, a force at least as strong as friction must be continuously applied. Objects stop moving because friction or another force acts on them.

Key Concept Check

7. Summarize What effect does inertia have on the motion of an object?

_____

_____

Think it Over

8. Apply Why does a rolling ball eventually stop rolling?

_____

_____

_____

FOLDABLES

Make a chart with six columns and six rows to define and show how this lesson's vocabulary words are related.

	Net Force	Balanced Forces	Unbalanced Forces	Newton's First Law	Inertia
Net Force					
Balanced Forces					
Unbalanced Forces					
Newton's First Law					
Inertia					

Mini Glossary

balanced forces: forces acting on an object that combine and form a net force of zero

inertia (ihn UR shuh): the tendency of an object to resist a change in its motion

net force: the combination of all the forces acting on an object

Newton's first law of motion: the law that states that if the net force on an object is zero, the motion of the object does not change

unbalanced forces: forces acting on an object that combine and form a net force that is not zero

1. Review the terms and their definitions in the Mini Glossary. Write a sentence that explains how balanced forces affect an object at rest.

_____

_____

_____

2. Use the information in the diagram to complete the equation and determine the net force on the object. The reference direction is "to the right."

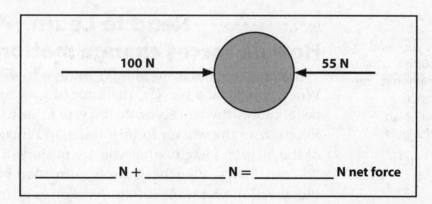

100 N   55 N

_____ N + _____ N = _____ N net force

3. How did underlining the main ideas help you review Newton's first law of motion?

_____

_____

_____

What do you think NOW?

Reread the statements at the beginning of the lesson. Fill in the After column with an A if you agree with the statement or a D if you disagree. Did you change your mind?

Log on to ConnectED.mcgraw-hill.com and access your textbook to find this lesson's resources.

END OF LESSON

The Laws of Motion

Newton's Second Law

Key Concepts

- What is Newton's second law of motion?
- How does centripetal force affect circular motion?

Copyright © Glencoe/McGraw-Hill, a division of The McGraw-Hill Companies, Inc.

·············· **Before You Read** ··············

What do you think? Read the two statements below and decide whether you agree or disagree with them. Place an A in the Before column if you agree with the statement or a D if you disagree. After you've read this lesson, reread the statements to see if you have changed your mind.

Before	Statement	After
	5. When an object's speed increases, the object accelerates.	
	6. If an object's mass increases, its acceleration also increases if the net force acting on the object stays the same.	

Study Coach

Create a Quiz As you read this lesson, write quiz questions based on what you have read under each heading. After you finish reading, answer the quiz questions.

·············· **Read to Learn** ··············

How do forces change motion?

Forces can change an object's motion in different ways. When you pedal a bicycle, the force of your foot on the pedal causes the wheels of the bicycle to turn. Increasing the force causes the wheels to turn faster and increases the speed of the bicycle. Imagine that you are pushing a wheelbarrow. You can change the wheelbarrow's direction by pushing it in the direction you want it to move. Forces change an object's motion by changing its speed of motion, its direction of motion, or both its speed and its direction of motion.

Unbalanced Forces and Velocity

Velocity is the speed of an object in a certain direction. Only unbalanced forces change an object's velocity. A bicycle's speed will not increase unless the force of the person's foot on the pedal is greater than friction that slows the wheels. If someone pushes the wheelbarrow with the same force but in the opposite direction that you are pushing, the wheelbarrow's direction will not change. In this lesson, you will read about how unbalanced forces affect the velocity of an object.

💡 Think it Over

1. Identify Name three ways that forces can change the motion of an object.

_____

_____

_____

Unbalanced Forces on an Object at Rest

Unbalanced forces affect an object at rest. If you hold a ball in your hand, the ball does not move. Your hand holds the ball up against the downward pull of gravity. The forces acting on the ball are balanced. When your hand moves out of the way, the ball falls. You know that the forces on the ball are now unbalanced because the ball's motion changed. The ball moves in the direction of the net force. When unbalanced forces act on an object at rest, the object begins moving in the direction of the net force.

Unbalanced Forces on an Object in Motion

Unbalanced forces change the velocity of an object that is moving. Recall that one way to change an object's velocity is to change its speed.

Speeding Up If a net force acts on a moving object in the direction that the object is moving, the object will speed up. For example, imagine that you are pushing someone on a sled. If you push in the direction that the sled is already moving, the sled will speed up.

Slowing Down If the direction of the net force on an object is opposite to the direction the object is moving, the object will slow down. If you are riding on a sled and push your foot against the ground, friction acts in the direction opposite to the motion of the sled. Because the net force is in the direction opposite to the sled's motion, the sled's speed decreases.

Changes in Direction of Motion

Unbalanced forces can also change an object's velocity by changing its direction. The ball shown in the figure moves at a constant velocity until it hits the tree. The tree exerts a force on the ball, which makes the ball change direction.

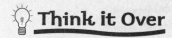

2. Predict What will happen to an object at rest if unbalanced forces act upon it?

_____

_____

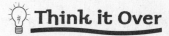 **Think it Over**

3. Apply What will happen to the speed of a wagon rolling to the right if a net force pushes it to the right?

_____

_____

✔ **Visual Check**

4. Interpret Why does the velocity of the ball change when it hits the tree?

_____

_____

_____

_____

Make a half-book to organize your notes on Newton's second law.

Key Concept Check

5. Define What is Newton's second law of motion?

_____

_____

_____

Math Skills ✕ ÷

You throw a 0.5-kg basketball with a force of 10 N. What is the acceleration of the ball?

mass: $m = 0.5$ kg

force: $F = 10$ N
or 10 kg·m/s^2

acceleration: a

Use this formula: $a = \frac{F}{m}$

Substitute the values for F and m into the formula and divide:

$$a = \frac{10 \text{ N}}{0.5 \text{ kg}} = \frac{20 \text{ kg·m/s}^2}{\text{kg}} = 20 \text{ m/s}^2$$

Acceleration = 20 m/s^2

6. Solve for Force

A 24-N net force acts on an 8-kg rock. What is the acceleration of the rock?

_____

Unbalanced Forces and Acceleration

You have read how unbalanced forces can change an object's velocity by changing its speed, its direction, or both. Recall that another name for a change in velocity over time is acceleration. When you push a sled forward, the sled accelerates because its speed changes. When the soccer ball in the figure hit the tree, the ball accelerated because its direction changed. Unbalanced forces can make an object accelerate by changing its speed, its direction, or both.

Newton's Second Law of Motion

Newton's second law of motion describes the relationship between an object's acceleration and the net force that acts on the object. *According to* **Newton's second law of motion,** *the acceleration of an object is equal to the net force acting on the object divided by the object's mass.* The direction of acceleration is the same as the direction of the net force. 🔑

Newton's Second Law Equation

$$\text{acceleration (in m/s}^2) = \frac{\text{net force (in N)}}{\text{mass (in kg)}}$$

$$a = \frac{F}{m}$$

SI units are included in the equation. Acceleration is expressed in meters per second squared (m/s^2), mass in kilograms (kg), and force in newtons (N). From this equation, it follows that a newton is the same as kg·m/s^2.

Circular Motion

Newton's second law of motion describes the relationship between an object's change in velocity over time, or acceleration, and unbalanced forces acting on the object. You learned how this relationship applies to motion along a line. **Circular motion** *is any motion in which an object is moving along a curved path.* Velocity and acceleration also apply to centripetal force.

Centripetal Force

You can tie a string to a ball and swing it around above your head. The ball has a tendency to move along a straight path. Inertia—not a force—causes this motion. The ball's path is curved, however, because the string pulls the ball inward. *In circular motion, a force that acts perpendicular to the direction of motion, toward the center of the curve, is* **centripetal** (sen TRIH puh tuhl) **force.** The ball accelerates in the direction of the centripetal force.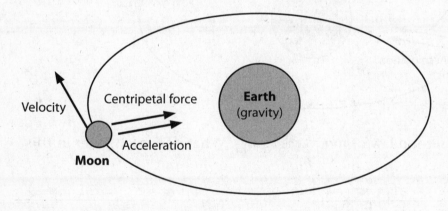

The Motion of Satellites and Planets

A satellite is another object that is acted on by centripetal force. A satellite is any object in space that orbits a larger object. Like the ball described above, a satellite tends to move in a straight path because of inertia. But just as the string pulls the ball inward, gravity pulls a satellite inward.

Gravity is the centripetal force that keeps a satellite in orbit by changing its direction. Look at the figure of Earth and the Moon below. The Moon is a satellite of Earth. Earth's gravity changes the Moon's direction. The inertia of the Moon and Earth's gravity determine the circular motion of the Moon's orbit around Earth. Similarly, the Sun's gravity changes the direction of its satellites, including Earth.

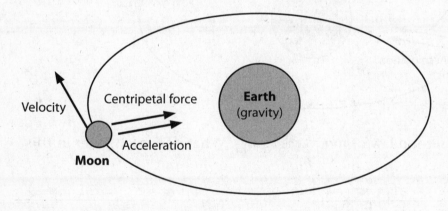

✔ Key Concept Check

7. Describe How does centripetal force affect circular motion?

_____

_____

_____

💡 Think it Over

8. Explain What prevents Earth from leaving its orbit and flying out into space?

_____

_____

_____

✔ Visual Check

9. Discover How does the direction of the velocity of a satellite differ from the direction of its acceleration?

_____

_____

_____

Mini Glossary

centripetal (sen TRIH puh tuhl) force: in circular motion, a force that acts perpendicular to the direction of motion, toward the center of the curve

circular motion: any motion in which an object is moving along a curved path

Newton's second law of motion: the law that states that the acceleration of an object is equal to the net force acting on the object divided by the object's mass

1. Review the terms and their definitions in the Mini Glossary. Write a sentence that describes how centripetal force affects circular motion.

_____

_____

2. Identify the force that keeps the Moon in orbit around Earth.

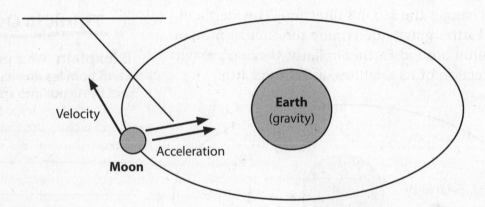

Velocity

Acceleration

Moon

Earth (gravity)

3. The equation for Newton's second law of motion is $a = \frac{F}{m}$. What does each letter in this formula stand for?

_____

_____

What do you think NOW?

Reread the statements at the beginning of the lesson. Fill in the After column with an A if you agree with the statement or a D if you disagree. Did you change your mind?

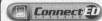

Log on to ConnectED.mcgraw-hill.com and access your textbook to find this lesson's resources.

END OF LESSON

The Laws of Motion

Newton's Third Law

········· **Before You Read** ·············

Before	Statement	After
	What do you think? Read the two statements below and decide whether you agree or disagree with them. Place an A in the Before column if you agree with the statement or a D if you disagree. After you've read this lesson, reread the statements to see if you have changed your mind.	
	7. If objects collide, the object with more mass applies more force.	
	8. Momentum is a measure of how hard it is to stop a moving object.	

············ **Read to Learn** ················

Opposite Forces

If you are wearing skates and push against a wall, you will move away from the wall. What force causes you to move? You might think that the force of your muscles moves you away from the wall. Think about the direction of your push. Your push is against the wall in the opposite direction from your movement. In fact, when you push against the wall, the wall pushes back in the opposite direction. The push of the wall causes you to accelerate away from the wall. When an object applies a force on another object, the second object applies a force of the same strength on the first object, but the force is in the opposite direction.

Newton's Third Law of Motion

Newton's first two laws of motion describe the effects of balanced and unbalanced forces on one object. Newton's third law relates forces between two objects. *According to* **Newton's third law of motion,** *when one object exerts a force on a second object, the second object exerts a force of the same size but in the opposite direction on the first object.* An example of forces described in Newton's third law of motion is a gymnast pushing against the floor during a flip. When the gymnast applies force against the floor, the floor applies force back.

Key Concepts 🔑

- What is Newton's third law of motion?
- Why don't the forces in a force pair cancel each other?
- What is the law of conservation of momentum?

Study Coach

Outline Main Ideas As you read, make an outline to summarize the information in the lesson. Use the main headings in the lesson as the main headings in the outline. Complete the outline with the information under each heading. Review the outline to help you learn the material in this lesson.

🔑 **Key Concept Check**
1. Define What is Newton's third law of motion?

_____

_____

_____

Force Pairs

The forces described by Newton's third law depend on each other. *A* **force pair** *is the forces two objects apply to each other.* Recall that you can add forces to calculate the net force. If the forces of a force pair always act in opposite directions and are always the same strength, why don't they cancel each other? The reason is that each force acts on a different object. Adding forces results in a net force of zero only if the forces act on the same object.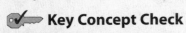

Action and Reaction

In a force pair, one force is the action force and the other is the reaction force. Swimmers diving from a boat apply an action force against the boat. The boat applies a reaction force on the swimmers. For every action force, there is a reaction force of equal strength but in the opposite direction.

Using Newton's Third Law of Motion

When you push against an object, the force you apply is the action force. The object then pushes back against you. The force applied by the object is the reaction force. According to Newton's second law, when the reaction force results in an unbalanced force, there is a net force, and the object accelerates. Newton's third law explains many common activities, such as those described in the table.

Key Concept Check

2. Explain Why don't the forces in a force pair cancel each other?

_____

_____

_____

FOLDABLES®

Make a half-book to summarize how Newton's third law explains the motion of a variety of common activities.

Using Newton's Third Law

Action and Reaction Forces		
Swimming When you push your arms against the water to swim, the water pushes back in the opposite (forward) direction. If you push with enough force, the water's reaction force becomes greater than the force of fluid friction. You accelerate in the direction of the net force and swim forward.	**Jumping** When you jump, you push down on the ground, and the ground pushes up on you. The upward force of the ground combines with the downward force of gravity to form the net force acting on you. If you push down hard enough, the upward force becomes greater than the downward force of gravity. The net force is upward, and you accelerate in the direction of the net force.	**Rocket Motion** The burning fuel in a rocket engine produces a hot gas. The engine pushes the hot gas out in a downward direction. The gas pushes upward on the engine. When the upward force of the gas pushing on the engine becomes greater than the downward force of gravity on the rocket, the net force is upward. The rocket then accelerates upward.

Interpreting Tables

3. Specify On what part of a swimmer's body does the water's reaction force push?

_____

_____

Momentum

Because action and reaction forces do not cancel each other, they can change the motion of objects. **Momentum** *is a measure of how hard it is to stop a moving object.* It is the product of an object's mass and velocity. The momentum equation appears on the top of the opposite page. An object's momentum is in the same direction as its velocity.

Momentum Equation

momentum (in kg·m/s) = mass (in kg) × velocity (in m/s)

$$p = m \times v$$

Momentum and Mass If a large truck and a car are moving at the same speed, the truck is harder to stop. Because the truck has more mass, it has more momentum. If cars of equal mass move at different speeds, the faster car has more momentum and is more difficult to stop.

Newton's Laws and Momentum According to Newton's first law, if the net force on an object is zero, its velocity does not change. This means its momentum does not change. Newton's second law states that the net force on an object is the product of its mass and its change in velocity. Because momentum is the product of mass and velocity, the force on an object equals its change in momentum.

Conservation of Momentum

In a game of billiards, when the moving cue ball hits a ball that is not moving, the motion of both balls changes. The cue ball has momentum because it has mass and is moving. When it hits the other ball, the cue ball's velocity and momentum decrease. The other ball starts moving. Because this ball then has mass and velocity, it also has momentum.

The Law of Conservation of Momentum

In any collision, momentum transfers from one object to another. The billiard ball gains the momentum lost by the cue ball. The total momentum, however, does not change. *According to the* **law of conservation of momentum,** *the total momentum of a group of objects stays the same unless outside forces act on the objects.* Outside forces include friction. Friction between the balls and the billiard table decreases their velocities, and they lose momentum. 🔑

Types of Collisions

Objects collide with each other in different ways. When colliding objects bounce off each other, an elastic collision occurs. If objects collide and stick together, such as when one football player tackles another, the collision is inelastic. No matter the type of collision, the total momentum will be the same before and after the collision.

What is the momentum of a 12-kg bicycle moving at 5.5 m/s?

mass: $m = 12$ kg

velocity: $v = 5.5$ m/s

momentum: p

Use this formula:

$$p = m \times v$$

Substitute the values for m and v into the formula and multiply:

$p = 12$ kg $\times 5.5$ m/s
$= 66$ kg·m/s

Momentum = 66 kg·m/s in the direction of the velocity.

4. Solve for Momentum
What is the momentum of a 1.5-kg ball rolling at 3.0 m/s?

_____

_____

🔑 **Key Concept Check**
5. Define What is the law of conservation of momentum?

_____

_____

_____

Mini Glossary

force pair: the forces two objects apply to each other

law of conservation of momentum: the law that states that the total momentum of a group of objects stays the same unless outside forces act on the objects

momentum: a measure of how hard it is to stop a moving object

Newton's third law of motion: the law that states that when one object exerts a force on a second object, the second object exerts a force of the same size but in the opposite direction on the first object

1. Review the terms and their definitions in the Mini Glossary. Write a sentence that summarizes Newton's third law of motion in your own words.

_____

_____

2. Circle the diagram below that shows an example of an inelastic collision.

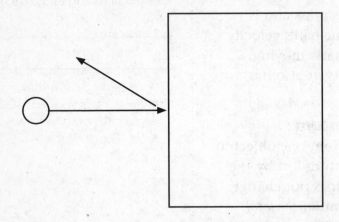 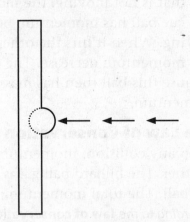

3. If a tennis ball and a bowling ball are rolling at the same speed, which ball is harder to stop? Explain why.

_____

_____

_____

What do you think NOW?

Reread the statements at the beginning of the lesson. Fill in the After column with an A if you agree with the statement or a D if you disagree. Did you change your mind?

 **Connect ED**

Log on to ConnectED.mcgraw-hill.com and access your textbook to find this lesson's resources.

END OF LESSON

Energy, Work, and Simple Machines

Types of Energy

·············· Before You Read ··············

What do you think? Read the two statements below and decide whether you agree or disagree with them. Place an A in the Before column if you agree with the statement or a D if you disagree. After you've read this lesson, reread the statements to see if you have changed your mind.

Before	Statement	After
	1. Energy is the ability to produce motion.	
	2. Waves transfer energy from place to place.	

·············· Read to Learn ··············

What is energy?

You probably have heard the word *energy* used on television, the radio, or the Internet. Commercials claim that the newest models of cars are energy efficient. What is energy? Scientists define **energy** as *the ability to cause a change.*

Using this definition, what does energy have to do with cars? Most cars use some type of fuel, such as gasoline or diesel, as their energy source. A car's engine transforms the energy stored in the fuel to a form of energy that moves the car. Compared to other cars, the car that is energy efficient uses less fuel to make the car move a certain distance.

Gasoline and diesel fuel are not the only sources of energy. Food is an energy source for your body. Also, satellites need a source of energy to run their systems and to stay in orbit. Solar panels provide energy for the *International Space Station*. As you will read, wind, coal, nuclear fuel, Earth's interior, and the Sun also are sources of energy.

Energy from each of these sources can be transformed into other forms of energy, such as electric energy. Every time you turn on a light, you use energy that was transformed from one form to another.

Key Concepts

- What is energy?
- What are the different forms of energy?
- How is energy used?

Study Coach

Create a Quiz Create a quiz about types of energy. Exchange quizzes with a partner. After taking the quizzes, discuss your answers. Read more about the topics you don't understand.

Key Concept Check

1. Define What is energy?

_____

_____

_____

✓ **Reading Check**

2. Assess What is one drawback of wind energy?

_____

_____

_____

✓ **Reading Check**

3. Explain What creates an electric current?

_____

_____

_____

Kinetic Energy

You just turned the page of this book. As the page was moving, it had **kinetic energy**—*the energy an object has because it is in motion.* Anything that is in motion has kinetic energy, including large objects that you can see as well as small particles, such as molecules, ions, atoms, and electrons.

Kinetic Energy of Objects

Huge wind turbines on wind farms use wind as a source of energy. When the wind blows, the large blades of the wind turbines turn. Because the blades are moving, they have kinetic energy. Kinetic energy depends on mass. If the turbine blades were smaller and had less mass, they would have less kinetic energy.

Kinetic energy also depends on speed. When the wind blows harder, the blades move faster. The faster the blades move, the more kinetic energy they have. When the wind stops, the blades stop. When the blades are not moving, the kinetic energy of the blades is zero.

One of the drawbacks of using wind-generated energy is that wind does not always blow. As a result, wind provides an inconsistent supply of energy. ✓

Electric Energy

When you turn on a lamp or use a cell phone, you are using a type of kinetic energy—electric energy. Recall that all objects are composed of atoms. Electrons move around the nucleus of an atom, and they move from one atom to another. When electrons move, they have kinetic energy and create an electric current. *The energy that an electric current carries is a form of kinetic energy called* **electric energy.** ✓

Electric energy can be produced by moving objects. When the blades of the wind turbines rotate, they turn a generator. The generator changes the kinetic energy of the moving blades into electric energy. Electric energy generated from the kinetic energy of wind is a clean source of energy because it creates no waste products.

Potential Energy

Suppose you hold up a piece of paper. When the paper is held above the ground, it has potential energy. **Potential energy** *is stored energy that depends on the interaction of objects, particles, or atoms.*

Gravitational Potential Energy

Gravitational potential energy is a type of potential energy stored in an object due to its height above Earth's surface. The water at the top of a dam has gravitational potential energy because it is higher above the surface of Earth than the water at the base of the dam. Gravitational potential energy depends on the mass of an object and its distance from Earth's surface. The more mass an object has and the greater its distance from Earth, the greater its gravitational potential energy.

In a hydroelectric energy plant, water above a dam flows through turbines as it falls. Generators connected to the spinning turbines convert the gravitational potential energy of the water into electric energy. Hydroelectric energy plants are a very clean source of energy. About 7 percent of all electric power in the United States is produced from hydroelectric energy. However, hydroelectric plants can interrupt the movement of animals in streams and rivers.

Chemical Energy

Most electric energy in the United States comes from fossil fuels such as petroleum, natural gas, and coal. Chemical bonds join the atoms that make up these fossil fuels. Chemical bonds have the potential to break apart. Therefore, chemical bonds have a form of potential energy called chemical energy. **Chemical energy** *is energy that is stored in and released from the bonds between atoms.*

When an energy plant burns fossil fuels, the chemical bonds between the atoms break apart. When this happens, chemical energy transforms to thermal energy. The plant uses this thermal energy to heat water and form steam. The steam turns a turbine, which is connected to a generator that generates electric energy. ✓

A drawback of fossil fuels is that they introduce harmful waste products, such as sulfur dioxide and carbon dioxide, into the environment. Sulfur dioxide in the air creates acid rain. Most scientists suspect that increased levels of carbon dioxide in the atmosphere contribute to climate change. Scientists are searching for replacement fuels.

Fossil fuels are not the only source of chemical energy. Chemical energy also is stored in the foods you eat. Your body converts the energy stored in chemical bonds in food into the kinetic energy of your moving muscles and into the electric energy that sends signals through your nerves to your brain.

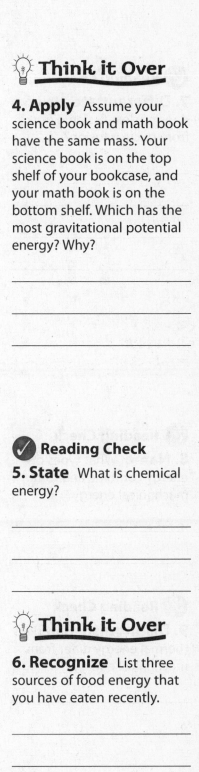

💡 Think it Over

4. Apply Assume your science book and math book have the same mass. Your science book is on the top shelf of your bookcase, and your math book is on the bottom shelf. Which has the most gravitational potential energy? Why?

_____

_____

✓ Reading Check

5. State What is chemical energy?

_____

_____

_____

💡 Think it Over

6. Recognize List three sources of food energy that you have eaten recently.

_____

_____

Nuclear Energy

Most energy on Earth comes from the Sun. A process in the Sun called nuclear fusion joins the nuclei of atoms. Nuclear fusion releases large amounts of energy.

On Earth, nuclear energy plants use the potential energy stored in the nuclei of atoms to produce electric energy. In these plants, the nuclei of certain atoms break apart using a process called nuclear fission. Both nuclear fusion and nuclear fission release **nuclear energy**— *energy stored in and released from the nucleus of an atom.* ✓

Nuclear fission produces a large amount of energy from just a small amount of fuel. However, the process produces radioactive waste that is hazardous and difficult to dispose of safely.

Kinetic and Potential Energies Combined

Recall that a moving object has kinetic energy. Objects such as wind turbine blades and particles, such as molecules, ions, atoms, and electrons, often have kinetic and potential energies.

Mechanical Energy

The sum of potential energy and kinetic energy in a system of objects is **mechanical energy.** Mechanical energy is the energy a system has because of the movement of its parts (kinetic energy) and because of the position of its parts (potential energy). ✓

An object, such as a wind turbine, has mechanical energy because the parts that make up the system have both potential energy and kinetic energy. A rotating blade has kinetic energy because of its motion. It also has gravitational potential energy because of its distance from Earth's surface.

Thermal Energy

The particles that make a wind turbine also have thermal energy. **Thermal energy** *is the sum of the kinetic energy and potential energy of the particles that make up an object.* Although you cannot see the individual particles move, they vibrate back and forth in place. This movement gives the particles kinetic energy. The particles also have potential energy because of the distance between particles and the charge of the particles. ✓

✓ **Reading Check**

7. Differentiate How does nuclear fission differ from nuclear fusion?

_____

_____

_____

_____

✓ **Reading Check**

8. Name What types of energy combine and form mechanical energy?

_____

_____

✓ **Reading Check**

9. Distinguish How does thermal energy differ from mechanical energy?

_____

_____

_____

Geothermal Energy

The particles in Earth's interior contain great amounts of thermal energy. This energy is called geothermal energy. Geothermal energy plants use this thermal energy to heat water and turn it to steam. The steam turns turbines in electric generators, converting the geothermal energy to electric energy. Geothermal energy produces almost no pollution. However, geothermal plants must be built in places where molten rock is close to Earth's surface. ✔

Energy from Waves

Have you ever seen waves crash on a beach? When a big wave crashes, you hear the sound of the impact. The movement and the sound result from the energy carried by the wave. Waves are disturbances that carry energy from one place to another. Waves move only energy, not matter.

Sound Energy

Clapping your hands together creates a sound wave in the air. Sound waves move through matter. **Sound energy** *is energy carried by sound waves.* Some animals, such as bats, emit sound waves to find their prey. The length of time it takes sound waves to travel to a bat's prey and then echo back tells the bat the location of its prey.

Seismic Energy

You probably have seen news reports showing photographs of damage caused by earthquakes. Earthquakes occur when Earth's tectonic plates, or large portions of Earth's crust, suddenly shift position. The kinetic energy of the plate movement is carried through the ground by seismic waves. **Seismic energy** *is the energy transferred by waves moving through the ground.* Seismic energy can destroy buildings and roads.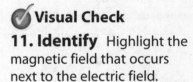

Radiant Energy

When you listen to the radio, use a lamp, or talk on your cell phone, do you think of waves? Electromagnetic waves are electric and magnetic waves that move perpendicular to each other, as shown in the figure at right.

Electromagnetic Waves

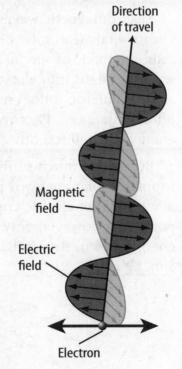

Direction of travel

Magnetic field

Electric field

Electron

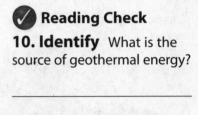

Reading Check

10. Identify What is the source of geothermal energy?

_____

_____

_____

Visual Check

11. Identify Highlight the magnetic field that occurs next to the electric field.

Key Concept Check

12. Name What are the different forms of energy?

_____

_____

_____

_____

_____

Forms of Electromagnetic Waves

Visible light waves

Radio waves

Microwaves

Solar cell (transforms radiant energy to electric energy)

Infrared waves (used by computer to read compact disc)

Infrared waves (thermal energy emitted from body as waves)

Microwaves

Visual Check

13. Name What type of electromagnetic waves does a cell phone use?

_____

_____

Key Concept Check

14. Summarize How is radiant energy used?

_____

_____

_____

_____

_____

Radio waves, light waves, and microwaves are all electromagnetic waves, as shown in the figure above. Some electromagnetic waves can travel through solids, liquids, gases, and vacuums. _The energy carried by electromagnetic waves is_ **radiant energy.**

Electromagnetic waves transmit the Sun's energy to Earth. Photovoltaic (foh toh vohl TAY ihk) cells, also called solar cells, are made of special material that transforms the radiant energy of light into electric energy. You might have used a solar calculator. It does not need batteries because it has a photovoltaic cell. Photovoltaic cells also are used to provide energy to satellites, offices, and homes.

Because so much sunlight hits the surface of Earth, the supply of solar energy is plentiful. Also, using solar energy as a source for electric energy produces almost no waste or pollution. However, only about 0.1 percent of the electric energy used in the United States comes directly from the Sun.

Mini Glossary

chemical energy: energy that is stored in and released from the bonds between atoms

electric energy: a form of kinetic energy that an electric current carries

energy: the ability to cause a change

kinetic energy: the energy an object has because it is in motion

mechanical energy: the sum of potential energy and kinetic energy in a system of objects

nuclear energy: energy stored in and released from the nucleus of an atom

potential energy: stored energy that depends on the interaction of objects, particles, or atoms

radiant energy: the energy carried by electromagnetic waves

seismic energy: the energy transferred by waves moving through the ground

sound energy: energy carried by sound waves

thermal energy: the sum of the kinetic energy and potential energy of the particles that make up an object

1. Review the terms and their definitions in the Mini Glossary. Write a sentence describing the source of seismic energy.

_____

2. Write the type of energy next to each clue in the graphic organizer.

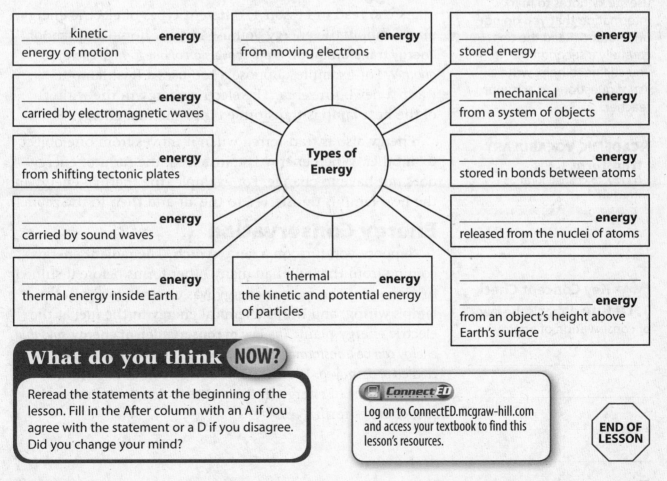

kinetic **energy**	**energy**	**energy**
energy of motion	from moving electrons	stored energy
energy	mechanical **energy**
carried by electromagnetic waves	from a system of objects
energy	**energy**
from shifting tectonic plates	stored in bonds between atoms
energy	**energy**
carried by sound waves	released from the nuclei of atoms
energy	thermal **energy**	**energy**
thermal energy inside Earth	the kinetic and potential energy of particles	from an object's height above Earth's surface

Types of Energy

What do you think NOW?

Reread the statements at the beginning of the lesson. Fill in the After column with an A if you agree with the statement or a D if you disagree. Did you change your mind?

ConnectED

Log on to ConnectED.mcgraw-hill.com and access your textbook to find this lesson's resources.

END OF LESSON

Energy, Work, and Simple Machines

Energy Transformations and Work

Copyright © Glencoe/McGraw-Hill, a division of The McGraw-Hill Companies, Inc.

Key Concepts 🔑

- What is the law of conservation of energy?
- In what ways can energy be transformed?
- How are energy and work related?

Mark the Text ▶

Sticky Notes As you read, use sticky notes to mark information that you do not understand. Read the text carefully a second time. If you still need help, write a list of questions to ask your teacher.

ACADEMIC VOCABULARY

transform

(verb) to change form or structure

🔑 **Key Concept Check**

1. Define What is the law of conservation of energy?

_____

_____

_____

·············· **Before You Read** ··············

What do you think? Read the two statements below and decide whether you agree or disagree with them. Place an A in the Before column if you agree with the statement or a D if you disagree. After you've read this lesson, reread the statements to see if you have changed your mind.

Before	Statement	After
	3. Energy cannot be created or destroyed, but it can be transformed.	
	4. Work describes how much energy it takes for a force to push or to pull an object.	

·············· **Read to Learn** ··············

Energy Transformations

As you read in Lesson 1, different types of electric energy plants supply the energy you use in your home and school. **Energy transformation** *is the conversion of one form of energy to another.* For example, suppose a zoo uses a heat lamp to warm a newborn zebra. The electric energy in the wiring of the heat lamp is <u>transformed</u> into thermal energy.

Energy also is transferred when it moves from one object to another. When energy is transferred, the form of energy does not have to change. For example, the thermal energy in the heat lamp is transferred to the air and then to the zebra.

Energy Conservation

Suppose you turn on a lamp switch. The radiant energy coming from the bulb had many other forms before it shined in your eyes. The radiant energy was electric energy in the lamp's wiring, and it was chemical energy in the fuel at the electric energy plant. *The* **law of conservation of energy** *says that energy can be transformed from one form to another, but it cannot be created or destroyed.* Even though energy can change forms, the total amount of energy in the universe does not change. It just changes form. 🔑

Roller Coasters

Have you ever thought about the energy transformations that occur on a roller coaster? Most roller coasters first pull your car to the top of a big hill. At this point, the distance between you and Earth increases. Your potential energy also increases. Next, your car races down the hill. You move faster and faster. The gravitational potential energy is transformed to kinetic energy. At the bottom of the hill, your gravitational potential energy is small, but you have a lot of kinetic energy. This kinetic energy is transformed back to gravitational potential energy as your car moves up the next hill. ✔

Plants and the Body

When a plant, such as broccoli, carries on photosynthesis, it transforms radiant energy from the Sun into chemical energy. The chemical energy is stored in the bonds of the plant's molecules. When you eat the broccoli, your body breaks apart the chemical bonds in the molecules that make up the broccoli. This releases chemical energy. Your body transforms the chemical energy to energy your body needs for movement, temperature control, and other life processes.

Electric Energy Power Plants

Plants that lived millions of years ago carried on photosynthesis just like the plants that are alive today do. These ancient plants stored radiant energy from the Sun as chemical energy in their molecular bonds. After the plants died, they were buried under sediment. After much time, pressure from the sediments that covered the plant remains turned the plants into fossil fuels.

When fossil fuels burn, the chemical energy, which was stored in the molecules of plants that lived millions of years ago, transforms. At electric energy power plants, that chemical energy is transformed to the electric energy that you use in your home and school. As you read in Lesson 1, other forms of energy also can be transformed at electric energy plants. At these facilities, forms of energy—solar, wind, geothermal, and hydroelectric energy—are transformed to electric energy. 🗝

Energy and Work

When you study for a test, it might feel like a lot of work. But science does not define studying as work. **Work** *is the transfer of energy that occurs when a force makes an object move in the direction of the force while the force acts on the object.*

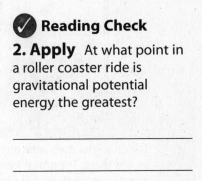

✔ Reading Check

2. Apply At what point in a roller coaster ride is gravitational potential energy the greatest?

_____

_____

FOLDABLES

Create a vertical half-book to summarize, in your own words, the relationship between work and energy.

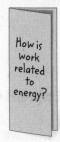

How is work related to energy?

🗝 Key Concept Check

3. Identify three energy transformations that occur to make electric energy.

_____

_____

_____

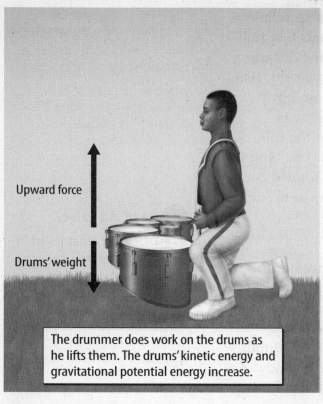

Upward force

Drums' weight

The drummer does work on the drums as he lifts them. The drums' kinetic energy and gravitational potential energy increase.

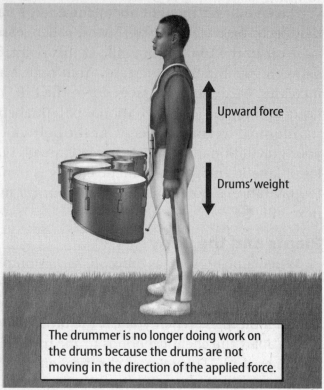

Upward force

Drums' weight

The drummer is no longer doing work on the drums because the drums are not moving in the direction of the applied force.

✔ Visual Check

4. Recognize What energy transformations occur as the drums are lifted?

_____

_____

_____

🔑 Key Concept Check

5. Describe If you do work on an object, how will its energy change?

_____

_____

_____

Recall that forces are pushes or pulls. When you lift an object, you transfer energy from your body to the object. As the boy lifts the drums in the figure above, the drums move and have kinetic energy. As the drums move higher above Earth's surface, they gain gravitational potential energy. The boy has done work on the drums.

On the right in the figure above, the boy is standing still with his drums lifted in place. Because he is not moving the drums, he is not doing work. To do work on an object, an object must move in the direction of the force. Work is done only while the force is moving the object. 🔑

Doing Work

How much work do you do when you lift your backpack off the ground? The amount of work depends on the force you apply. If you lift a backpack with a force of 20 N, you do less work than if you lift with a force of 40 N.

Work also depends on the distance the object moves during the time the force is applied. If you lift a backpack 1 m, you do less work than if you lift it 2 m. Suppose you toss a backpack in the air. When you release it, it continues moving upward. Even though the backpack is still moving when you let go, no work is being done. This is because you are no longer applying a force to the backpack while it is in the air.

Calculating Work

The equation for work is shown below. *Force* is the force applied to the object. *Distance* is the distance the object moves in the direction of the force while the force is acting on it.

Work Equation

work (in joules) = force (in newtons) × distance (in meters)

$$W = Fd$$

The force in the equation is in newtons (N), and distance is in meters (m). The product of newtons and meters is newton-meter (N·m). A newton-meter is a joule (J).

Energy and Heat

Have you ever heard the phrase *burning rubber*? The tires of race cars are made of rubber. The tires and the road are in contact, and they move past each other quickly. Recall that friction is a force between two surfaces in contact with each other. The direction of friction is in the opposite direction of the motion.

Friction between a car's tires and the road causes some of the kinetic energy of the tires to transform into thermal energy. If race cars are moving very quickly, thermal energy in the tires causes the rubber to give off a burnt odor.

As a race car moves, energy is transformed and transferred. For example, a car's engine transforms energy from fuel and transfers it to the wheel axle. In every energy transformation and every energy transfer, some energy is transformed into thermal energy. This thermal energy is transferred to the surroundings. Thermal energy moving from a region of higher temperature to a region of lower temperature is called heat. Scientists sometimes call this heat *waste energy* because it is not easily used to do useful work.

Math Skills ✕⋅÷

A student lifts a bag from the floor to his or her shoulder 1.2 m above the floor, using a force of 50 N. How much work does the student do on the bag?

a. This is what you know:

force: $F = 50$ N

distance: $d = 1.2$ m

b. You need to find:

work: W

c. Use this formula:

$W = Fd$

d. Substitute the values for F and d into the formula and multiply:

$W = (50$ N$) \times (1.2$ m$)$

$= 60$ N·m $= 60$ J

Answer: The amount of work done is 60 J.

6. Solve for Work

A student pulls out his or her chair in order to sit down. The student pulls the chair 0.75 m with a force of 20 N. How much work does he or she do on the chair?

Mini Glossary

energy transformation: the conversion of one form of energy to another

law of conservation of energy: a principle that says that energy can be transformed from one form to another, but it cannot be created or destroyed

work: the transfer of energy that occurs when a force makes an object move in the direction of the force while the force acts on the object

1. Review the terms and their definitions in the Mini Glossary. Write a sentence that describes an energy transformation that occurs during photosynthesis.

   _____

   _____

2. The figure below depicts a hill in the middle of a roller coaster ride. On each line, identify the energy transformation that is occurring at that point in the ride.

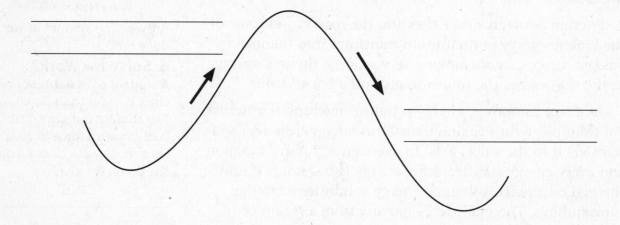

3. Write a question about something you marked with a sticky note. Then answer the question from what you learned by rereading the lesson or asking your teacher.

   _____

   _____

   _____

What do you think NOW?

Reread the statements at the beginning of the lesson. Fill in the After column with an A if you agree with the statement or a D if you disagree. Did you change your mind?

 **Connect ED**

Log on to ConnectED.mcgraw-hill.com and access your textbook to find this lesson's resources.

 END OF LESSON

Energy, Work, and Simple Machines

Machines

············· **Before You Read** ·············

What do you think? Read the two statements below and decide whether you agree or disagree with them. Place an A in the Before column if you agree with the statement or a D if you disagree. After you've read this lesson, reread the statements to see if you have changed your mind.

Before	Statement	After
	5. All machines are 100 percent efficient.	
	6. Simple machines do work using one motion.	

Key Concepts

- What are simple machines?
- In what ways can machines make work easier?

················· **Read to Learn** ·················

Machines Transfer Mechanical Energy

If you use a bottle opener, you can easily pry the top off a bottle. A bottle opener is a machine. Many machines transfer mechanical energy from one object to another. The bottle opener transfers mechanical energy from your hand to the bottle cap. In this lesson, you will read about ways that machines transfer mechanical energy to other objects.

Simple Machines

When you walk up a ramp or cut food with a knife, you are using a simple machine. **Simple machines** *are machines that do work using one movement.* A simple machine can be an inclined <u>plane</u>, a screw, a wedge, a lever, a pulley, or a wheel and axle. Simple machines do not change the amount of work required to do a task. They only change the way work is done. ✓

Inclined Plane
Movers often use a ramp to move heavy furniture into their truck. It is easier to slide a sofa up a ramp than to lift it straight up into the truck. An **inclined plane,** such as a ramp, *is a flat, sloped surface.*

A ramp with a gentle slope requires less force to move an object than a ramp with a steeper slope, but the object will travel a greater distance on the ramp with a gentle slope.

Mark the Text

Identify the Main Ideas To help you learn about machines, highlight each heading in one color. Then highlight the details that support and explain it in a different color. Refer to this highlighted text as you study the lesson.

REVIEW VOCABULARY
plane
a flat, level surface

✓ **Reading Check**
1. Define What is a simple machine?

_____

_____

_____

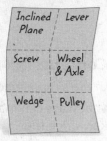
Screw A screw, such as a screw-top bottle, is a special type of inclined plane. *A **screw** is an inclined plane wrapped around a cylinder.* A screw changes the direction of the force from one that acts in a straight line to one that rotates.

Wedge Like all knives, pizza cutters are a special type of inclined plane. *A **wedge** is an inclined plane that moves.* A wedge changes the direction of the input force. As you push the pizza cutter into the pizza, the cutter changes the downward force from your hand to a sideways force that splits the pizza.

Lever The tab on a beverage can is a **lever,** *which is a simple machine that pivots around a fixed point.* The fixed point on a beverage can is where the finger tab attaches to the can. Bottle openers, scissors, seesaws, tennis rackets, and wheelbarrows are examples of levers. Levers decrease the amount of force required to complete a task, but the force must be applied over a longer distance.

Wheel and Axle Screwdrivers are a type of simple machine called a **wheel and axle**—*a shaft attached to a wheel of a larger diameter so that both rotate together.* The wheel and axle are usually circular objects. Because the wheel (the screwdriver handle) has a larger diameter than the axle (the screwdriver shaft), you apply a small input force over a large distance to the wheel (screwdriver handle). This causes the axle (screwdriver shaft) to rotate a smaller distance with a greater output force.

Pulley Have you ever raised a flag on a flagpole? The rope that you pull goes through a **pulley,** *which is a grooved wheel with a rope or cable wrapped around it.* A single pulley, such as the kind on a flagpole, changes the direction of a force. It changes your downward pull on the rope to an upward pull on the flag. A series of pulleys decreases the force you need to lift an object because the number of ropes or cables supporting the object increases.

Complex Machines

Bicycles are made up of many different simple machines. The pedal stem is a lever. The pedal and gears together act as a wheel and axle. The chain around the gear acts as a pulley system. *Two or more simple machines working together are a **complex machine.*** Complex machines use more than one motion to accomplish tasks.

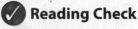

Key Concept Check

2. Describe What are examples of simple machines?

_____

_____

_____

Reading Check

3. Contrast How is a complex machine different from a simple machine?

_____

_____

_____

Machines and Work

Think of a window washer like the one in the figure below. It takes a great amount of work for the washer to lift his own weight plus the weight of buckets of water, tools, and the platform. The washer is able to do all this work because the pulley system that lifts him and the platform make the work easier. Because two ropes are supporting the platform, the force required is half. The work you do on a machine is called the input work. The work the machine does on an object is the output work. Recall that work is the product of force and distance. Machines make work easier by changing the distance the object moves or the force required to do work on an object. ✓

Changing Distance and Force

To pull himself toward the top of the building, the window washer pulls down on a rope. The rope runs through a pulley system. The distance the window washer must pull the rope (the input distance) is much greater than the distance he moves (the output distance). The force the window washer has to use to lift the platform (the input force) is much less than the force the pulley exerts on the platform (the output force). When the input distance of a machine is larger than the output distance, the output force is larger than the input force. This is true for all simple machines. The input force is decreased, but the distance it is applied is increased.

Changing Direction

Machines also can change the direction of a force. A window washer pulls down on the rope. The pulley system changes the direction of the force, which pulls the platform up. 🗝

A Pulley System

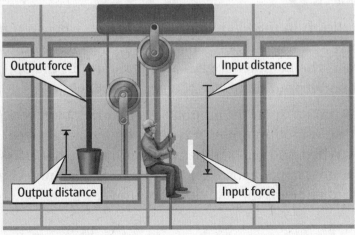

Output force | Input distance | Output distance | Input force

✓ Reading Check

4. Recognize What is input work?

_____

_____

_____

🗝 Key Concept Check

5. State How can machines make work easier?

_____

_____

_____

✓ Visual Check

6. Explain How does the pulley make raising the platform easier for the window washer?

_____

_____

_____

Efficiency

Suppose the window washer wants to buy a new pulley system. One way to compare machines is to calculate their **efficiency,** *the ratio of output work to input work.* In other words, efficiency is a measure of how much work put into the machine is changed into useful output work. Input and output work are measured in joules (J). Efficiency is expressed as a percentage by multiplying the ratio by 100%.

Efficiency Equation

$$\text{efficiency (in \%)} = \frac{\text{output work (in J)}}{\text{input work (in J)}} \times 100\% = \frac{W_{out}}{W_{in}} \times 100\%$$

The window washer considers two systems that require 100 J of input work. The first one does 90 J of output work on his platform. The other pulley system does 95 J of output work. The efficiency of the first pulley system is (90 J/100 J) × 100% = 90%. The efficiency of the second one is (95 J/100 J) × 100% = 95%. The window washer decides to buy the second pulley system.

The efficiency of a machine is never 100%. Some work is always transformed into wasted thermal energy because of friction. One way to improve the efficiency of a machine is to lubricate the moving parts by applying a substance, such as oil. Lubrication reduces the friction between the moving parts so that less input work is transformed to waste energy. ✓

Newton's Laws and Simple Machines

Recall that Newton's laws of motion tell you how forces change the motion of objects. As you have read, machines apply forces on objects. For example, Newton's third law says that if one object applies a force on a second object, the second object applies an equal and opposite force on the first object. ✓

As shown in the figure on the next page, when you use a hammer as a lever to pull a nail, you apply a force on the hammer. The hammer applies an equal force in the opposite direction on your hand.

Newton's first law of motion says that an object remains at rest or in motion unless it is acted on by unbalanced forces. When you pull on the hammer handle, the claws of the hammer apply a force on the nail. However, unless you pull hard enough, the nail does not move.

Think it Over

7. Solve What is the efficiency of a machine that requires 80 J of input work and exerts 60 J of output work?

_____

✓ **Reading Check**

8. Explain Why is the efficiency of a machine never 100%?

_____

_____

✓ **Reading Check**

9. Define What is Newton's third law?

_____

_____

_____

Newton's Laws Applied to a Simple Machine

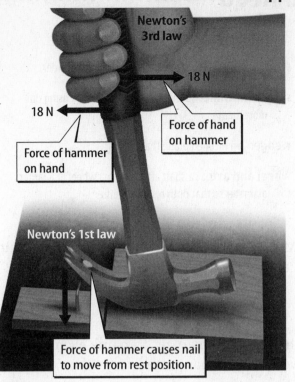

Newton's 3rd law

18 N

18 N

Force of hand on hammer

Force of hammer on hand

Newton's 1st law

Force of hammer causes nail to move from rest position.

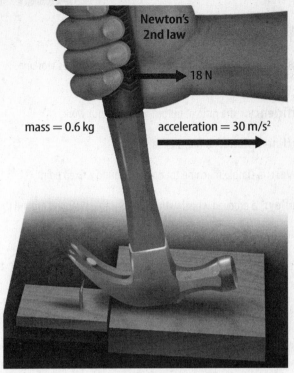

Newton's 2nd law

18 N

mass = 0.6 kg

acceleration = 30 m/s²

The nail does not move because there is another force acting on the nail—the force due to friction between the nail and the wood. Unless you pull hard enough, the force of friction balances the force the hammer exerts on the nail. As a result, the motion of the nail does not change—the nail does not move.

If you pull hard enough, then the upward force the hammer applies on the nail is greater than the force of friction on the nail, as shown on the right in the figure above. Then the forces on the nail are unbalanced and the motion of the nail changes. The nail moves upward.

According to Newton's second law of motion, the change in motion of an object is in the same direction as the total, or net, force on the object. The nail moves upward because the net force on the nail is upward.

Visual Check

10. Apply See the left side of the figure above. Suppose the hand pulls the hammer to the right with a force of 20 N. Based on Newton's third law, how much force will the hammer exert on the hand and in what direction?

_____

Mini Glossary

complex machine: two or more simple machines working together

efficiency: the ratio of output work to input work

inclined plane: a flat, sloped surface

lever: a simple machine that pivots around a fixed point

pulley: a grooved wheel with a rope or cable wrapped around it

screw: an inclined plane wrapped around a cylinder

simple machine: a machine that does work using one movement

wedge: an inclined plane that moves

wheel and axle: a shaft attached to a wheel of a larger diameter so that both rotate together

1. Review the terms and their definitions in the Mini Glossary. Write a sentence that explains what makes one machine more efficient than another machine.

_____

2. Write the words below on the blank lines of the graphic organizer to name each type of simple machine. You may need to use some words more than once.

wheel and axle   **lever**   **wedge**   **screw**   **pulley**

_____ raises flag on a flagpole

_____ wheelbarrow

_____ a seesaw

Simple Machines

_____ a screwdriver

_____ a knife

_____ a screw-top bottle

What do you think NOW?

Reread the statements at the beginning of the lesson. Fill in the After column with an A if you agree with the statement or a D if you disagree. Did you change your mind?

 Connect ED

Log on to ConnectED.mcgraw-hill.com and access your textbook to find this lesson's resources.

 END OF LESSON

Sound and Light

Sound

·············· **Before You Read** ··············

What do you think? Read the two statements below and decide whether you agree or disagree with them. Place an A in the Before column if you agree with the statement or a D if you disagree. After you've read this lesson, reread the statements to see if you have changed your mind.

Before	Statement	After
	1. Vibrating objects make sound waves.	
	2. Human ears are sensitive to more sound frequencies than any other animal's ears.	

·············· **Read to Learn** ··············

What is sound?

Have you ever walked down a busy city street and noticed all the sounds? They all have one thing in common. The sounds travel from one place to another as sound waves. *A* **sound wave** *is a* <u>longitudinal wave</u> *that can travel only through matter.* Sound waves can travel through solids, liquids, and gases. The sounds you hear now are traveling through air—a mixture of solids and gases.

You might have dived under water and heard someone call you. Those sound waves traveled through a liquid. Sound waves travel through a solid when you knock on a door. Your knock makes the door vibrate. Vibrating objects produce sound waves.

Vibrations and Sound

Some objects, such as doors or drums, vibrate when you hit them. When you hit a drum, the drumhead moves up and down, or vibrates. These vibrations produce sound waves by moving molecules in air.

Compressions and Rarefactions

As the drumhead moves up, it pushes the molecules in the air above it closer together. The region where molecules are closer together is a compression.

Key Concepts

- How are sound waves produced?
- Why does the speed of sound waves vary in different materials?
- How do your ears enable you to hear sounds?

> **Mark the Text**

Identify Main Ideas As you read, underline the main ideas under each heading. After you finish reading, review the main ideas that you underlined.

REVIEW VOCABULARY
longitudinal wave
a wave in which particles in a material move along the same direction that the wave travels

When the drumhead moves down, it makes a rarefaction. This is a region where the molecules in the air are farther apart. As the drumhead vibrates up and down, it produces a series of compressions and rarefactions, as shown in the figure above, that travels away from the drumhead. This series of compressions and rarefactions is a sound wave.

The vibrating drumhead causes molecules in the air to move closer together and then farther apart. The molecules move back and forth in the same direction that the sound wave travels. As a result, a sound wave is a longitudinal wave.

Wavelength and Frequency

Wavelength is the distance between a point on a wave and the nearest point just like it. The figure below shows that the wavelength is the distance between one compression and the next compression or the distance between a rarefaction and the next rarefaction.

The frequency of a sound wave is the number of wavelengths that pass a given point in one second. The faster an object vibrates, the higher the frequency of the sound wave it produces. Frequency is measured in hertz (Hz).

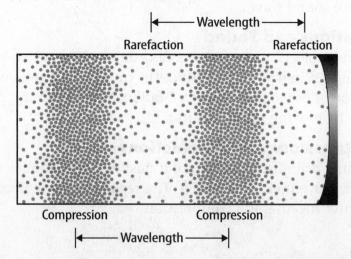

Visual Check

1. Point Out Highlight the air molecules above each drumhead that are part of the compression.

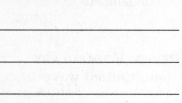

Key Concept Check

2. Explain How do vibrating objects produce sound waves?

_____

_____

_____

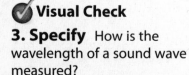

Visual Check

3. Specify How is the wavelength of a sound wave measured?

_____

_____

_____

Speeds of Sound Waves

Sound waves traveling through air cause most of the sounds you hear every day. Recall that sound waves can also travel through liquids and solids. Like all types of waves, the speed of a sound wave depends on the material in which it travels.

The Speed of Sound Waves in Different Materials					
Gases (0°C)		Liquids (25°C)		Solids	
Material	Speed (m/s)	Material	Speed (m/s)	Material	Speed (m/s)
Carbon dioxide	259	Ethanol	1,207	Brick	3,480
Dry air	331	Mercury	1,450	Ice	3,850
Water vapor	405	Water	1,500	Aluminum	6,420
Helium	965	Glycerine	1,904	Diamond	17,500

Sound in Gases, Liquids, and Solids

Sound waves travel at different speeds in different materials. The table above lists the speed of sound waves in different materials. The more dense the material is, the faster a sound wave can move through it. Solids and liquids are usually more dense than gases. Sound waves move fastest through solids and slowest through gases.

A sound wave's speed also depends on the strength of the forces between the particles—atoms or molecules—in the material. The stronger these forces, the faster a sound wave can move through the material.

These forces are usually strongest in solids and weakest in gases. Overall, sound waves usually travel faster in solids than in liquids or gases. 🗝️

Temperature and Sound Waves

The temperature of a material also affects the speed of a sound wave. The speed of a sound wave in a material increases as the temperature of the material increases.

For example, the speed of a sound wave in dry air increases from 331 m/s to 343 m/s as the air temperature increases from 0°C to 20°C. Therefore, sound waves in air travel faster on a warm, summer day than on a cold, winter day.

Math Skills

Speed (s) is equal to the distance (d) something travels divided by the time (t) it takes to cover that distance:

$$s = \frac{d}{t}$$

You can use this equation to calculate the speed of sound waves. For example, if a sound wave travels a distance of 662 meters in 2 seconds in air, its speed is:

$$s = \frac{d}{t} = \frac{662 \text{ m}}{2 \text{ s}} = 331 \text{ m/s}$$

4. Use a Simple Equation How fast is a sound wave traveling if it travels 5,000 m in 5 s?

_____

Interpreting a Table

5. Compare Through which material do sound waves move fastest? (Circle the correct answer.)

a. dry air

b. water

c. ice

🗝️ **Key Concept Check**

6. Explain Why is the speed of sound waves faster in solids than in liquids or gases?

_____

_____

_____

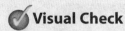

7. Identify In which part of the ear is the cochlea located?

_____

_____

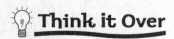 **Think it Over**

8. Analyze Bat-eared foxes have very large outer ears. How do large outer ears benefit these foxes?

_____

_____

_____

_____

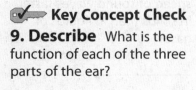 **Key Concept Check**

9. Describe What is the function of each of the three parts of the ear?

_____

_____

_____

_____

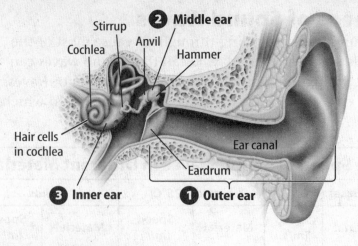

The Human Ear

When you think about your ears, you probably think only about the structure on each side of your head. However, the human ear has three parts—the outer ear, the middle ear, and the inner ear. These parts collect and amplify sound waves and convert the waves into nerve signals. The parts of the ear are shown in the figure above.

1. The Outer Ear

The outer ear collects sound waves. The structure on each side of your head and the ear canal are included in the outer ear. The visible part of the outer ear collects sound waves and funnels them into the ear canal. The ear canal channels sound waves into the middle ear.

2. The Middle Ear

The middle ear amplifies, or strengthens, sound waves. As shown in the figure above, the middle ear includes the eardrum and three tiny bones—the hammer, the anvil, and the stirrup. The eardrum is a thin membrane that stretches across the ear canal. When a sound wave hits the eardrum, it causes the eardrum to vibrate. The vibrations travel to the three tiny bones, which amplify the sound wave.

3. The Inner Ear

The inner ear converts, or changes, vibrations into nerve signals that travel to the brain. The inner ear has a small chamber called the cochlea (KOH klee uh). The cochlea is filled with fluid. Tiny hairlike cells line the inside of the cochlea. These cells are sensitive to vibrations. As a sound wave passes into the cochlea, it causes some hair cells to vibrate. The movements of these cells produce nerve signals that travel to the brain.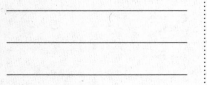

Frequencies and the Human Ear

Recall that frequency—vibrations per second—is measured in hertz (Hz). The table shows that humans hear sounds with frequencies between about 20 Hz and 20,000 Hz. Some mammals can hear sounds with frequencies greater than 100,000 Hz.

Frequencies Different Mammals Can Hear	
Creature	**Frequency Range (Hz)**
Human	20–20,000
Dog	67–45,000
Cat	45–64,000
Bat	2,000–110,000
Beluga whale	1,000–123,000
Porpoise	75–150,000

Sound and Pitch

If you pluck a guitar string, you hear a note. A thick guitar string makes a low note. A thin guitar string makes a higher note. The sound a thick string makes has a lower pitch than the sound a thin string makes. *The **pitch** of a sound is the perception of how high or low a sound seems*. A sound wave with a higher frequency has a higher pitch. A sound wave with a lower frequency has a lower pitch.

You use your vocal cords to make sounds of different pitches. As shown below, vocal cords are two membranes in your neck above your windpipe, or trachea (TRAY kee uh). When you speak, you force air from your lungs through the space between the vocal cords. Your vocal cords then vibrate, making sound waves that people hear. This is your voice.

You change the pitch of your voice by using the muscles connected to your vocal cords. When these muscles contract, they pull on your vocal cords. This stretches the vocal cords, and they become longer and thinner. The pitch of your voice is then higher, just as a thinner guitar string produces a higher pitch. When these muscles relax, the vocal cords become shorter and thicker, and the pitch of your voice is lower.

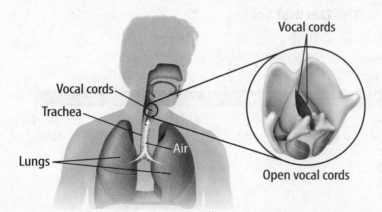

Vocal cords
Trachea
Lungs
Air
Vocal cords
Open vocal cords

Copyright © Glencoe/McGraw-Hill, a division of The McGraw-Hill Companies, Inc.

Interpreting a Table

10. Compare Which mammals listed in the table can hear a sound with a frequency of 55 Hz?

_____

_____

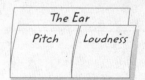

Make a two-tab concept-map book to organize information about pitch and loudness.

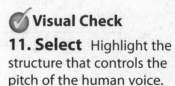

The Ear
Pitch   Loudness

Visual Check

11. Select Highlight the structure that controls the pitch of the human voice.

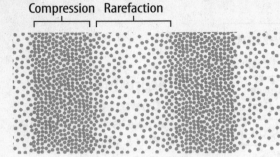
Compression   Rarefaction

Low-amplitude sound wave

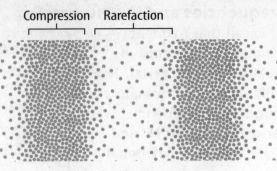
Compression   Rarefaction

High-amplitude sound wave

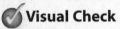 **Visual Check**

12. Contrast How do distances between particles differ in high- and low-amplitude sound waves?

_____

_____

_____

Sound and Loudness

Loudness is the human sensation of how much energy a sound wave carries. Sound waves made by a shout carry more energy than sound waves made by a whisper. Because a shout carries more energy, it sounds louder than a whisper.

Amplitude and Energy

The amplitude of a wave depends on the amount of energy the wave carries. The more energy the wave has, the greater the amplitude.

The figure above shows the difference between a high-amplitude sound wave and a low-amplitude sound wave. High-amplitude sound waves have particles that are closer together in the compressions and farther apart in the rarefactions.

The Decibel Scale

The decibel scale is one way to compare the loudness of sounds. The figure below shows the decibel measurements for some sounds.

The softest sound a person can hear is about 0 decibels (dB). Normal conversation is about 50 dB. A sound wave that is 10 dB higher than another sound wave carries ten times more energy. However, people hear the higher-energy sound wave as being only twice as loud.

 **Visual Check**

13. Calculate What is the difference in decibels between a vacuum cleaner and a jet plane taking off?

_____

The Decibel Scale

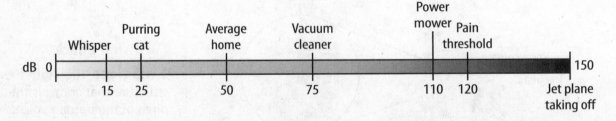

Sonar System

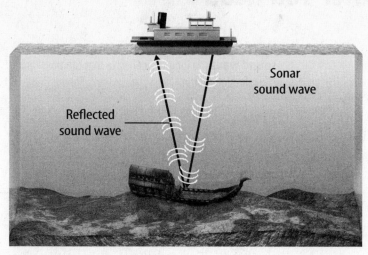

Sonar sound wave

Reflected sound wave

Visual Check
14. Identify What is the echo in the figure at the left?

_____

_____

_____

Using Sound Waves

If you have ever shouted in a cave or a big, empty room, you might have heard an echo of your voice. *An* **echo** *is a reflected sound wave.* You probably can't tell how far away a wall is by hearing an echo. However, sonar systems and some animals use reflected sound waves to determine how far away objects are.

Sonar and Echolocation

Sonar systems use reflected sound waves to locate objects under water, as shown in the figure above. The sonar system sends a sound wave that reflects off an underwater object. The sonar system calculates the distance to the object by measuring the time difference between when the sound leaves the ship and when the sound returns to the ship. Sonar is used to map the ocean floor and to detect submarines, schools of fish, and other objects under water. ✓

Reading Check
15. Explain How do sonar systems use sound waves?

_____

_____

_____

_____

Some animals use echolocation to hunt or to find their way. Echolocation is a type of sonar. Bats and dolphins make high-pitched sounds and interpret the echoes reflected from objects. Echolocation makes it possible for bats and dolphins to locate prey and detect objects.

Ultrasound

Ultrasound scanners use high-frequency sound waves to make images of internal body parts. The sound waves reflect from structures within the body. The scanner analyzes the reflected waves and produces images, called sonograms, of body structures. The images can help doctors diagnose disease or other medical conditions.

Think it Over

16. Apply When a bat flies in darkness, why is it able to avoid objects in its path?

_____

_____

_____

_____

Mini Glossary

echo: a reflected sound wave

pitch: the perception of how high or low a sound seems

sound wave: a longitudinal wave that can travel only through matter

1. Review the terms and their definitions in the Mini Glossary. Write a sentence that describes how animals use sound waves.

_____

_____

2. Sounds waves travel at different speeds through different types of matter. Place the terms *liquid, gas,* and *solid* in the correct boxes below according to the speed that sound waves travel through each. Then find two examples of each type of matter in the lesson and record them below. Also record the speed that sound waves travel through each.

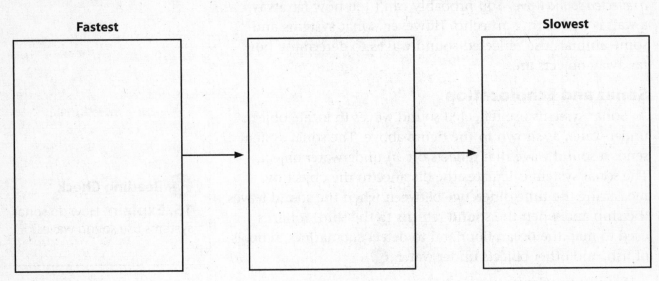

Fastest   Slowest

3. How did underlining the main ideas in the lesson help you learn about sound?

_____

_____

What do you think 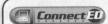 NOW?

Reread the statements at the beginning of the lesson. Fill in the After column with an A if you agree with the statement or a D if you disagree. Did you change your mind?

 Connect ED

Log on to ConnectED.mcgraw-hill.com and access your textbook to find this lesson's resources.

END OF LESSON

Copyright © Glencoe/McGraw-Hill, a division of The McGraw-Hill Companies, Inc.

Sound and Light

Light

What do you think? Read the two statements below and decide whether you agree or disagree with them. Place an A in the Before column if you agree with the statement or a D if you disagree. After you've read this lesson, reread the statements to see if you have changed your mind.

Before	Statement	After
	3. Unlike sound waves, light waves can travel through a vacuum.	
	4. Light waves always travel at the same speed.	

··············· **Read to Learn** ···············

What is light?

What are your eyes detecting as you read the words on this page? When you see the words, your eyes are sensing light waves. You see the words in a book or books on a desk when light waves reflect off these objects and enter your eyes. Some objects also emit, or send out, light waves. You see a candle flame or a glowing lightbulb because the light waves they emit enter your eyes.

Light—An Electromagnetic Wave

Light is a type of wave called an electromagnetic wave. Like sound waves, electromagnetic waves can travel through matter. But electromagnetic waves can also travel through a vacuum where no matter is present. For example, light can travel through the space between Earth and the Sun.

Light waves travel fastest through a vacuum. The speed of light waves in a vacuum is about 300,000 km/s. Light waves travel more slowly when they move through matter. Light waves travel at different speeds in different materials. They move fastest in gases and slowest in solids.

Light waves travel much faster than sound waves. The speed of light is about 900,000 times faster than the speed of sound.

Key Concepts 🔑

- How are light waves different from sound waves?
- How do waves in the electromagnetic spectrum differ?
- What happens to light waves when they interact with matter?

◀ **Study Coach**

Create a Quiz Write a quiz question for each paragraph. Answer the question with information from the paragraph. Then work with a partner to quiz each other.

🔑 **Key Concept Check**
1. Contrast How are light waves different from sound waves?

_____

_____

_____

_____

_____

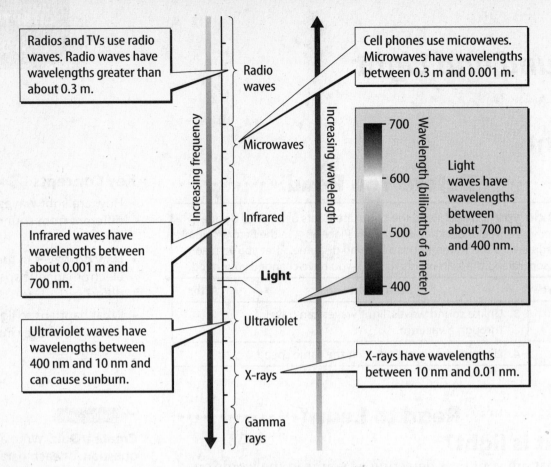

Radios and TVs use radio waves. Radio waves have wavelengths greater than about 0.3 m.

Cell phones use microwaves. Microwaves have wavelengths between 0.3 m and 0.001 m.

Infrared waves have wavelengths between about 0.001 m and 700 nm.

Light waves have wavelengths between about 700 nm and 400 nm.

Ultraviolet waves have wavelengths between 400 nm and 10 nm and can cause sunburn.

X-rays have wavelengths between 10 nm and 0.01 nm.

Increasing frequency

Increasing wavelength

Wavelength (billionths of a meter)

Radio waves

Microwaves

Infrared

Light

Ultraviolet

X-rays

Gamma rays

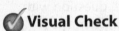 **Visual Check**

2. Identify Which type of electromagnetic waves has the longest wavelengths?

_____

_____

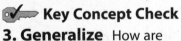 **Key Concept Check**

3. Generalize How are waves in the electromagnetic spectrum different?

_____

_____

_____

The Electromagnetic Spectrum

Visible light waves are one type of electromagnetic wave. There are other types, as shown in the figure above.

Scientists classify electromagnetic waves into groups based on their wavelengths. The main groups are radio waves, microwaves, infrared waves, light waves, ultraviolet waves, X-rays, and gamma rays. The whole range of electromagnetic waves is called the electromagnetic spectrum.

Light waves are only a small part of the electromagnetic spectrum. The wavelengths of light waves are very short. Because they are so short, they are usually measured in nanometers (nm). One nanometer equals one-billionth of a meter.

The wavelengths of light waves range from about 700 nm to about 400 nm. This is about one-hundredth the width of a human hair. When different wavelengths of light waves enter your eyes, you see them as different colors.

Light-Emitting Objects

When you turn on a light, the lightbulb produces light waves that travel away from the bulb in all directions. A **light source** *is something that emits light.* In order to emit light, the lightbulb transforms electric energy into light energy. The Sun is a light source that transforms nuclear energy into light energy. A burning candle transforms chemical energy into light energy. In general, light sources transform other forms of energy into light energy.

Light Rays

You have read that light waves spread out in all directions from a light source. You also can think of light in terms of light rays. A **light ray** *is a narrow beam of light that travels in a straight line.* Light rays travel in straight lines until they hit a surface or pass through a different material. ✓

Light Reflection

Light sources emit light. Other objects, like books, reflect light. In order to see an object that is not a light source, light waves must reflect from the object and enter your eyes.

Seeing Objects

When you see a light source, light rays travel directly from the light source into your eyes. When you see an object that is not a light source, light waves reflect from the object in many directions.

The lamp in the figure below is a light source. The book is not a light source. The lamp emits light waves in many directions. The boy sees the book when some of the light waves reflect off the book and enter his eyes.

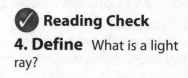

Reading Check

4. Define What is a light ray?

_____

_____

_____

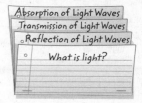

Make a layered book to summarize information about light and how light waves interact with matter.

| Absorption of Light Waves |
| Transmission of Light Waves |
| Reflection of Light Waves |
| What is light? |

Visual Check

5. Explain Why is the boy able to see the book?

_____

_____

The Interaction of Light and Matter

Like all waves, when light waves interact with matter, they can be reflected, transmitted, or absorbed.

- Reflection occurs when light waves strike the surface of a material and bounce off.

- Transmission occurs when light waves travel through a material.

- Absorption occurs when interactions with a material convert light energy into other forms of energy.

In some materials, reflection, transmission, and absorption occur at the same time. For example, the tinted glass of an office building reflects some light, transmits some light, and absorbs some light.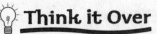

Materials can be classified as transparent, translucent, or opaque. *A material is* **transparent** *if it allows almost all light that strikes it to pass through and forms a clear image.* Window glass is transparent. *A material is* **translucent** *if it allows most of the light that strikes it to pass but forms a blurry image.* For example, frosted glass is translucent. *A material is* **opaque** *if light does not pass through it.* Heavy curtains that block light are opaque.

The Reflection of Light Waves

The figure below shows what happens when a surface reflects light waves. All waves, including light waves, obey the law of reflection. In the figure below, the line that is perpendicular to the surface is called the normal. The angle between the incoming light ray and the normal is the angle of incidence. The angle between the reflected light ray and the normal is the angle of reflection. The law of reflection states that the angle of incidence equals the angle of reflection.

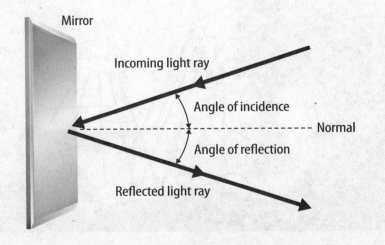

Mirror

Incoming light ray

Angle of incidence

Normal

Angle of reflection

Reflected light ray

Key Concept Check
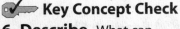

6. Describe What can happen to light waves when they interact with matter?

_____

_____

_____

Think it Over

7. Apply Give an example of an object that is opaque.

_____

_____

_____

Visual Check

8. Interpret How will the angle of reflection change if the angle of incidence increases?

_____

_____

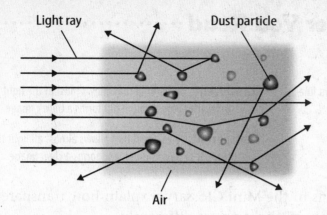

Light ray   Dust particle

Air

Visual Check
9. Discover What can you observe about the dust particles in the figure?

_____

_____

_____

Scattering

The figure above illustrates a beam of sunlight reflecting off tiny particles of dust floating in air. Dust particles have different shapes. As a result, the particles reflect the light waves in many different directions. This is an example of scattering. Scattering occurs when waves traveling in one direction are made to travel in many different directions. The dust particles scatter the light waves in the sunbeam.

The Refraction of Light Waves

Like all types of waves, light waves can change direction when they travel from one material to another. The figure on the right shows how a beam of light changes direction as it moves from air, through glass, into water, and back again. A wave that changes direction as it travels from one material to another is refracting.

Visual Check
10. Show Extend the pencil to show how it would appear if there were no water in the glass.

Refraction occurs when a wave changes speed. Because waves move at different speeds through different materials, they change direction when they travel into a different material. The pencil looks broken because light waves are refracted as they change speed when they pass through the different materials. The greater the change in speed, the more the light wave refracts or changes direction.

Reading Check
11. Explain When does refraction occur?

_____

_____

_____

Mini Glossary

light ray: a narrow beam of light that travels in a straight line

light source: something that emits light

opaque: a material that light does not pass through

translucent: a material that allows most of the light that strikes it to pass through but forms a blurry image

transparent: a material that allows almost all light that strikes it to pass through and forms a clear image

1. Review the terms and their definitions in the Mini Glossary. Explain how transparent, translucent, and opaque materials transmit light waves differently.

_____

_____

_____

2. Fill in the table to describe the different ways light waves interact with matter.

Interaction	Description
Absorption	
	Light waves strike the surface of a material and bounce off.
Transmission	

3. Describe what can happen when a light wave strikes an object.

_____

_____

_____

What do you think NOW?

Reread the statements at the beginning of the lesson. Fill in the After column with an A if you agree with the statement or a D if you disagree. Did you change your mind?

Log on to ConnectED.mcgraw-hill.com and access your textbook to find this lesson's resources.

END OF LESSON

Sound and Light

Mirrors, Lenses, and the Eye

· · · · · · · · · · · · Before You Read · · · · · · · · · · · ·

What do you think? Read the two statements below and decide whether you agree or disagree with them. Place an A in the Before column if you agree with the statement or a D if you disagree. After you've read this lesson, reread the statements to see if you have changed your mind.

Before	Statement	After
	5. All mirrors form images that appear identical to the object itself.	
	6. Lenses always magnify objects.	

· · · · · · · · · · · · Read to Learn · · · · · · · · · · · ·

Why are some surfaces mirrors?

When you look at a smooth lake, you can see a sharp image of yourself reflected off the water's surface. If you look at the lake on a windy day, you do not see a sharp image. Why are these images different? A smooth surface reflects light rays traveling in the same direction at the same angle. This is called regular reflection, as shown in the figure below. Because the light rays travel parallel to each other before and after they reflect from the surface, the reflected light rays form a sharp image.

When a surface is not smooth, light rays still follow the law of reflection. When light rays traveling in the same direction hit the rough surface, they hit at different angles. The reflected light rays travel in many different directions. This is called diffuse reflection. Diffuse reflection does not form a clear image. Light rays in diffuse reflection are shown in the figure above.

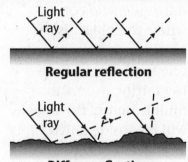

Regular reflection

Diffuse reflection

Key Concepts 🔑

- What is the difference between regular and diffuse reflection?

- What types of images are formed by mirrors and lenses?

- How does the human eye enable a person to see?

◄ **Mark the Text**

Summarize Write a phrase beside each paragraph that summarizes the main point of the paragraph. Use the phrases to review the lesson.

✓ **Visual Check**

1. Explain In diffuse reflection, why are light rays reflected in different directions?

_____

_____

_____

Types of Mirrors

When you look at a wall mirror, the image you see is about the same size that you are and right-side up. *A **mirror** is any reflecting surface that forms an image by regular reflection.* The shape of the mirror's surface determines how the image in the mirror looks.

Plane Mirrors

A plane mirror has a flat reflecting surface. The reflected image looks just like the object, except that the reflection is reversed left to right.

The size of the image in a plane mirror depends on how far the object is from the mirror. The image gets smaller as the object gets farther from the mirror.

Concave Mirrors

Concave mirrors, shown in the figures on the right, are reflecting surfaces that are curved inward. Notice the optical axis in the top figure. Light rays that are parallel to the optical axis reflect through one point—the focal point. The distance from the mirror to the focal point is the focal length.

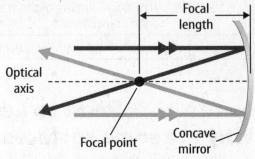

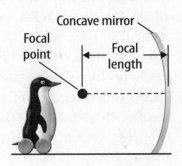

The type of image that forms in a concave mirror depends on where the object is. If an object is more than one focal length from the mirror, as in the middle figure, the image will be upside down. If an object is closer than one focal length to the concave mirror, as in the bottom figure, the image will be right-side up. If the object is exactly at the focal point, no image will form.

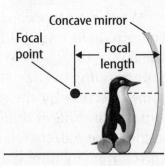

Copyright © Glencoe/McGraw-Hill, a division of The McGraw-Hill Companies, Inc.

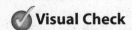

Think it Over

2. Predict How would the word *pen* appear if you looked at it in a plane mirror? Write the word as it would appear in the mirror.

_____

_____

Visual Check

3. Recognize Look at the toy penguin in the middle figure. Where is the toy penguin in relation to the focal point?

_____

_____

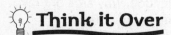

Think it Over

4. Apply The focal length of a concave mirror is 3 m. If you place an object 3 m from that mirror, how will this object appear in the mirror?

_____

_____

_____

Convex Mirrors

A convex mirror has a reflecting surface that is curved outward. The image in a convex mirror is always right-side up and smaller than the object itself. Store security mirrors and passenger-side car mirrors are usually convex mirrors.

Types of Lenses

Magnifying lenses, eyeglasses, and binoculars all use lenses to change the way an image of an object forms. *A **lens** is a transparent object with at least one curved side that causes light to change direction.* The more curved the sides of a lens, the more the light changes direction as it passes through the lens.

Convex Lenses

A convex lens is curved outward on at least one side. It is thicker in the middle than at its edges.

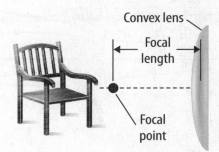

Like a concave mirror, a convex lens has a focal point and a focal length. The more curved the lens is, the shorter the focal length. A convex lens is shown in the figure above.

The image formed by a convex lens depends on where the object is, just like it does for a concave mirror. See the figure to the right. When an object is farther than one focal length from a convex lens, the image is upside down. When the chair in the figure is viewed through the lens, it will appear upside down.

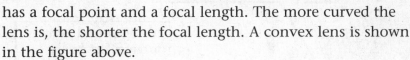

When an object is closer than one focal length to a convex lens, the image is larger and right-side up. The image of the dollar bill in the figure to the right will appear larger and right-side up in the lens. Both a magnifying lens and a camera lens are convex lenses.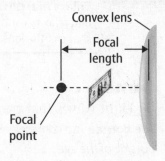

Copyright © Glencoe/McGraw-Hill, a division of The McGraw-Hill Companies, Inc.

Key Concept Check

5. Describe How do the images formed by plane mirrors, concave mirrors, and convex mirrors depend on the distance of an object from the mirror?

_____

_____

_____

_____

✓ **Visual Check**

6. Describe Look at the bottom two figures to the left. Circle the object that will appear right-side up in the lens.

_____

_____

_____

Key Concept Check

7. Consider How does the image formed by a convex lens depend on the distance of the object from the lens?

_____

_____

_____

_____

_____

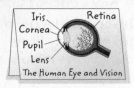

Make a half book to identify the parts of the eye and their function.

ACADEMIC VOCABULARY
convert
(verb) to change from one form into another

Concave Lenses

A concave lens is curved inward on at least one side. It is thicker at its edges than it is in the middle. A concave lens forms an image that is upright and smaller than the object. Concave lenses along with other lenses are usually used in microscopes and telescopes.

Light and the Human Eye

The human eye contains lenses, as well as other parts, that enable a person to see. The structure of the human eye is shown in the figure below. To see an object, light waves from the object must travel through two convex lenses in the eye. The first lens is called the cornea. The second is simply called the lens. At the back of the eye is a thin layer of tissue called the retina. The lenses form an image of the object on the retina. Special cells in the retina convert the image into electrical signals. Nerves carry these signals to the brain.

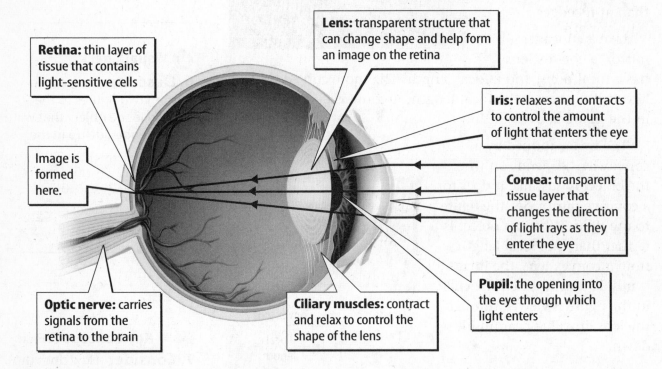

Retina: thin layer of tissue that contains light-sensitive cells

Image is formed here.

Optic nerve: carries signals from the retina to the brain

Lens: transparent structure that can change shape and help form an image on the retina

Iris: relaxes and contracts to control the amount of light that enters the eye

Cornea: transparent tissue layer that changes the direction of light rays as they enter the eye

Pupil: the opening into the eye through which light enters

Ciliary muscles: contract and relax to control the shape of the lens

Visual Check

8. Identify On which part of the eye is an image formed?

Cornea

Light waves first travel through the cornea (KOR nee uh). *The **cornea** is a convex lens made of transparent tissue located on the outside of the eye.*

Most of the change of direction in light rays occurs in the cornea. Doctors can correct some vision problems by changing the cornea's shape. Locate the cornea in the figure above.

Iris and Pupil

The **iris** *is the colored part of the eye. The* **pupil** *is an opening into the interior of the eye at the center of the iris.* See the figure below. When the iris changes size, the amount of light that enters the eye changes. In bright light, the iris relaxes. The iris gets larger and the pupil gets smaller. Then less light enters the eye. In dim light, the iris contracts, or gets smaller, and the pupil becomes larger. Then more light enters the eye.

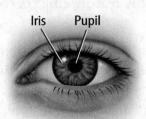 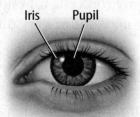

Iris   Pupil   Iris   Pupil

The iris relaxes   The iris contracts
in bright light.   in dim light.

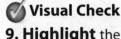 **Visual Check**

9. Highlight the part of the eye that controls the amount of light that enters the pupil.

Lens

Behind the iris is the lens. The lens is flexible, transparent tissue. The lens enables the eye to form a sharp image of nearby and distant objects. The muscles surrounding the lens change the lens's shape. To focus on nearby objects, these muscles relax and the lens becomes more curved. To focus on distant objects, these muscles pull on the lens and make it flatter. The figure below shows how the lens in the eye changes shape.

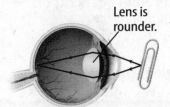 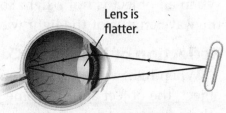

Lens is   Lens is
rounder.   flatter.

Lens becomes rounder and a sharp   Lens becomes flatter and a sharp
image forms of a nearby object.   image forms of a distant object.

Visual Check

10. Describe How does the shape of the lens change when the muscles relax?

_____

_____

_____

Retina

The **retina** *is a layer of special light-sensitive cells in the back of the eye.* After light travels through the lens, an image forms on the retina.

On the retina, chemical reactions produce nerve signals. The optic nerve sends these signals to the brain. The retina has two types of light-sensitive cells—rod cells and cone cells.

Key Concept Check

11. Identify the parts of the eye that form a sharp image of an object and the parts that convert an image into electrical signals.

_____

_____

_____

_____

Copyright © Glencoe/McGraw-Hill, a division of The McGraw-Hill Companies, Inc.

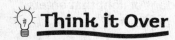
12. Contrast How are cone cells different from rod cells?

_____

_____

_____

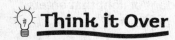 **Think it Over**

13. Identify What property of light waves determines the type of cone that will respond to the light?

_____

_____

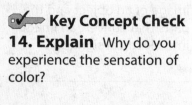 **Key Concept Check**

14. Explain Why do you experience the sensation of color?

_____

_____

Rod Cells There are more than 100 million rod cells in a human retina. Rod cells are sensitive to low levels of light. These cells enable you to see objects in dim light. However, the signals that rod cells send to the brain do not enable you to see colors.

Cone Cells A retina contains over 6 million cone cells. Cone cells make it possible for you to see colors. However, cone cells need brighter light to work than rod cells do. In dim light, only rod cells function. For this reason, objects seem to have no color in very dim light.

The responses of cone cells to light waves with different wavelengths enable you to see different colors. The retina has three types of cone cells. Each type responds to a different range of wavelengths. This means that different wavelengths of light cause each type of cone cell to send different signals to the brain. Your brain interprets the different combinations of signals from the three types of cone cells as different colors.

In some people, not all types of cone cells work properly. These people cannot detect certain colors. This condition is commonly known as color blindness.

The Colors of Objects

The objects you see around you are different colors. Why do you see a banana as yellow instead of red? Most objects do not give off, or emit, light. Instead, they reflect light. When an object is not a light source, its colors depend on the wavelengths of the light waves it reflects.

Reflection of Light and Color

When light waves of different wavelengths strike an object, the object absorbs some light waves and reflects others. The materials that make up the object determine the wavelengths of light that the object absorbs or reflects.

For example, a red rose reflects light waves with certain wavelengths and absorbs all other wavelengths of light. When the reflected light waves enter your eye, they cause cone cells in your retina to send certain nerve signals to your brain. These signals cause you to see the rose as red.

A banana absorbs and reflects different wavelengths of light than a red rose does. The reflected wavelengths cause cone cells to send different signals to your brain. These signals cause you to see the banana as yellow instead of red.

Light waves have no color. Color is a sensation produced by your brain when light waves enter your eyes.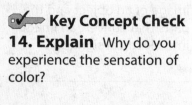

The Color of Objects that Emit Light

Some objects, such as the Sun, lightbulbs, and neon lights, emit light. The color of an object that emits light depends on the wavelengths of the light waves it emits. For example, a red neon light emits light waves with wavelengths that you see as red.

White Light—A Combination of Light Waves

You might have noticed at a concert that the colors of objects on stage depend on the colors of the spotlights. A shirt might appear blue when a blue or white spotlight shines on it. The same shirt might appear black when a red spotlight shines on it.

Light that you see as white is actually a combination of light waves of many different wavelengths. When white light travels through a prism, light waves with different wavelengths spread out after passing through the prism and form a color spectrum. They spread out because the different wavelengths of light change direction by different amounts when they move into and out of the prism.

Changing Colors

The color of an object depends on the wavelengths of light it reflects. A blue shirt will appear to be different colors when different spotlights shine on it. When white light strikes the shirt, the shirt reflects only the wavelengths that you see as blue. It absorbs all other wavelengths of light. The shirt appears blue under a blue spotlight because the shirt reflects the blue light.

But when red light strikes the same shirt, the shirt absorbs nearly all of the light. Almost no light is reflected. This causes the shirt to appear black. An object appears black when it absorbs almost all light waves that strike it.

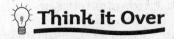

Think it Over

15. Contrast How is light that is white different from light that is red or blue?

_____

_____

_____

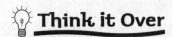

Think it Over

16. Describe If an object absorbs all the wavelengths of light that strike it, how does the object look?

_____

Mini Glossary

cornea (KOR nee uh): a convex lens made of transparent tissue located on the outside of the eye

iris: the colored part of the eye

lens: a transparent object with at least one curved side that causes light to change direction

mirror: any reflecting surface that forms an image by regular reflection

pupil: an opening into the interior of the eye at the center of the iris

retina: a layer of special light-sensitive cells in the back of the eye

1. Review the terms and their definitions in the Mini Glossary. Write one or two sentences that describe where light waves go after they enter your eye.

_____

_____

_____

_____

2. The diagram below shows the path of reflected light from a concave mirror. How will this person appear in the mirror? Explain why.

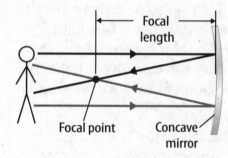

_____

_____

3. Contrast regular and diffuse reflection.

_____

_____

What do you think NOW?

Reread the statements at the beginning of the lesson. Fill in the After column with an A if you agree with the statement or a D if you disagree. Did you change your mind?

 Connect **ED**

Log on to ConnectED.mcgraw-hill.com and access your textbook to find this lesson's resources.

END OF LESSON

Thermal Energy

Thermal Energy, Temperature, and Heat

·············· **Before You Read** ··············

What do you think? Read the two statements below and decide whether you agree or disagree with them. Place an A in the Before column if you agree with the statement or a D if you disagree. After you've read this lesson, reread the statements to see if you have changed your mind.

Before	Statement	After
	1. Temperature is the same as thermal energy.	
	2. Heat is the movement of thermal energy from a hotter object to a cooler object.	

············· **Read to Learn** ···············

Kinetic and Potential Energy

What do a soaring soccer ball and the particles that make up hot maple syrup have in common? They have energy, or the ability to cause change. What type of energy does a moving soccer ball have? Recall that any moving object has kinetic energy. When an athlete kicks a soccer ball and puts it in motion, the ball has kinetic energy.

In addition to having kinetic energy when it is in the air, the soccer ball also has potential energy. Potential energy is stored energy due to the interaction between two objects. For example, think of Earth as one object and the ball as another. When the ball is in the air, it is attracted to Earth due to gravity. This attraction is called gravitational potential energy. In other words, because the ball has the potential to change, it has potential energy. And, the higher the ball is in the air, the greater the potential energy of the ball.

You also might recall that the potential energy plus the kinetic energy of an object is the mechanical energy of the object. When a soccer ball is flying through the air, you could describe the mechanical energy of the ball by describing its kinetic and potential energy. On the next page, you will read about how the particles that make up maple syrup have energy, just like a soaring soccer ball.

Key Concepts

- How are temperature and kinetic energy related?
- How do heat and thermal energy differ?

◀ Study Coach

Building Vocabulary Work with another student to write a question about each vocabulary term in this lesson. Answer the questions and compare your answers. Reread the text to clarify the meaning of the terms.

FOLDABLES

Make a three-column chart book to organize your notes on the properties of heat, temperature, and thermal energy.

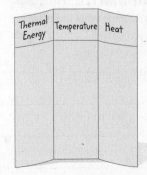

Copyright © Glencoe/McGraw-Hill, a division of The McGraw-Hill Companies, Inc.

What is thermal energy?

Every solid, liquid, and gas is made up of trillions of particles that are constantly moving. The particles that make up your book, or any solid, vibrate in place. The particles that make up the air around you, or any gas, spread out and move freely and quickly. Because the particles are in motion, they have kinetic energy. The faster particles move, the more kinetic energy they have. ✓

The particles that make up matter also have potential energy because they interact with and are attracted to one another. The particles that make up solids usually are held very close together by attractive forces. The particles that make up a liquid are slightly farther apart than those that make up a solid. And, the particles that make up a gas are much more spread out than those that make up either a solid or a liquid. The greater the average distance between particles, the greater the potential energy of the particles.

Recall that a flying soccer ball has mechanical energy, which is the sum of its potential energy and its kinetic energy. The particles that make up the ball, or any material, have thermal energy. **Thermal energy** *is the sum of the kinetic energy and the potential energy of the particles that make up a material.* Thermal energy describes the energy of the particles that make up a solid, a liquid, or a gas. ✓

What is temperature?

You probably think of temperature as a measurement of how warm or cold something is. However, scientists define temperature in terms of kinetic energy.

Average Kinetic Energy and Temperature

The particles that make up the air inside and outside a house on a cold night are moving. However, the particles are not all moving at the same speed. The air particles inside the warm house move faster and have more kinetic energy than the air particles outside. **Temperature** *represents the average kinetic energy of the particles that make up a material.*

The greater the average kinetic energy of particles, the greater the temperature is. The temperature of the air in the house is higher because the particles that make up the air inside the house have greater average kinetic energy than the particles outside. The particles of air inside the house are moving at a greater average speed than those outside. Because temperature represents the average kinetic energy of particles, the temperature of the outside air is lower. 🗝

Thermal Energy and Temperature

Temperature and thermal energy are related, but they are not the same thing. For example, as a frozen pond melts, ice and water are present and they have the same temperature. The particles of ice and water have the same average kinetic energy, or speed. The particles do not have the same thermal energy. This is because the average distance of the particles that make up liquid water and ice are different. The particles that make up the liquid water and the solid water have different potential energies and thermal energies. ✓

Measuring Temperature

How can you measure temperature? It would be impossible to measure the kinetic energy of individual particles and then calculate their average kinetic energy to determine the temperature. Instead, you can use thermometers, such as the ones in the figure below, to measure temperature.

A bulb thermometer is a common type of thermometer. It is a glass tube connected to a bulb that contains a liquid such as alcohol. When the temperature of the alcohol increases, the alcohol expands and rises in the glass tube. When the temperature of the alcohol decreases, the alcohol contracts back into the bulb. The height of the alcohol in the tube indicates the temperature. An electronic thermometer measures changes in the resistance of an electric circuit. It converts this measurement to a temperature. ✓

Measuring Temperature

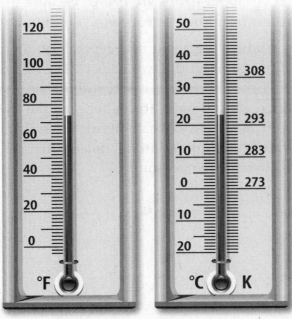

Copyright © Glencoe/McGraw-Hill, a division of The McGraw-Hill Companies, Inc.

Reading Check

4. Explain Why do particles in liquid water have greater potential energy than particles in ice?

_____

_____

_____

_____

Reading Check

5. Differentiate between a bulb thermometer and an electronic thermometer.

_____

_____

_____

_____

Visual Check

6. Identify Are these bulb thermometers or electronic thermometers?

_____

_____

To convert Fahrenheit to Celsius, use the following equation:

$$°C = \frac{(°F - 32)}{1.8}$$

For example, to convert 176°F to Celsius:

a. Always perform the operation in parentheses first.

$$176 - 32 = 144$$

b. Divide the answer from Step a by 1.8.

$$\frac{144}{1.8} = 80°C$$

To convert Celsius to Fahrenheit, follow the same steps using the following equation:

$$°F = (°C \times 1.8) + 32$$

7. Convert Between Temperature Scales

Convert 86°F to Celsius.

_____

Convert 37°C to Fahrenheit.

_____

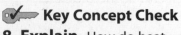 **Key Concept Check**

8. Explain How do heat and thermal energy differ?

_____

_____

_____

Temperature Scales

In a weather report, the temperature might be given in degrees Fahrenheit and degrees Celsius. On the Fahrenheit scale, water freezes at 32° and boils at 212°. On the Celsius scale, water freezes at 0° and boils at 100°. The Celsius scale is used by scientists worldwide.

Scientists also use the Kelvin scale. On the Kelvin scale, water freezes at 273 K and boils at 373 K. The lowest possible temperature for any material is 0 K. This is known as absolute zero. If a material were at 0 K, the particles in that material would not be moving and would no longer have kinetic energy. Scientists have not been able to cool any material to 0 K.

What is heat?

Have you ever held a cup of hot cocoa on a cold day? Hot cocoa has a high temperature. Thermal energy is transferred from the cup to its surroundings. As you hold the cup, thermal energy moves from the warm cup to the air and to your hands. *The movement of thermal energy from a warmer object to a cooler object is called* **heat.** Another way to say this is that thermal energy from the cup heats your hands, or the cup is heating your hands.

Just as temperature and thermal energy are not the same thing, neither are heat and thermal energy. All objects have thermal energy. However, something is heated when thermal energy transfers from one object to another. When you hold the cup of cocoa, your hands are heated because thermal energy transfers from the hot cocoa to your hands.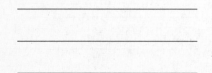

The rate at which heating occurs depends on the difference in temperatures between the two objects. The difference in temperatures between the hot cocoa and the air is greater than the difference in temperatures between the hot cocoa and the cup. The hot cocoa heats the air more than it heats the cup. Heating continues until all objects that are in contact are the same temperature.

Mini Glossary

heat: the movement of thermal energy from a warmer object to a cooler object

temperature: the average kinetic energy of the particles that make up a material

thermal energy: the sum of the kinetic energy and the potential energy in the particles that make up a material

1. Review the terms and their definitions in the Mini Glossary. Write a sentence explaining the difference between heat and temperature.

_____

_____

2. Imagine that you are preparing for a snowball fight with a friend. Identify each activity listed in the diagram by one of the terms listed below. An activity might involve more than one of these terms.

potential energy   heat   kinetic energy

Activity	Example of . . .
You cup some snow in your hands, and it melts enough to hold the ball of snow together.	
The finished snowball sits on top of your pile of snowy ammunition.	
The contest begins, and you pick up the top snowball and fire it at your friend.	

3. Have a classmate select one of the questions you wrote as you read the lesson. Without checking against the text, answer the question in the space below.

_____

_____

What do you think NOW?

Reread the statements at the beginning of the lesson. Fill in the After column with an A if you agree with the statement or a D if you disagree. Did you change your mind?

Log on to ConnectED.mcgraw-hill.com and access your textbook to find this lesson's resources.

END OF LESSON

Thermal Energy

Thermal Energy Transfers

Copyright © Glencoe/McGraw-Hill, a division of The McGraw-Hill Companies, Inc.

Key Concepts 🔑
- What is the effect of having a small specific heat?
- What happens to a material when it is heated?
- In what ways can thermal energy be transferred?

Mark the Text

Main Ideas and Details
Highlight the main idea of each paragraph. Highlight two details that support each main idea with a different color. Use your highlighted copy to review what you studied in this lesson.

✔ Reading Check
1. Define What is radiation?

_____

_____

_____

_____

·············· **Before You Read** ··············

Before	Statement	After
	3. It takes a large amount of energy to significantly change the temperature of an object with a low specific heat.	
	4. The thermal energy of an object can never be increased or decreased.	

What do you think? Read the two statements below and decide whether you agree or disagree with them. Place an A in the Before column if you agree with the statement or a D if you disagree. After you've read this lesson, reread the statements to see if you have changed your mind.

·············· **Read to Learn** ··············

How is thermal energy transferred?

Have you ever gotten into a car on a hot summer day? You can guess that the inside of the car is hot even before you touch the door handle. You open the door and hot air seems to pour out of the car. When you touch the metal safety-belt buckle, it is hot. How is thermal energy transferred between objects? Thermal energy is transferred in three ways—by radiation, by conduction, and by convection.

Radiation

The transfer of thermal energy from one material to another by electromagnetic waves is called **radiation.** All matter, including the Sun, fire, you, and even ice, transfers thermal energy by radiation. Warm objects emit more radiation than cold objects do. You feel the transfer of thermal energy by radiation less when you place your hands near a block of ice than when you place your hands near a fire. ✔

Thermal energy from the Sun heats the inside of a car by radiation. Radiation is the only way thermal energy can travel from the Sun to Earth because space is a vacuum. However, radiation also transfers thermal energy through solids, liquids, and gases.

Conduction

Suppose it's a hot summer day and you are outside drinking a glass of cold lemonade. The lemonade has a lower temperature than the surrounding air. Therefore, the particles that make up the lemonade have less kinetic energy than the particles that make up the air. When particles with different kinetic energies collide, the particles with higher kinetic energy transfer energy to particles with lower kinetic energy.

In this case, the particles that make up the air collide with and transfer kinetic energy to the particles that make up the lemonade. As a result, the average kinetic energy, or temperature, of the particles that make up the lemonade increases. The hot air transfers thermal energy to, or heats, the cool lemonade.

Because kinetic energy is being transferred, thermal energy is being transferred. *The transfer of thermal energy between materials by the collisions of particles is called* **conduction.** Conduction continues until the thermal energy of all particles that are in contact is equal.

Thermal Conductors and Insulators

On a hot day, a metal safety-belt buckle in a car feels hotter than the cloth safety belt. The buckle and safety belt receive the same amount of thermal energy from the Sun. So why does the buckle feel hotter? The reason is that the metal that makes up the buckle is a good thermal conductor.

A **thermal conductor** *is a material through which thermal energy flows easily.* Atoms in good thermal conductors have electrons that move easily. These electrons transfer kinetic energy when they collide with other electrons and atoms. Metals (like those in safety-belt buckles) are better thermal conductors than nonmetals (like the materials in safety-belt straps).

By contrast, the material that makes up a safety belt is a good thermal insulator. *A* **thermal insulator** *is a material through which thermal energy does not flow easily.* The electrons in the atoms of a good thermal insulator do not move easily. These materials do not transfer thermal energy easily because fewer collisions occur between electrons and atoms. ✓

FOLDABLES

Make a three-column chart book to describe the ways thermal energy is transferred.

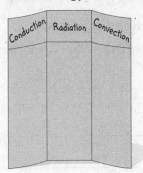

💡 Think it Over

2. Predict Suppose you have a cup of hot tea outside on a cold day. What will happen? (Circle the correct answer.)

a. Cold air particles will move into the hot tea.

b. Hot tea particles will transfer energy to the cold air particles.

c. The average kinetic energy of the tea particles will increase.

✓ Reading Check

3. Contrast What is the difference between a thermal conductor and a thermal insulator?

_____

_____

_____

_____

Specific Heat

The amount of thermal energy required to increase the temperature of 1 kg of a material by 1°C is called its **specific heat.** Every material has a specific heat. The temperature of a material with a low specific heat changes easily. The temperature of a material with a high specific heat does not change easily. ✓

Thermal conductors, such as metal safety-belt buckles, have a lower specific heat than thermal insulators, such as cloth safety belts and seat covers. This means it takes less thermal energy to increase a buckle's temperature than it takes to increase the temperature of a cloth safety belt or seat cover by the same amount. Thermal conductors and thermal insulators are shown in the figure below.

The specific heat of water is especially high. It takes a large amount of energy to increase or decrease the temperature of water. The high specific heat of water has many beneficial effects. For example, much of your body is water. Water's high specific heat helps prevent your body from overheating. The high specific heat of water is one of the reasons why pools, lakes, and oceans stay cool in summer. Water's high specific heat also makes it ideal for cooling machinery, such as car engines and rock-cutting saws. ✓

Specific Heat, Thermal Conductors, and Thermal Insulators

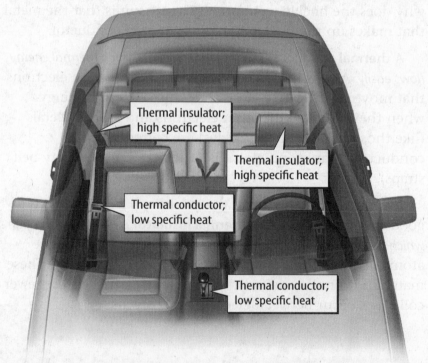

Thermal insulator; high specific heat

Thermal insulator; high specific heat

Thermal conductor; low specific heat

Thermal conductor; low specific heat

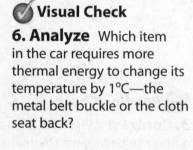 **Key Concept Check**

5. Summarize What does it mean if a material has a low specific heat?

_____

_____

✓ **Visual Check**

6. Analyze Which item in the car requires more thermal energy to change its temperature by 1°C—the metal belt buckle or the cloth seat back?

_____

_____

Thermal Expansion and Contraction

What happens if you take an inflated balloon outside on a cold day? Thermal energy transfers from the particles that make up the air inside the balloon to the particles that make up the balloon material and then to the cold outside air. As the particles that make up the air in the balloon lose thermal energy, which included kinetic energy, they slow down and move closer together. This cause the volume of the balloon to decrease. **Thermal contraction** *is a decrease in a material's volume when its temperature decreases.*

How could you reinflate the balloon? You could heat the air inside the balloon with a hair dryer. The particles that make up the hot air coming out of the hair dryer transfer thermal energy, which includes kinetic energy, to the particles that make up the air inside the balloon.

As the average kinetic energy of the particles increases, the air temperature increases. Also, as the average kinetic energy of the particles increases, they speed up and spread out. This increases the volume of the air inside the balloon. **Thermal expansion** *is an increase in a material's volume when its temperature increases.*

Thermal expansion and contraction are most noticeable in gases and less noticeable in liquids. They are least noticeable in solids.

Sidewalk Gaps

In may locations, the air temperatures are very hot in the summer. The high temperatures cause thermal expansion in structures, such as concrete sidewalks. If the concrete expands too much or expands unevenly, it could crack. Therefore, control joints are cut into sidewalks. If the sidewalk does crack, it should crack smoothly at the control joint. Sidewalks can withstand thermal expansion and contraction because of control joints. ✔

Key Concept Check

7. Explain What happens to the volume of a gas when it is heated?

_____

💡 **Think it Over**

8. Evaluate Thermal expansion is most noticeable in ___. (Circle the correct answer.)

a. water

b. rock

c. oxygen

✔ **Reading Check**

9. Consider What process occurs in a sidewalk when the temperature decreases?

_____

_____

Hot-Air Balloons

Hot-air balloons float because a burner heats the air in the balloon, causing thermal expansion. The particles that make up the air inside the balloon move faster and faster. The particles collide, and some are forced outside the balloon through the opening at the bottom. Now there are fewer particles in the balloon than in the same volume of air outside the balloon. The balloon is less dense and it begins to rise through denser outside air. ✓

To land a hot-air balloon, the balloonist allows the air inside the balloon to gradually cool. The air undergoes thermal contraction. But the balloon does not contract. Instead, denser air from outside the balloon fills the space inside. As the balloon's density increases, it slowly descends.

Ovenproof Glass

If you put an ordinary drinking glass into a hot oven, the glass might break or shatter. But a hot oven does not damage an ovenproof glass dish. Why? Different parts of ordinary glass expand at different rates when heated. This causes it to crack or shatter. But ovenproof glass is designed to expand less than ordinary glass when heated. This means that it usually does not crack in the oven. ✓

Convection

When you heat a pan of water on the stove, the burner heats the pan by conduction. This process involves the movement of thermal energy within a fluid. Particles that make up liquids and gases move around easily, transferring thermal energy from one location to another. **Convection** *is the transfer of thermal energy by the movement of particles from one part of a material to another.* Convection occurs only in fluids, such as water, air, magma, and maple syrup. ✓

Density, Thermal Expansion, and Thermal Contraction

If you heat a beaker of water on a burner, the burner transfers thermal energy to the beaker, which transfers thermal energy to the water. Thermal expansion occurs in water nearest the bottom of the beaker. Heating increases the water's volume, making it less dense. At the same time, water molecules at the water's surface transfer thermal energy to the air. This causes cooling and thermal contraction of the surface water. The denser surface water sinks to the bottom, forcing the less-dense water upward. This cycle continues until all the water in the beaker is at the same temperature.

Reading Check

10. Identify What process causes a hot-air balloon to rise when the balloonist turns on the burner?

_____

_____

Reading Check

11. Explain Why doesn't ovenproof glass shatter in a hot oven?

_____

_____

_____

Key Concept Check

12. Name What are the three processes that transfer thermal energy?

_____

_____

_____

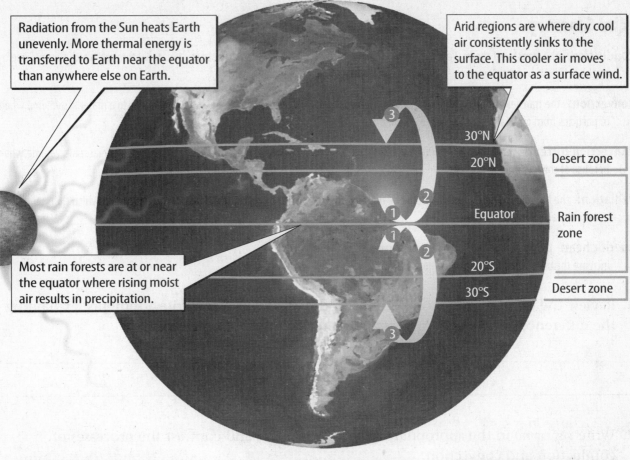

Radiation from the Sun heats Earth unevenly. More thermal energy is transferred to Earth near the equator than anywhere else on Earth.

Arid regions are where dry cool air consistently sinks to the surface. This cooler air moves to the equator as a surface wind.

Most rain forests are at or near the equator where rising moist air results in precipitation.

30°N

20°N — Desert zone

Equator — Rain forest zone

20°S

30°S — Desert zone

1 The higher amount of thermal energy at the equator heats the air. The air becomes less dense and rises.

2 Water vapor in the rising air condenses as the air rises and cools. The water falls back to Earth as rain.

3 Cooler air sinks back to Earth's surface where it moves to the equator to replace the less dense, rising air.

Convection Currents in Earth's Atmosphere

The movement of fluids in a cycle because of convection is a **convection current.** Convection currents circulate the water in Earth's oceans and other bodies of water. They also circulate the air in a room and the materials in Earth's interior. Convection currents also move matter and thermal energy from inside the Sun to its surface.

On Earth, convection currents move air between the equator and latitudes near 30°N and 30°S. This plays an important role in Earth's climates, as shown in the figure above. The locations of rain forests and deserts are influenced by convection currents.

✓ Visual Check

13. Locate At what latitudes are most of Earth's rain forests located?

_____

_____

Mini Glossary

conduction: the transfer of thermal energy between materials by the collisions of particles

convection: the transfer of thermal energy by the movement of particles from one part of a material to another

convection current: the movement of fluids in a cycle because of convection

radiation: the transfer of thermal energy from one material to another by electromagnetic waves

specific heat: the amount of thermal energy required to increase the temperature of 1 kg of a material by 1°C

thermal conductor: a material through which thermal energy flows easily

thermal contraction: a decrease in a material's volume when its temperature decreases

thermal expansion: an increase in a material's volume when its temperature increases

thermal insulator: a material through which thermal energy does not flow easily

1. Review the terms and their definitions in the Mini Glossary. Write a sentence explaining the difference between thermal contraction and thermal expansion.

_____

_____

2. Write *yes* or *no* in the appropriate spaces to compare and contrast the processes of conduction and convection.

	Conduction	Convection
Transfers thermal energy	yes	yes
Transfers thermal energy only in fluids		
Transfers thermal energy from warmer to cooler materials		
Transfers thermal energy between objects that touch		

3. Use your knowledge of specific heat to explain why you might burn your hand if you touched a copper pan on a hot burner while the water inside the pan was still only warm.

_____

What do you think NOW?

Reread the statements at the beginning of the lesson. Fill in the After column with an A if you agree with the statement or a D if you disagree. Did you change your mind?

 Connect ED

Log on to ConnectED.mcgraw-hill.com and access your textbook to find this lesson's resources.

 END OF LESSON

Thermal Energy

Using Thermal Energy

•••••••••••••• **Before You Read** ••••••••••••••

What do you think? Read the two statements below and decide whether you agree or disagree with them. Place an A in the Before column if you agree with the statement or a D if you disagree. After you've read this lesson, reread the statements to see if you have changed your mind.

Before	Statement	After
	5. Car engines create energy.	
	6. Refrigerators cool food by moving thermal energy from inside the refrigerator to the outside.	

•••••••••••••• **Read to Learn** ••••••••••••••

Thermal Energy Transformations

Burning wood heats the air. A toaster gets hot when you turn it on. You can convert other forms of energy into thermal energy. You also can convert thermal energy into other forms of energy. Thermostats switch heaters on and off, transforming thermal energy into mechanical energy. When you convert energy from one form to another, you can use the energy to perform useful tasks. Energy cannot be created or destroyed. Many devices transform energy from one form to another or transfer energy from one place to another. However, the total amount of energy does not change.

Heating Appliances

A device that converts electric energy into thermal energy is a **heating appliances.** Curling irons and coffeemakers are heating appliances. Computers and cell phones also become warm when you use them. This is because some electric energy always converts to thermal energy in an electronic device. However, the thermal energy that most electronic devices generate is not used for any purpose.

Thermostats

A **thermostat** *is a device that regulates the temperature of a system.* Refrigerators, toasters, and ovens have thermostats.

Key Concepts

- How does a thermostat work?
- How does a refrigerator keep food cold?
- What are the energy transformations in a car engine?

▸ **Mark the Text**

Identify the Main Ideas
Write a phrase beside each paragraph that summarizes the main point of the paragraph. Use the phrases to review the lesson.

FOLDABLES

Make a four-tab book to explain the energy transformation that occurs in each device.

Copyright © Glencoe/McGraw-Hill, a division of The McGraw-Hill Companies, Inc.

Turning a Furnace Off Most thermostats in home heating systems contain a bimetallic coil. A bimetallic coil is made of two types of metal that are joined together and bent into a coil. The metal on the inside of the coil expands and contracts more with changes in temperature than the metal on the outside of the coil. After a room warms, the thermal energy in the air causes the bimetallic coil to uncurl slightly. This tilts a switch that turns off the furnace.

Turning a Furnace On As the air in the room cools, the metal on the inside of the coil contracts more than the metal on the outside. This curls the coil tighter, which tilts the switch in the other direction, turning on the furnace.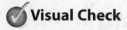

Refrigerators

A device that uses electric energy to transfer thermal energy from a cooler location to a warmer location is called a **refrigerator.** Recall that thermal energy naturally flows from a warmer area to a cooler area. A refrigerator does the opposite. It moves thermal energy from from its cold inside to the warm air outside. As shown below, pipes that surround the refrigerator are filled with a fluid, called a coolant, that flows through the pipes. Thermal energy inside the refrigerator transfers to the coolant, keeping the inside of the refrigerator cold. ✔

Vaporizing the Coolant

A coolant is a substance that evaporates at a low temperature. A coolant is pumped through the pipes inside and outside of the refrigerator. The coolant, which begins as a liquid, passes through an expansion valve and cools. The cold gas flows through pipes inside the refrigerator, absorbs thermal energy, and vaporizes. The coolant gas becomes warmer, and the inside of the refrigerator becomes cooler.

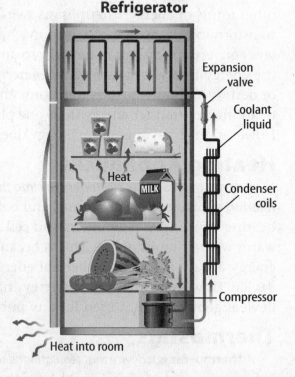

Refrigerator

Expansion valve

Coolant liquid

Heat

MILK

Condenser coils

Compressor

Heat into room

Key Concept Check

1. Explain How does the bimetallic coil in a thermostat respond to heating and cooling?

_____

_____

_____

_____

Reading Check

2. Summarize What type of energy does the coolant in a refrigerator move?

_____

_____

Visual Check

3. Explain What is happening as the coolant gas becomes warmer?

_____

_____

_____

_____

Condensing the Coolant

The coolant flows to an electric compressor. Here, the coolant is compressed, or forced into a smaller space, which increases its thermal energy. Then, the gas is pumped through condenser coils. There, the thermal energy of the gas is greater than that of the surrounding air. This causes thermal energy to flow from the coolant gas to the air. As thermal energy is removed from the gas, it condenses, or becomes a liquid. The liquid coolant is pumped up through the expansion valve. The cycle repeats.

Heat Engines

A car engine is a heat engine. *A heat engine is a machine that converts thermal energy into mechanical energy.* This mechanical energy then moves the car. Most vehicles use a type of heat engine called an internal combustion engine. The figure below shows how one type of internal combustion engine converts thermal energy into mechanical energy.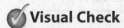

A cylinder is a tube with a piston that moves up and down. At one end of the cylinder, a spark ignites a fuel-air mixture. The ignited mixture expands and pushes the piston down. The fuel's chemical energy converts to thermal energy and some of the thermal energy immediately converts to mechanical energy. A heat engine is not efficient. Most car engines convert only about 20 percent of the chemical energy in gasoline into mechanical energy. The remaining energy from the gasoline is lost to the environment.

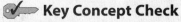
Key Concept Check
4. Identify What is one form of energy that is output from a heat engine?

_____

_____

Visual Check
5. State What device ignites the fuel-air mixture in an internal combustion engine?

_____

_____

Internal Combustion Engine

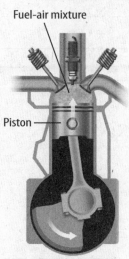
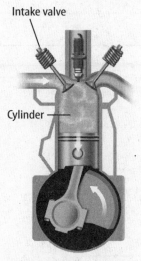
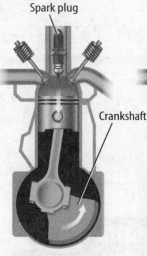
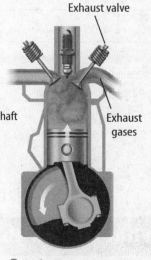

❶ The intake valve opens as the piston moves downward, drawing a mixture of gasoline and air into the cylinder.

❷ The intake valve closes as the piston moves upward, compressing the fuel-air mixture.

❸ A spark plug ignites the fuel-air mixture. As the mixture burns, hot gases expand, pushing the piston down.

❹ As the piston moves up, the exhaust valve opens, and the hot gases are pushed out of the cylinder.

Mini Glossary

heat engine: a machine that converts thermal energy into mechanical energy

heating appliance: a device that converts electric energy into thermal energy

refrigerator: a device that uses electric energy to transfer thermal energy from a cooler location to a warmer location

thermostat: a device that regulates the temperature of a system

1. Review the terms and their definitions in the Mini Glossary. Write a sentence explaining how a refrigerator keeps food cold.

   _____

   _____

2. Complete the graphic organizer to show the types of energy a heat engine converts and produces.

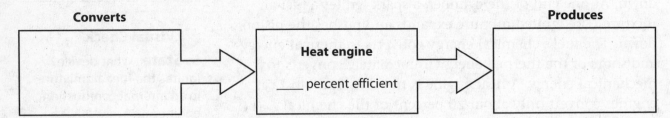

Converts   Heat engine _____ percent efficient   Produces

3. You use some logs to build a campfire. Have you created energy? Destroyed energy? Explain your answer.

   _____

   _____

   _____

What do you think NOW?

Reread the statements at the beginning of the lesson. Fill in the After column with an A if you agree with the statement or a D if you disagree. Did you change your mind?

 Log on to ConnectED.mcgraw-hill.com and access your textbook to find this lesson's resources.

 END OF LESSON

States of Matter

Solids, Liquids, and Gases

·············· Before You Read ··············

What do you think? Read the two statements below and decide whether you agree or disagree with them. Place an A in the Before column if you agree with the statement or a D if you disagree. After you've read this lesson, reread the statements to see if you have changed your mind.

Before	Statement	After
	1. Particles moving at the same speed make up all matter.	
	2. The particles in a solid do not move.	

·············· Read to Learn ··············

Describing Matter

Picture yourself blowing bubbles by the seaside. Do you see matter in this scene? The three most common forms, or states, of matter on Earth are solids, liquids, and gases. The bubbles you blow hold air, which is a mixture of gases. The soap mixture used to make the bubbles and the ocean water are liquids. The sand, your shoes, and nearby seashells are a few of the solids you might see by the seaside.

There is a fourth state of matter, plasma. Plasma is high-energy matter made up of particles that have positive and negative charges. Plasma is the most common state of matter in space. Plasma also is in lightning flashes, fluorescent lights, and stars, such as the Sun. ✔

Matter can be described in many ways. You can describe matter using your senses. You can describe its state, color, texture, and smell. You also can describe matter using measurements, such as mass, volume, and density. Mass is the amount of matter in an object. The units for mass are often grams (g) or kilograms (kg). Volume is the amount of space that a sample of matter takes up. The units for liquid volume are usually liters (L) or milliliters (mL). The units for solid volume are usually cubic centimeters (cm^3) or cubic meters (m^3). Density is a quantity calculated by dividing an object's mass by its volume. The units of density are usually g/cm^3 or g/mL.

Key Concepts

- How do particles move in solids, liquids, and gases?
- How are the forces between particles different in solids, liquids, and gases?

Study Coach

Make a Table with three columns to contrast solids, liquids, and gases. Label one column *Particle Motion and Forces*. Label the second column *Definite Shape?* Label the third column *Definite Volume?* Complete the table as you read this lesson.

✔ Reading Check

1. Name the four states of matter.

_____

_____

_____

_____

Solid

Liquid

Gas

Think it Over

2. Relate How does particle speed relate to the distance between particles?

_____

_____

_____

Visual Check

3. Draw Circle the particles that show the weakest attractive forces between them.

Particles in Motion

Have you ever wondered what makes something a solid, a liquid, or a gas? Two main factors that determine the state of matter are particle motion and particle forces.

Atoms, ions, or molecules make up all matter. These particles can move in different ways. In some matter, they are close together and vibrate back and forth. In other matter, the particles are farther apart. Sometimes, they slide past each other. At other times, they move freely and spread out. It does not matter how close the particles are to each other. All particles have random motion. Random motion is movement in all directions and at different speeds. If particles are free to move, they move in straight lines until they collide with something. Collisions usually change the speed and direction of the particles' movements.

Forces Between Particles

Recall that atoms that make up matter have positively charged protons and negatively charged electrons. These opposite charges attract each other. They create attractive forces between any two particles. Attractive forces pull particles together.

Strong attractive forces hold slow-moving particles close together, as shown in the figure below. As the motion of particles gets faster, particles move farther apart. When they get farther apart, the attractive forces between particles have a weaker effect. The spaces between them increase. This bigger space lets other particles slip past. As the motion of particles gets even faster, particles move even farther apart. In time, the distance between particles is so great that there is little or no attractive force between them. The particles move randomly and spread out.

Particle Motion

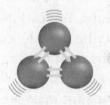

Particles move slowly and can only vibrate in place. Therefore, the attractive forces between particles are strong.

Particles move faster and slip past each other. The distance between particles increases. Therefore, the attractive forces between particles are weaker.

Particles move fast. The distance between the particles is great, and therefore, the attractive forces between particles are very weak.

Solids

If a skateboard moves from one place to another, its shape and volume do not change. A skateboard's shape and volume do not change because a skateboard is a solid. *A **solid** is matter that has a definite shape and a definite volume.*

Particles in a Solid

Why doesn't a solid change shape or volume? Remember that the particles in a solid are close together. The particles are touching neighboring particles. The attractive forces between them are strong. Their strong attractive forces and slow motion hold the particles tightly in their positions. The particles still move, but they do not get away from each other. They simply vibrate back and forth in place. This arrangement gives solids a definite shape and volume. 🔑

Types of Solids

All solids are not the same. For example, a diamond and a piece of charcoal do not look alike. However, they are both solids made of carbon atoms. They both have particles that strongly attract each other and vibrate in place. What makes them different is the arrangement of their particles. A diamond is a crystalline solid. It has particles arranged in a specific, repeating order. Charcoal is an amorphous solid. It has particles that are arranged randomly. Different particle arrangements give these materials different properties. For example, a diamond is a hard material. Charcoal is brittle. ✔

Liquids

Have you ever seen a waterfall flowing into a riverbed? Water is a liquid. *A **liquid** is matter with a definite volume but no definite shape.* Liquids flow and can take the shape of their containers. Water from a waterfall takes the shape of the riverbed that it fills.

Particles in a Liquid

How can liquids change shape? The particle motion in liquids is faster than the particle motion in solids. This faster motion causes the particles to move slightly farther apart. As they move farther apart, the effect of the attractive forces between them decreases. The faster motion also causes gaps to form between the particles. The gaps allow particles to slip past each other. The slightly weaker attractive forces and gaps between particles let liquids flow and take the shape of their containers.

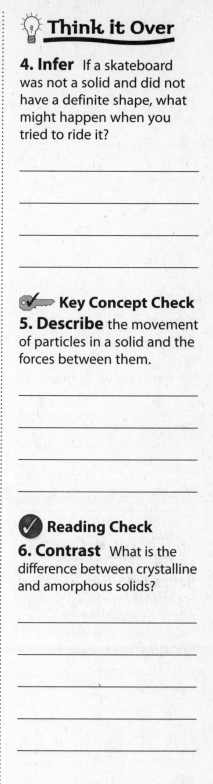

💡 Think it Over

4. Infer If a skateboard was not a solid and did not have a definite shape, what might happen when you tried to ride it?

_____

_____

_____

_____

🔑 Key Concept Check

5. Describe the movement of particles in a solid and the forces between them.

_____

_____

_____

_____

✔ Reading Check

6. Contrast What is the difference between crystalline and amorphous solids?

_____

_____

_____

_____

_____

Viscosity

If you have ever poured or dipped honey, you know what a liquid with a high viscosity is like. **Viscosity** (vihs KAW sih tee) *is a measurement of a liquid's resistance to flow.* Honey has high viscosity. Water, on the other hand, has low viscosity. This property of a liquid is due to the strength of attraction between particles, particle mass, and particle shape.

- Strong forces between particles slow particle movement as particles slip past each other. In general, the stronger the forces are between particles, the higher the viscosity. For many liquids, viscosity decreases as the liquid becomes warmer.

- The mass of a particle also affects its ability to slip past other particles. More massive particles tend to move more slowly.

- Particles with complex shapes, such as long chains, also have high viscosity. Such long particles have difficulty slipping past other particles.

Surface Tension

Have you ever seen an insect that can walk on water? Believe it or not, some insects can do this because of the forces between molecules.

Water molecules below the surface of water are surrounded on all sides by other water molecules. Therefore, they have attractive forces, or pulls, in all directions. The attraction between similar molecules, such as water molecules, is called cohesion.

Water molecules at the surface of a liquid do not have liquid water molecules above them. As a result, there is a greater downward pull on the molecules. This downward pull causes the surface particles of water to become tightly stretched like the head of a drum. Molecules at the surface of a liquid have **surface tension,** *the uneven forces acting on the particles on the surface of a liquid.* Surface tension makes it possible for some insects to walk on water. In general, the stronger the attractive forces are between particles, the greater the surface tension of the liquid.

Think about the bubbles you were blowing in the imaginary scene at the beginning of the lesson. The thin water-soap film surrounding the bubbles formed because of surface tension between the particles. ✓

Copyright © Glencoe/McGraw-Hill, a division of The McGraw-Hill Companies, Inc.

Key Concept Check

7. Describe the movement of particles in a liquid and the forces between them.

_____

_____

_____

_____

Think it Over

8. Predict whether an insect could walk on a liquid such as rubbing alcohol, which has a smaller attraction between molecules than water.

_____

_____

_____

Reading Check

9. Explain What causes surface tension?

_____

_____

_____

_____

Gases

Think about a beach ball. To make it big and round, you have to fill it with a gas. *A gas is matter that has no definite volume and no definite shape.*

It is not easy to identify the gas in a beach ball because you cannot see it. In fact, there are gas particles inside and outside a beach ball. Air is all around us all the time. Air is a mixture of gases, including nitrogen, oxygen, argon, and carbon dioxide. ✓

Particles in a Gas

Why don't gases have definite volumes or definite shapes? Compared to the particles in the solid and the liquid states, the particles in gases are very far apart.

The distances between the particles in a solid are small. The particles touch each other. The distances between the particles in a liquid are greater, and the particles can slip past each other. The distances between the particles in a gas differ from those in solids and liquids. In a gas, the forces of attraction between the particles are not strong enough to keep the particles close together. As a result, distances between particles are greater in the gas state. ✓

Forces Between Particles

Particles in the gas state have greater motion than the same particles in the solid or liquid state. Because the particles are moving quickly, the distances between particles increase. As the distances increase, the attractive forces between particles have less of an effect.

The distances are so great and the effect of the attractive forces so small that gas particles act like they have little or no attraction to each other. As a result, the particles spread out to fill their container. Gases have no definite shape or volume.

Vapor

Have you ever heard the term *vapor? The gas state of a substance that is normally a solid or a liquid at room temperature is called* **vapor.** For example, water is normally a liquid at room temperature. When it is a gas, such as in air, it is called water vapor. Other substances that can form a vapor are rubbing alcohol, grain alcohol, iodine, mercury, and gasoline. 🗝

✓ **Reading Check**

10. Identify What is a gas, and what is another object that contains a gas?

_____

_____

_____

✓ **Reading Check**

11. Generalize Which state of matter has the greatest distance between particles?

_____

🗝 **Key Concept Check**

12. Describe How do particles move and interact in a gas?

_____

_____

_____

_____

Mini Glossary

gas: matter that has no definite volume and no definite shape

liquid: matter with a definite volume but no definite shape

solid: matter that has a definite shape and a definite volume

surface tension: the uneven forces acting on the particles on the surface of a liquid

vapor: the gas state of a substance that is normally a solid or a liquid at room temperature

viscosity (vihs KAW sih tee): a measurement of a liquid's resistance to flow

1. Review the terms and their definitions in the Mini Glossary. Write a sentence that explains the differences between liquids and gases.

_____

_____

2. Using what you have learned about the states of matter, draw lines to connect two characteristics (shown in ovals) to each state of matter (shown in boxes).

| Definite Shape | Definite Volume | No Definite Shape | No Definite Volume |

| SOLID | LIQUID | GAS |

3. Which state of matter is a vapor? From what states of matter does a vapor form?

_____

_____

What do you think NOW?

Reread the statements at the beginning of the lesson. Fill in the After column with an A if you agree with the statement or a D if you disagree. Did you change your mind?

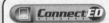 Connect ED

Log on to ConnectED.mcgraw-hill.com and access your textbook to find this lesson's resources.

 END OF LESSON

States of Matter

Changes in State

······· **Before You Read** ·······

What do you think? Read the two statements below and decide whether you agree or disagree with them. Place an A in the Before column if you agree with the statement or a D if you disagree. After you've read this lesson, reread the statements to see if you have changed your mind.

Before	Statement	After
	3. Particles of matter have both potential energy and kinetic energy.	
	4. When a solid melts, thermal energy is removed from the solid.	

······· **Read to Learn** ·······

Kinetic and Potential Energy

When snow melts after a snowstorm, all three states of water are present. The snow is a solid, the melted snow is a liquid, and the air above the snow and ice contains water vapor, a gas. What causes particles to change state?

Kinetic Energy

Recall that the particles that make up matter are always moving. These particles have **kinetic energy,** *the energy an object has due to its motion.* The faster particles move, the more kinetic energy they have. Within a given substance, such as water, particles in the solid state have the least amount of kinetic energy. This is because they only vibrate in place. Particles in the liquid state move faster than particles in the solid state. Therefore, they have more kinetic energy. Particles in the gaseous state move quickly. They have the most kinetic energy of particles of a given substance.

Temperature *is a measure of the average kinetic energy of all the particles in an object.* Within a given substance, a rise in temperature means that the particles, on average, are moving at greater speeds. Therefore, the particles have more kinetic energy. For example, water molecules at 25°C are moving faster and have more kinetic energy than water molecules at 10°C.

Key Concepts

- How is temperature related to particle motion?
- How are temperature and thermal energy different?
- What happens to thermal energy when matter changes from one state to another?

▶ **Mark the Text**

Building Vocabulary Skim this lesson and circle any words you do not know. If you still do not understand a word after reading the lesson, look it up in the dictionary. Keep a list of these words and definitions to refer to when you study other chapters.

☑⚷ **Key Concept Check**
1. Relate How is temperature related to particle motion?

_____

_____

_____

_____

Potential Energy

In addition to kinetic energy, particles have potential energy. Recall that potential energy is stored energy due to the interactions between particles or objects. Think about holding a basketball and then letting it go. The gravitational force between the ball and Earth causes the ball to fall toward Earth. Before you let the ball go, it has potential, or stored, energy.

Potential energy typically increases when objects get farther apart. It decreases when objects get closer together. When you hold up a basketball, it is farther off the ground than when it is falling from your hands. It has a higher potential energy than when it is falling. When the basketball is touching the ground, it has no more potential energy. The farther an object is from Earth's surface, the greater its gravitational potential energy is. As the ball gets closer to the ground, its potential energy decreases. ✓

You can think of the potential energy of particles in a similar way. The chemical potential energy of particles is due to their position relative to other particles. The chemical potential energy of particles increases and decreases as the distances between particles increase or decrease. Thus, particles that are farther apart have greater chemical potential energy than particles that are closer together.

Thermal Energy

Changes in state are caused by changes in thermal energy. **Thermal energy** *is the total potential and kinetic energies of an object.* You can change an object's state of matter by adding or removing thermal energy. When you add thermal energy to an object, these things can happen:

- Particles move faster (increased kinetic energy).

- Particles get farther apart (increased potential energy).

- Particles get faster and move farther apart (increased kinetic and potential energy).

The opposite is true when you remove thermal energy:

- Particles move slower (less kinetic energy).

- Particles get closer together (less potential energy).

- Particles move slower and closer together (less kinetic and potential energy).

If enough thermal energy is added or removed, a change of state can occur. ✓

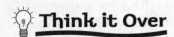

Reading Check

2. Apply Which has more potential energy: a baseball held 1 m above the ground or a baseball held 2 m above the ground?

_____

_____

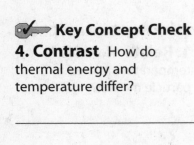

Think it Over

3. Model Imagine two balls connected by a spring. In which arrangement do the balls have more potential energy: when they are pulled apart or when they are closer together?

_____

_____

Key Concept Check

4. Contrast How do thermal energy and temperature differ?

_____

_____

_____

Solid to Liquid or Liquid to Solid

After you drink a beverage from an aluminum can, do you recycle the can? Aluminum recycling is one example of a process that involves matter changing from one state to another by adding or removing thermal energy.

Melting

The first part of the recycling process involves melting aluminum cans. To change matter from a solid to a liquid, thermal energy must be added. The graph below shows the relationship between increasing temperature and increasing thermal energy (potential energy + kinetic energy).

At first, the thermal energy and the temperature increase. The temperature stops rising when it reaches the melting point of the matter. The melting point is the temperature at which the solid changes to a liquid. As aluminum changes from solid to liquid, the temperature does not change. However, energy changes still occur.

Energy Changes

What happens when a solid reaches its melting point? Notice that the line on the graph below is horizontal. This means that the temperature, or average kinetic energy, stops increasing. However, the amount of thermal energy continues to increase. How is this possible?

Once a solid reaches the melting point, additional thermal energy causes the particles to overcome their attractive forces. The particles move farther apart and potential energy increases. Once a solid completely melts, the addition of thermal energy will cause the kinetic energy of the particles to increase again, as shown by a temperature increase.

Copyright © Glencoe/McGraw-Hill, a division of The McGraw-Hill Companies, Inc.

Reading Check

5. Infer What must be added to matter to change it from a solid to a liquid?

_____

_____

Visual Check

6. Analyze During melting, which factor remains constant?

_____

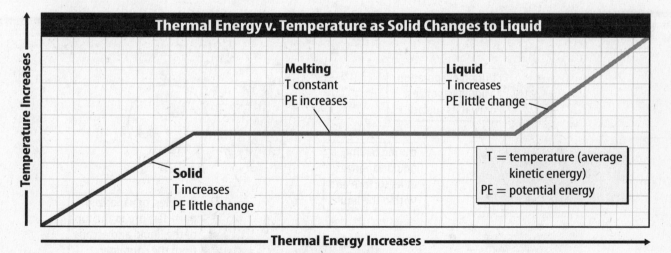

Thermal Energy v. Temperature as Solid Changes to Liquid

Temperature Increases

Thermal Energy Increases

Solid
T increases
PE little change

Melting
T constant
PE increases

Liquid
T increases
PE little change

T = temperature (average kinetic energy)
PE = potential energy

Freezing

After the aluminum melts, it is poured into molds to cool. As the aluminum cools, thermal energy leaves it. If enough energy is removed, the aluminum will freeze. Freezing is a process that is the opposite of melting—liquid changes to solid. The temperature at which matter changes from the liquid state to the solid state is its freezing point. You can look at the graph of melting on the previous page to follow the process of freezing as thermal energy is removed. To observe the temperature and thermal energy changes that take place as liquid aluminum forms solid blocks, move from right to left on the graph on the previous page. ✓

Liquid to Gas or Gas to Liquid

When you heat water, do you ever notice how bubbles begin to form at the bottom and rise to the surface? The bubbles contain water vapor, a gas. As the water heats, it changes from the liquid state to the gaseous state. *The change in state of a liquid into a gas is* **vaporization.** The figure below shows two types of vaporization—evaporation and boiling. The two types of vaporization differ in where they take place in the liquid.

Boiling

Vaporization that occurs within a liquid is called boiling. During boiling, vaporization takes place throughout the liquid. The temperature at which boiling occurs in a liquid is called its boiling point.

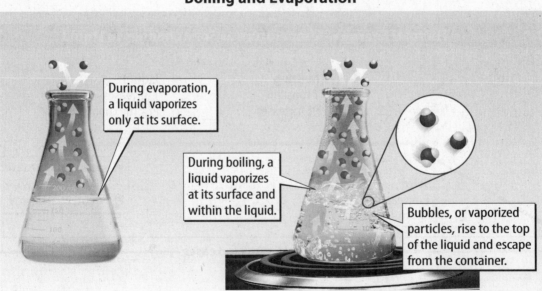

Boiling and Evaporation

During evaporation, a liquid vaporizes only at its surface.

During boiling, a liquid vaporizes at its surface and within the liquid.

Bubbles, or vaporized particles, rise to the top of the liquid and escape from the container.

Visual Check

8. Explain Why doesn't the evaporation flask have bubbles below the surface?

_____

_____

_____

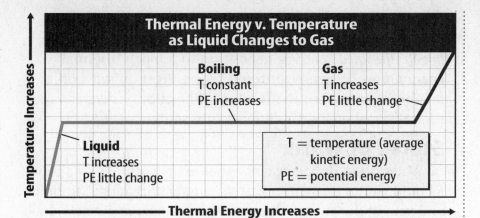

Thermal Energy v. Temperature as Liquid Changes to Gas

Temperature Increases →

Boiling
T constant
PE increases

Gas
T increases
PE little change

Liquid
T increases
PE little change

T = temperature (average kinetic energy)
PE = potential energy

Thermal Energy Increases →

Copyright © Glencoe/McGraw-Hill, a division of The McGraw-Hill Companies, Inc.

In the graph above, notice the energy changes that occur as thermal energy is added. The kinetic energy of particles increases until the liquid reaches its boiling point. At the boiling point, the potential energy of particles begins increasing. The particles move farther apart until the attractive forces no longer hold them together. At this point, the liquid changes to a gas. When boiling ends, if thermal energy continues to be added, the kinetic energy of the gas particles begins to increase again. Therefore, the temperature begins to increase again as shown on the graph above.

Evaporation

Unlike boiling, **evaporation** *is vaporization that occurs only at the surface of a liquid.* A small amount of liquid in an open container will disappear after several days due to evaporation.

Condensation

Boiling and evaporation are processes that change a liquid to a gas. The opposite process also occurs. When a gas loses enough thermal energy, the gas changes to a liquid, or condenses. *The change of state from a gas to a liquid is called* **condensation.** Overnight, water vapor often condenses on blades of grass and forms dew.

Solid to Gas or Gas to Solid

A solid can become a gas without turning into a liquid. Also, a gas can become a solid without turning into a liquid.

Solid to Gas Dry ice is solid carbon dioxide. It turns immediately into a gas when thermal energy is added to it. The process is called sublimation. **Sublimation** *is the change of state from a solid to a gas without going through the liquid state.* As dry ice sublimes, it cools and condenses the water vapor in the surrounding air, creating a thick fog.

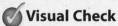 **Visual Check**

9. Explain Why does the liquid change to a gas?

_____

_____

_____

FOLDABLES

Make a four-tab Foldable and record what you learn about each term under the tabs.

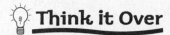

Think it Over

10. Apply Clouds can form when water vapor in the air condenses. Clouds are what state of matter?

_____

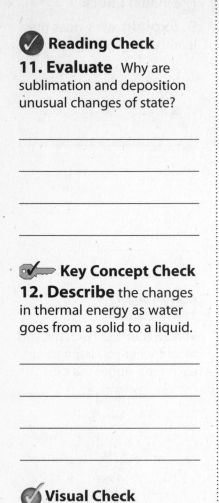

11. Evaluate Why are sublimation and deposition unusual changes of state?

_____

_____

_____

_____

Key Concept Check

12. Describe the changes in thermal energy as water goes from a solid to a liquid.

_____

_____

_____

_____

Visual Check

13. Locate Circle the location on the graph where water reaches its boiling point.

Gas to Solid The opposite of sublimation is deposition. **Deposition** _is the change of state of a gas to a solid without going through the liquid state._ For deposition to happen, thermal energy must be removed from the gas. Frost on grass on a fall morning is often the result of deposition. As water vapor loses thermal energy, it changes into solid frost. ✓

States of Water

Water is the only substance that exists naturally as a solid, a liquid, and a gas within Earth's temperature range. To better understand the energy changes during a change in state, look at the heating curve of water shown in the graph below.

Adding Thermal Energy

Suppose you place a beaker of ice on a hot plate. The hot plate moves thermal energy to the beaker and the ice. The temperature of the ice increases. Recall that this means the average kinetic energy of the water molecules increases.

At 0°C, the melting point of water, the water molecules vibrate so rapidly that they begin to move out of their places. At this point, any added thermal energy causes the particles to overcome their attractive forces, and melting occurs. Once melting is complete, the kinetic energy of the particles begins to increase again as more thermal energy is added. Then the temperature begins to increase, too.

When water reaches 100°C, its boiling point, liquid water begins to change to water vapor. Again, kinetic energy stays the same as vaporization occurs. When the change of state is complete, the kinetic energy of molecules increases again, and so does the temperature. ✓

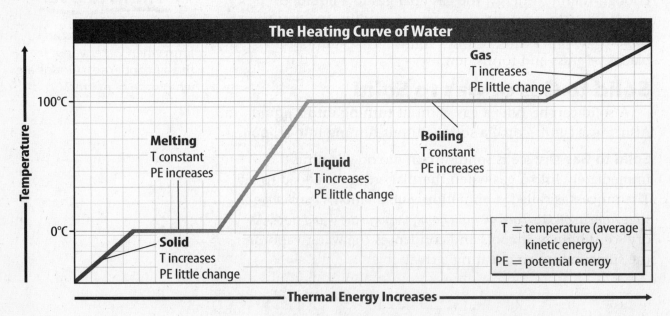

The Heating Curve of Water

Gas
T increases
PE little change

Melting
T constant
PE increases

Liquid
T increases
PE little change

Boiling
T constant
PE increases

Solid
T increases
PE little change

T = temperature (average kinetic energy)
PE = potential energy

Temperature — 100°C — 0°C

Thermal Energy Increases →

Removing Thermal Energy

The removal of thermal energy is the reverse of the process shown in the heating curve of water. You can follow what happens to water vapor as you remove thermal energy by following the graph on the previous page from right to left. Cooling water vapor changes the gas to a liquid. Cooling the water further changes it to ice. ✓

Conservation of Mass and Energy

The diagram below shows the energy changes that take place as thermal energy is added or removed from matter. Notice that there are three sets of opposite processes:

- melting and freezing
- vaporization and condensation
- sublimation and deposition

During all of these changes of state, matter and energy are always conserved.

Sometimes, such as when water vaporizes, it seems to have disappeared. However, it has just formed an invisible gas. If the gas were captured and its mass added to the remaining mass of the liquid, you would see that matter is conserved.

The same is true for energy. Surrounding matter often absorbs thermal energy. If you measured thermal energy in the matter and the surrounding matter, you would find that energy is also conserved.

Copyright © Glencoe/McGraw-Hill, a division of The McGraw-Hill Companies, Inc.

✓ **Reading Check**

14. Describe what happens to water vapor when thermal energy is removed from it.

_____

_____

_____

✓ **Visual Check**

15. Draw Circle the state of matter that results when thermal energy is added to a liquid.

Changes of State

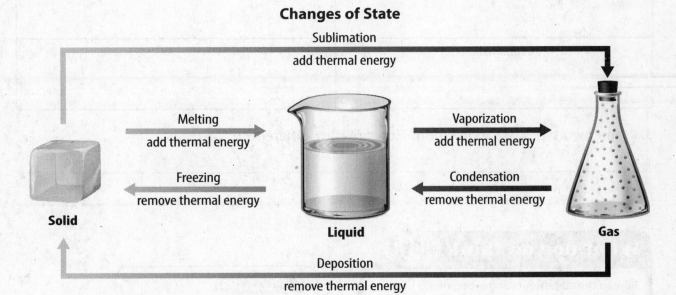

Sublimation
add thermal energy

Melting
add thermal energy

Freezing
remove thermal energy

Solid

Liquid

Vaporization
add thermal energy

Condensation
remove thermal energy

Gas

Deposition
remove thermal energy

Mini Glossary

condensation: the change of state from a gas to a liquid

deposition: the change of state of a gas to a solid without going through the liquid state

evaporation: vaporization that occurs only at the surface of a liquid

kinetic energy: the energy an object has due to its motion

sublimation: the change of state from a solid to a gas without going through the liquid state

temperature: a measure of the average kinetic energy of all the particles in an object

thermal energy: the total potential and kinetic energies of an object

vaporization: the change in state of a liquid into a gas

1. Review the terms and their definitions in the Mini Glossary. Write a sentence that includes an example of one change of state and its opposite process.

_____

_____

2. Write the correct term next to its opposite term in each row of the table below.

Happens When Thermal Energy Is Added	Happens When Thermal Energy Is Removed
melting	
	condensation
sublimation	

3. Name two types of vaporization. How are they different?

_____

_____

What do you think NOW?

Reread the statements at the beginning of the lesson. Fill in the After column with an A if you agree with the statement or a D if you disagree. Did you change your mind?

 **ConnectED**

Log on to ConnectED.mcgraw-hill.com and access your textbook to find this lesson's resources.

 END OF LESSON

States of Matter

The Behavior of Gases

··········· **Before You Read** ···········

What do you think? Read the two statements below and decide whether you agree or disagree with them. Place an A in the Before column if you agree with the statement or a D if you disagree. After you've read this lesson, reread the statements to see if you have changed your mind.

Before	Statement	After
	5. Changes in temperature and pressure affect gas behavior.	
	6. If the pressure on a gas increases, the volume of the gas also increases.	

··············· **Read to Learn** ···············

Understanding Gas Behavior

Pilots do not worry as much about solids and liquids at high altitudes as they do gases. That is because gases behave differently than solids and liquids. Changes in temperature, pressure, and volume affect the behavior of gases more than they affect the behavior of solids and liquids.

In the previous two lessons, you learned about how the particles of matter in different states behave—how they are arranged, how they move, how forces affect them, and how changes in energy affect them. The explanation of particle behavior in solids, liquids, and gases is based on the kinetic molecular <u>theory</u>. The **kinetic molecular theory** *is an explanation of how particles in matter behave.* Some basic ideas in this theory are

- small particles make up all matter;
- these particles are in constant, random motion;
- the particles collide with other particles, other objects, and the walls of their container;
- when particles collide, no energy is lost.

You read about most of these when you learned about all three states of matter. However, the last two statements are important in explaining how gases behave.

Key Concepts 🔑

- How does the kinetic molecular theory describe the behavior of a gas?
- How are temperature, pressure, and volume related in Boyle's law?
- How is Boyle's law different from Charles's law?

Study Coach

Make an outline as you read to summarize the information in the lesson. Use the main headings in the lesson as the main headings in your outline. Use your outline to review the lesson.

ACADEMIC VOCABULARY

theory
(noun) an explanation of things or events that is based on knowledge gained from many observations and investigations

🔑 Key Concept Check

1. Explain How does the kinetic molecular theory describe the behavior of a gas?

_____

_____

_____

Pressure Increases as Volume Decreases

| Greatest volume, least pressure | Less volume, more pressure | Least volume, most pressure |

✓ Visual Check

2. Confirm that the number of gas particles is the same in each cylinder by counting them and writing the number below each one.

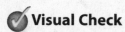

FOLDABLES®

Fold a sheet of notebook paper to make a three-tab Foldable to compare two important gas laws.

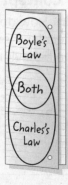

What is pressure?

Particles in gases are always moving. As a result of this movement, gas particles are always bumping into other particles and into their container. When particles collide with their container, pressure results. **Pressure** *is the amount of force applied per unit of area*. For example, a cylinder like the one shown in the figure above might hold trillions of gas particles. These particles exert forces on the cylinder each time they hit it. These forces make up the pressure exerted by the gas.

The figure above shows that gases are compressible, which means they can be squeezed into a smaller volume. When the plunger moves down because of the weight on it, the space between gas particles decreases. The gas is compressed. The empty spaces between particles make gases compressible.

Pressure and Volume

The figure above also shows the relationship between pressure and volume of gas at a constant temperature. Notice that when the volume is greater, as in the cylinder on the left, the particles have more room to move. This bigger space results in fewer collisions within the cylinder. There is less pressure. The gas particles in the middle cylinder have less volume and more pressure. In the cylinder on the right, the pressure is greatest because the particles have the least volume. The particles collide within the container more often and the pressure is greater.

Boyle's Law

You read that the pressure and volume of a gas are related. Robert Boyle (1627–1691), a British scientist, was the first to describe this property of gases. **Boyle's law** *states that pressure of a gas increases if the volume decreases and pressure of a gas decreases if the volume increases, when temperature is constant.* This law can be expressed mathematically as shown in Math Skills in the margin on the right. 🔑

Boyle's Law in Action

You have probably felt Boyle's law in action if you have ever traveled in an airplane. While on the ground, the air pressure inside your middle ear and the pressure of the air surrounding you are equal. As the airplane takes off and rises quickly, the air pressure of the air in the plane decreases. However, the air pressure inside your middle ear stays the same. Because the air pressure in your middle ear is greater than the pressure in the cabin, air in your middle ear increases in volume. The pressure on your eardrum can cause pain.

These pressure changes also occur as the plane is landing. During the flight, the pressure in your ears will have equalized with the cabin pressure. The cabin air pressure increases as the plane descends while the pressure in your middle ear remains the same. This can cause pain. You can equalize this pressure difference by yawning or chewing gum.

Graphing Boyle's Law

The graph below shows that as pressure increases, volume decreases when gas is at a constant temperature. Pressure is on the x-axis. Volume is on the y-axis. Notice that the line is a curve that decreases in value. As pressure increases along the x-axis, the volume decreases along the y-axis.

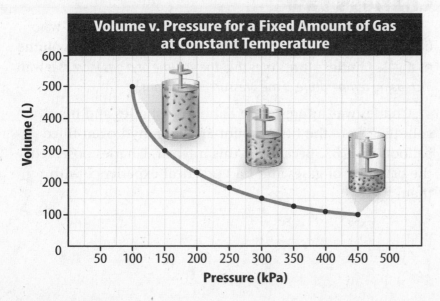

Volume v. Pressure for a Fixed Amount of Gas at Constant Temperature

Pressure (kPa) / *Volume (L)*

Key Concept Check

3. Relate What is the relationship between pressure and volume of a gas if temperature is constant?

_____

Math Skills

Boyle's law can be stated by the equation

$$V_2 = \frac{P_1 V_1}{P_2}$$

P_1 and V_1 represent the pressure and volume before a change. P_2 and V_2 are the pressure and volume after a change. Pressure is often measured in kilopascals (kPa). What is the final volume of a gas with an initial volume of 50.0 mL if the pressure increases from 600.0 kPa to 900.0 kPa?

a. Replace the terms in the equation with the values.

$$V_2 = \frac{(600.0 \text{ kPa}) (50.0 \text{ mL})}{(900.0 \text{ kPa})}$$

b. Cancel units, multiply, and then divide.

$$V_2 = \frac{(600.0 \text{ kPa}) (50.0 \text{ mL})}{(900.0 \text{ kPa})}$$

$$V_2 = 33.3 \text{ mL}$$

4. Solve Equations

What is the final volume of a gas with an initial volume of 100.0 mL if the pressure decreases from 500.0 kPa to 250.0 kPa?

✓ Visual Check

5. Label the left and right sides of the graph as showing low or high pressure and large or small volume.

Lower temperature, less volume

Higher temperature, greater volume

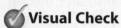 **Visual Check**

6. Illustrate Draw an ice cube or a snowflake next to the cylinder that is cooler. Draw a sun or a flame next to the cylinder that is warmer.

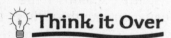 **Think it Over**

7. Apply Would the gases in car tires increase or decrease in volume on a hot day?

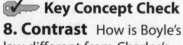 **Key Concept Check**

8. Contrast How is Boyle's law different from Charles's law?

Temperature and Volume

Pressure and volume changes are not the only factors that affect gas behavior. Changing the temperature of a gas also affects its behavior. The figure above shows two pictures of the same cylinder holding the same amount of gas. The picture on the left shows the gas in the cylinder at a low temperature. The average kinetic energy of the particles is low, and the particles move close together. The volume is small.

When thermal energy is added to the cylinder, as shown in the picture on the right, the gas particles move faster and spread farther apart. The temperature of the gas increases, which means the kinetic energy of its particles increases. The particles hit the cylinder more often. This pressure pushes up the plunger. The volume of the gas in the container is greater.

Charles's Law

Jacque Charles (1746–1823) was a French scientist who described the relationship between temperature and volume of a gas. **Charles's law** *states that the volume of a gas increases with increasing temperature, if the pressure is constant.*

Charles was interested in balloons. Charles and his colleague were the first to pilot and fly a hydrogen-filled balloon in 1783. Because of this interest, Charles observed the behavior of gases and had practical experience with them.

Charles's Law in Action

You have probably seen Charles's law in action if you have ever taken a balloon outside on a cold winter day. Why does a balloon seem to have less air in it when you take it from a warm place to a cold place? In fact, no air escapes from the balloon. When the balloon is in cold air, the temperature of the gas inside the balloon decreases. Recall that a decrease in temperature is a decrease in the average kinetic energy of particles. As a result, the gas particles slow down. They begin to get closer together. Fewer particles hit the inside of the balloon. The balloon seems to be only partway filled. If the balloon is returned to a warm place, the kinetic energy of the particles increases. More particles hit the inside of the balloon and push it out. The volume of the gas increases. ✅

Graphing Charles's Law

The relationship described in Charles's law is shown in the graph of several gases below. Temperature is on the *x*-axis. Volume is on the *y*-axis. The lines are straight. They slant upward, which means the values increase. The graph shows that the volume of a gas increases as the temperature of the gas increases at constant pressure.

Notice that each line in the graph is extrapolated to the same temperature: −273°C. (*Extrapolated* means the graph is extended beyond the data points that scientists have observed in the lab.) This temperature (−273°C) is referred to as 0 K (kelvin), or absolute zero. It is theoretically the lowest possible temperature that matter can have. At absolute zero, all particles have the lowest possible energy they can have. They do not move at all. The particles contain a minimal amount of thermal energy (potential energy + kinetic energy). 🔑

✅ **Reading Check**

9. Predict What happens when you warm a balloon?

_____

_____

_____

🔑 **Key Concept Check**

10. Analyze Which factors must be constant in Boyle's law and in Charles's law?

_____

_____

_____

✅ **Visual Check**

11. Interpret What do the dashed lines mean?

_____

Temperature and Volume

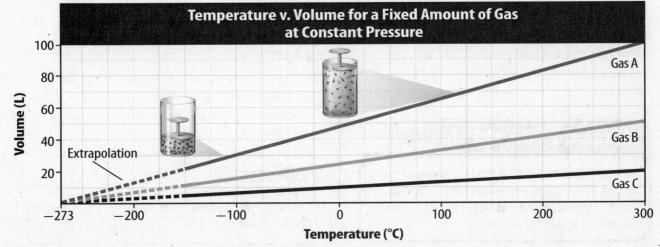

Temperature v. Volume for a Fixed Amount of Gas at Constant Pressure

Mini Glossary

Boyle's law: states that pressure of a gas increases if the volume decreases and pressure of a gas decreases if the volume increases, when temperature is constant

Charles's law: states that the volume of a gas increases with increasing temperature, if the pressure is constant

kinetic molecular theory: an explanation of how particles in matter behave

pressure: the amount of force applied per unit of area

1. Review the terms and their definitions in the Mini Glossary. Write a sentence that explains how the particles of a gas produce pressure.

_____

_____

_____

2. Use what you have learned about Boyle's law and Charles's law to complete the table.

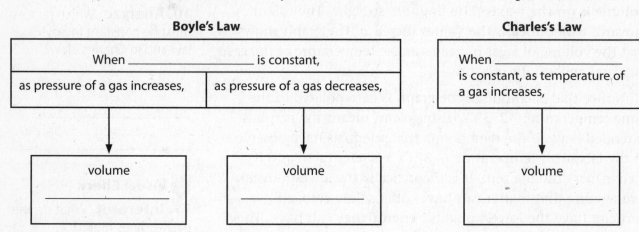

Boyle's Law

When _____ is constant,	
as pressure of a gas increases,	as pressure of a gas decreases,

volume
_____

volume
_____

Charles's Law

When _____ is constant, as temperature of a gas increases,

volume
_____

3. How did your outline of the lesson help you better understand the main ideas?

_____

_____

_____

What do you think **NOW?**

Reread the statements at the beginning of the lesson. Fill in the After column with an A if you agree with the statement or a D if you disagree. Did you change your mind?

 **Connect ED**

Log on to ConnectED.mcgraw-hill.com and access your textbook to find this lesson's resources.

END OF LESSON

Understanding the Atom

Discovering Parts of an Atom

························ **Before You Read** ··············

What do you think? Read the three statements below and decide whether you agree or disagree with them. Place an A in the Before column if you agree with the statement or a D if you disagree. After you've read this lesson, reread the statements to see if you have changed your mind.

Before	Statement	After
	1. The earliest model of an atom contained only protons and electrons.	
	2. Air fills most of an atom.	
	3. In the present-day model of the atom, the nucleus of the atom is at the center of an electron cloud.	

··············· **Read to Learn** ···············

Early Ideas About Matter

Look at your hands. What are they made of? You might answer that your hands are made of things such as skin, bone, muscle, and blood. Recall that each of these is made of even smaller structures called cells. Are cells made of even smaller parts? Imagine dividing something into smaller and smaller parts. What would the smallest part be?

Greek philosophers discussed and debated questions such as these more than 2,000 years ago. Most of them thought that all matter is made of only four elements—fire, water, air, and earth. However, they could not test their ideas. The scientific tools and methods for testing, such as experimentation, did not yet exist. The ideas proposed by the most influential philosophers usually were accepted over the ideas of less-influential philosophers. The popular idea of matter was challenged by Democritus (460–370 B.C.).

Democritus

The philosopher Democritus believed that matter is made of small, solid objects that cannot be divided, created, or destroyed. He called these objects *atomos*, from which the English word *atom* is derived. ✓

Key Concepts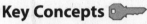
- What is an atom?
- How would you describe the size of an atom?
- How has the atomic model changed over time?

Study Coach

Create a Quiz Write five questions about discovering parts of the atom to create a quiz. Exchange quizzes with a partner. After taking the quizzes, discuss your answers. Reread the parts of the lesson that cover the topics you don't understand.

✓ **Reading Check**

1. Define What was Democritus's definition of an atom?

_____

_____

_____

_____

Atomic Theories		
Democritus	1. Atoms are small, solid objects that cannot be divided, created, or destroyed.	
	2. Atoms are constantly moving in empty space.	
	3. Different types of matter are made of different types of atoms.	
	4. The properties of the atoms determine the properties of matter.	
John Dalton	1. All matter is made of atoms that cannot be divided, created, or destroyed.	
	2. During a chemical reaction, atoms of one element cannot be converted into atoms of another element.	
	3. Atoms of one element are identical to each other but different from atoms of another element.	
	4. Atoms combine in specific ratios.	

Interpreting Tables

2. Identify Which philosopher in the table above proposed that atoms move in empty space?

_____

_____

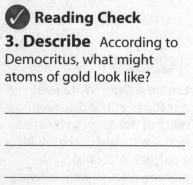 **Reading Check**

3. Describe According to Democritus, what might atoms of gold look like?

_____

_____

_____

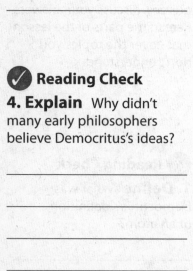 **Reading Check**

4. Explain Why didn't many early philosophers believe Democritus's ideas?

_____

_____

_____

_____

Democritus proposed that different types of matter are made from different types of atoms. For example, he said that smooth matter is made of smooth atoms. He also proposed that nothing was between these atoms except empty space. Democritus's ideas are summarized in the table above. ✓

Although Democritus had no way to test his ideas, many of his ideas are similar to the way scientists describe the atom today. Because Democritus's ideas did not conform to the popular opinion and could not be tested, they were open for debate. The philosopher Aristotle challenged Democritus's ideas.

Aristotle

Aristotle (384–322 B.C.) did not believe that empty space exists. Instead, he favored the more popular idea—that all matter is made of fire, water, air, and earth. Aristotle was highly respected. As a result, his ideas were accepted. Democritus's ideas about atoms were not studied again for more than 2,000 years. ✓

Dalton's Atomic Model

In the late 1700s, English schoolteacher and scientist John Dalton (1766–1844) looked again at the idea of atoms. Technology and scientific methods had advanced a great deal since Democritus's time. Dalton made careful observations and measurements of chemical reactions. He combined data from his own scientific research with data from the research of other scientists to propose the atomic theory. The table at the top of this page lists ways that Dalton's atomic theory supported some of the ideas of Democritus.

The Atom

Today, scientists agree that matter is made of atoms with empty space between and within them. What is an atom? Imagine dividing a piece of aluminum foil into smaller and smaller pieces. At first, you could cut the pieces with scissors. But eventually, the pieces would be too small to see. They would be much smaller than the smallest piece you could cut with scissors. This small piece is an aluminum atom. An aluminum atom cannot be divided into smaller aluminum pieces. An **atom** *is the smallest piece of an element that still represents that element.* �🔑➤

The Size of Atoms

Just how small is an atom? Atoms of different elements are different sizes. However, all are very, very small. You cannot see atoms even with most microscopes. Atoms are so small that about 7.5 trillion carbon atoms could fit into the period at the end of this sentence. 🔑➤

Seeing Atoms

Scientific experiments confirmed that matter is made of atoms long before scientists could see atoms. However, in 1981, a high-powered microscope, called a scanning tunneling microscope (STM), was invented. With this microscope, scientists could see individual atoms for the first time. An STM uses a tiny, metal tip to trace the surface of a piece of matter. The result is an image of atoms on the surface.

Even today, scientists still cannot see inside an atom. However, scientists have learned that atoms are not the smallest particles of matter. In fact, atoms are made of much smaller particles. What are these particles? How did scientists discover them if they could not see them?

Thomson—Discovering Electrons

Not long after Dalton's findings, another English scientist, named J.J. Thomson (1856–1940), made some important discoveries. Thomson and other scientists of that time worked with cathode ray tubes. If you have seen a neon sign, an older computer monitor, or the color display on an ATM screen, you have seen a cathode ray tube.

Thomson's cathode ray tube was a glass tube with pieces of metal, called electrodes, attached inside the tube. The electrodes were connected to wires. The wires were connected to a battery.

🔑➤ **Key Concept Check**

5. Apply What is a copper atom?

_____

_____

_____

🔑➤ **Key Concept Check**

6. Describe How would you describe the size of an atom?

_____

_____

_____

_____

FOLDABLES

Make a layered book to organize your notes and diagrams on the parts of an atom.

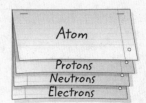

Thomson's Cathode Ray Tube Experiment

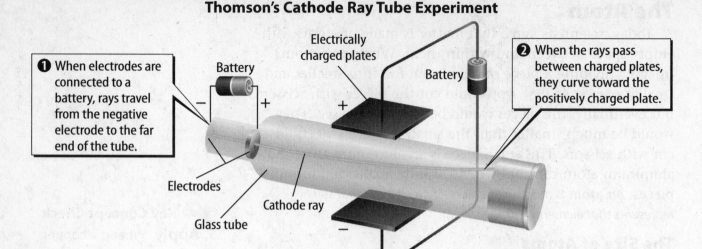

① When electrodes are connected to a battery, rays travel from the negative electrode to the far end of the tube.

Battery

Electrically charged plates

Battery

② When the rays pass between charged plates, they curve toward the positively charged plate.

Electrodes

Cathode ray

Glass tube

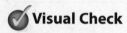 **Visual Check**

7. Recognize Did the ray in the experiment bend toward the plate with the positive charge or the plate with the negative charge?

_____

_____

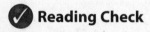 **Reading Check**

8. Analyze If the rays were positively charged, what would Thomson have observed as they passed between the plates?

_____

_____

_____

Thomson's cathode ray tube is shown above. Thomson removed most of the air from the tube. When he passed electricity through the wires, greenish-colored rays traveled from one electrode to the other end of the tube. What were these rays made of?

Negative Particles

Scientists called these rays cathode rays. Thomson wanted to know if these rays had an electric charge. To find out, he placed two plates on opposite sides of the tube. As shown in the figure above, one plate was positively charged. The other plate was negatively charged. As the cathode rays passed between the plates, the rays bent toward the positively charged plate and away from the negatively charged plate. Recall that opposite charges attract each other, and like charges repel each other. Thomson concluded that cathode rays are negatively charged. ✔

Parts of Atoms

Through more experiments, Thomson learned that these rays were made of particles that had mass. The mass of one of these particles was much smaller than the mass of the smallest atoms. This was surprising information to Thomson. Until then, scientists understood that an atom is the smallest particle of matter. But these rays were made of particles that were even smaller than atoms.

Metal Atoms Where did these small, negatively charged particles come from? Thomson proposed that these particles came from the metal atoms in the electrode. Thomson discovered that electrodes made of any kind of metal produced identical rays.

Charged Particles Putting these clues together, Thomson concluded that cathode rays were made of small, negatively charged particles. He called these particles electrons. *An* **electron** *is a particle with one negative charge (1–).* Atoms are neutral, or not electrically charged. Therefore, Thomson proposed that atoms also must contain a positive charge that balances the negatively charged electrons.

Thomson's Atomic Model

Thomson used this information to propose a new model of the atom. Instead of a solid, neutral sphere that was the same throughout, Thomson's model of the atom contained both positive and negative charges. He proposed that an atom was a sphere with a positive charge evenly spread throughout. Negatively charged electrons were mixed through the positive charge, similar to the way chocolate chips are mixed in cookie dough. The figure below shows this model. ✓

Thomson's Atomic Model

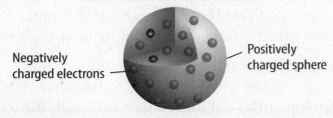

Negatively charged electrons

Positively charged sphere

Rutherford—Discovering the Nucleus

The discovery of electrons stunned scientists. Ernest Rutherford (1871–1937) was Thomson's student. He later had students of his own. Rutherford's students experimented with Thomson's model and discovered yet another surprise.

Rutherford's Predicted Result

Imagine throwing a baseball into a pile of table tennis balls. The baseball likely would knock the table tennis balls away and continue moving in a mostly straight line. This is similar to what Rutherford's students expected to see when they shot alpha particles into atoms. Alpha particles are dense and positively charged. Because they are so dense, only another dense particle could deflect the path of an alpha particle. According to Thomson's model, the positive charge of the atom was too spread out and not dense enough to change the path of an alpha particle. Electrons wouldn't affect the path of an alpha particle because electrons didn't have enough mass. Rutherford expected the alpha particles to travel straight without changing direction. ✓

✓ Reading Check

9. Differentiate How did Thomson's atomic model differ from Dalton's atomic model?

_____

_____

_____

_____

_____

✓ Visual Check

10. Describe How were the positive and negative charges arranged in Thomson's model?

_____

_____

_____

✓ Reading Check

11. Explain why Rutherford's students did not think an atom could change the path of an alpha particle.

_____

_____

_____

Rutherford's Predicted Result

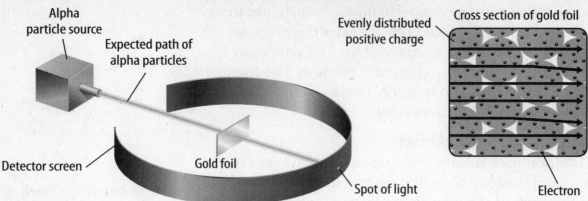

Alpha particle source

Expected path of alpha particles

Detector screen

Gold foil

Spot of light

Evenly distributed positive charge

Cross section of gold foil

Electron

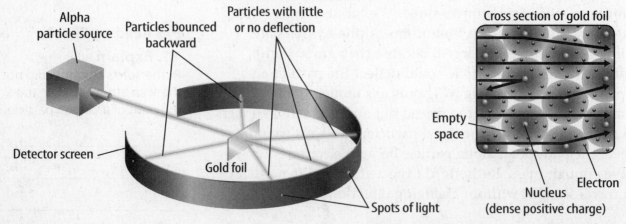

The figure above shows the result that Rutherford's students expected. They expected the positive alpha particles to travel straight through the foil without changing direction.

The Gold Foil Experiment

The students placed a source of alpha particles near a thin piece of gold foil made of gold atoms. A screen surrounded the gold foil. When an alpha particle struck the screen, it created a spot of light. The students could determine the path of the particles from the spots of light on the screen.

The Surprising Result

The figure below shows what the students observed. Most of the particles did indeed travel through the foil in a straight path. However, a few particles struck the foil and bounced off to the side. And one particle in 10,000 bounced straight back! Rutherford later said that this result was almost as surprising as if you fired a bullet at a piece of tissue paper and it came back and hit you. The alpha particles must have struck something dense and positively charged inside the atom. Thomson's model had to be refined.

Visual Check

12. Draw Highlight the expected path of the alpha particles.

Key Concept Check

13. Interpret Given the results of the gold foil experiment, how do you think an actual atom differs from Thomson's model?

_____

_____

_____

Visual Check

14. Recognize What do the dots on the screen indicate?

_____

_____

The Surprising Result

Alpha particle source

Particles bounced backward

Particles with little or no deflection

Detector screen

Gold foil

Spots of light

Cross section of gold foil

Empty space

Electron

Nucleus (dense positive charge)

Rutherford's Atomic Model

The result showed that most alpha particles traveled through the foil in a straight path. Therefore, Rutherford concluded that atoms are made mostly of empty space. The alpha particles that bounced backward must have hit a dense, positive mass. Rutherford concluded that *most of an atom's mass and positive charge is concentrated in a small area in the center of the atom called the* **nucleus.**

Rutherford's atomic model, shown below, contains a small, dense, positive nucleus. Further research showed that the nucleus was made up of positively charged particles called protons. *A* **proton** *is an atomic particle that has one positive charge (1+).* Negatively charged electrons move in the empty space surrounding the nucleus.

Rutherford's Atomic Model

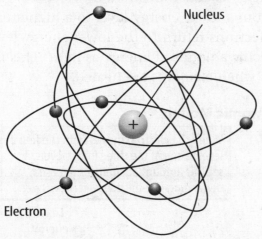

Nucleus

Electron

Discovering Neutrons

The modern model of the atom was beginning to take shape. James Chadwick (1891–1974) worked with Rutherford and also researched atoms. He discovered that in addition to protons, the nucleus contained neutrons. *A* **neutron** *is a neutral particle that exists in the nucleus of an atom.*

Bohr's Atomic Model

Rutherford's model explained much of his students' experimental evidence. However, the model could not explain several observations.

Colors of Light Scientists noticed that if they heated certain elements in a flame, the elements gave off specific colors of light. Each color of light had a specific amount of energy. Where did this light come from?

 Reading Check

15. Explain How did Rutherford explain the observation that some of the alpha particles bounced directly backward?

_____

_____

_____

 Visual Check

16. Identify In Rutherford's model, what makes up most of the area of an atom? (Circle the correct answer.)

a. the nucleus

b. electrons

c. empty space

Bohr's Experiments Niels Bohr (1885–1962), another student of Rutherford, proposed an answer to why certain elements heated in a flame give off light of specific colors. He studied hydrogen atoms because they contain only one electron.

Bohr experimented with adding electric energy to hydrogen and studying the energy that was released. His experiments led to a revised atomic model, shown in the figure below.

Electrons in the Bohr Model

Bohr proposed that electrons move in circular orbits, or energy levels, around the nucleus. Electrons in an energy level have a specific amount of energy. Electrons closer to the nucleus have less energy than electrons that are farther away from the nucleus.

When energy is added to an atom, electrons gain energy and move from a lower energy level to a higher energy level. When the electrons return to the lower energy level, they release a specific amount of energy as light. This is the light that appears when elements are heated.

✓ **Visual Check**

17. Predict According to Bohr's atomic model, what would happen to the electron in a hydrogen atom if you added energy to the atom?

_____

_____

_____

Bohr's Atomic Model

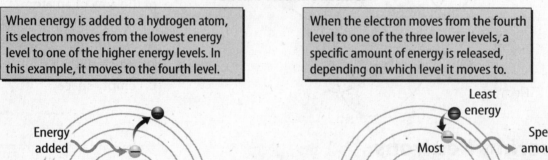

When energy is added to a hydrogen atom, its electron moves from the lowest energy level to one of the higher energy levels. In this example, it moves to the fourth level.

When the electron moves from the fourth level to one of the three lower levels, a specific amount of energy is released, depending on which level it moves to.

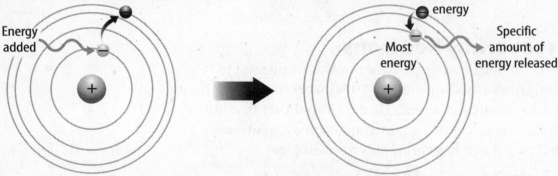

Energy added

Most energy

Least energy

Specific amount of energy released

Limitations of the Bohr Model

Bohr reasoned that if his model were accurate for atoms with one electron, it would be accurate for atoms with more than one electron. However, this was not the case.

More research confirmed that electrons do have specific amounts of energy, but energy levels are not arranged in circular orbits. How do electrons move in an atom? 🔑

🔑 **Key Concept Check**

18. Contrast How did Bohr's atomic model differ from Rutherford's?

_____

_____

_____

The Modern Atomic Model

In the modern atomic model, electrons form an electron cloud. *An **electron cloud** is an area around an atomic nucleus where an electron is most likely to be located.* Imagine taking a time-lapse photograph of bees around a hive. You might see a blurry cloud. The cloud might be denser near the hive than farther away because the bees spend more time near the hive.

In a similar way, electrons constantly move around the nucleus. It is impossible to know the speed and the exact location of an electron at a given moment. Instead, scientists only can predict the likelihood that an electron is in a particular location. The electron cloud, shown in the figure below, is mostly empty space. It represents the likelihood of finding an electron in a given area. The darker areas represent areas where electrons are more likely to be located.

Key Concept Check

19. Summarize How has the model of the atom changed over time?

_____

_____

_____

_____

The Modern Atomic Model

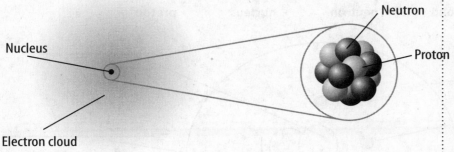

Nucleus

Electron cloud

Neutron

Proton

Visual Check

20. Consider Why do you think this model of the atom doesn't show the electrons?

_____

_____

_____

Quarks

You have read that atoms are made of smaller parts—protons, neutrons, and electrons. Are these particles made of even smaller parts? Scientists have discovered that electrons are not made of smaller parts. However, research has shown that protons and neutrons are made of smaller particles. Scientists call these particles quarks. Scientists theorize that there are six types of quarks. They named these quarks up, down, charm, strange, top, and bottom. Protons are made of two up quarks and one down quark. Neutrons are made of two down quarks and one up quark.

As you have read, the model of the atom has changed over time. The current model also might change with the invention of new technology that aids the discovery of new information.

Mini Glossary

atom: the smallest piece of an element that still represents that element

electron: a particle with one negative charge (1−)

electron cloud: an area around an atomic nucleus where an electron is most likely to be located

neutron: a neutral particle that exists in the nucleus of an atom

nucleus: a small, positively charged area in the center of an atom that contains most of the atom's mass

proton: an atomic particle that has one positive charge (1+)

1. Review the terms and their definitions in the Mini Glossary. Write a sentence that describes how neutrons and protons are related to a nucleus.

_____

_____

2. Name the parts of the modern atomic model in the diagram using the terms provided.

electron cloud   **neutron**   **nucleus**   **proton**

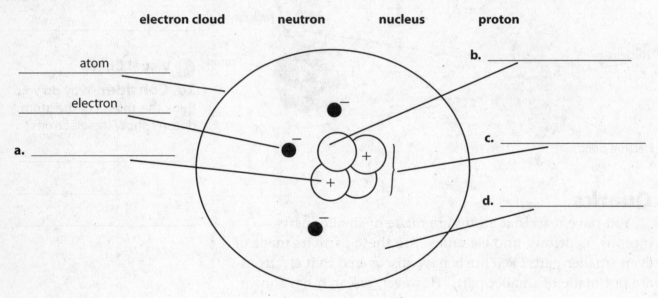

3. On the lines below, write one question from your partner's quiz that helped you learn an important concept about atoms. Then write the answer.

_____

_____

What do you think NOW?

Reread the statements at the beginning of the lesson. Fill in the After column with an A if you agree with the statement or a D if you disagree. Did you change your mind?

Log on to ConnectED.mcgraw-hill.com and access your textbook to find this lesson's resources.

END OF LESSON

Understanding the Atom

Protons, Neutrons, and Electrons—How Atoms Differ

············ **Before You Read** ············

What do you think? Read the three statements below and decide whether you agree or disagree with them. Place an A in the Before column if you agree with the statement or a D if you disagree. After you've read this lesson, reread the statements to see if you have changed your mind.

Before	Statement	After
	4. All atoms of the same element have the same number of protons.	
	5. Atoms of one element cannot be changed into atoms of another element.	
	6. Ions form when atoms lose or gain electrons.	

············ **Read to Learn** ············

The Parts of the Atom

Inside an atom is a very tiny nucleus surrounded by empty space. Positively charged protons and neutral neutrons are inside the nucleus. Negatively charged electrons whiz by in the empty space around the nucleus.

The table below compares the properties of protons, neutrons, and electrons. Protons and neutrons have about the same mass. Electrons have a much smaller mass than protons or neutrons. As a result, the nucleus contains most of an atom's mass. All atoms contain protons, neutrons, and electrons. However, different types of atoms have different numbers of these particles.

Properties of Protons, Neutrons, and Electrons			
	Electron	**Proton**	**Neutron**
Symbol	e–	p	n
Charge	1–	1+	0
Location	electron cloud around nucleus	nucleus	nucleus
Relative mass	1/1,840	1	1

Key Concepts

- What happens during nuclear decay?
- How does a neutral atom change when its number of protons, electrons, or neutrons changes?

Mark the Text

Building Vocabulary As you read, circle all the words you do not understand. Highlight the part of the text that helps you understand these words. Review the marked words and their definitions after you finish reading the lesson.

Interpreting Tables

1. Recognize What does the negative sign in the electron symbol mean?

_____

_____

_____

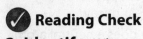
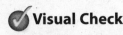
Reading Check

2. Identify What two numbers can be used to identify an element?

_____

_____

Visual Check

3. Apply Explain the difference between an oxygen atom and a carbon atom.

_____

_____

_____

Different Elements—Different Numbers of Protons

Look at the periodic table on the inside back cover of this book. Notice that more than 115 different elements have been identified. Recall that an element is a substance made from atoms that all have the same number of protons. For example, the element carbon is made from atoms that all have six protons. Likewise, all atoms that have six protons are carbon atoms. *The number of protons in an atom of an element is the element's* **atomic number.** The atomic number is the whole number listed with each element on the periodic table. ✓

What makes an atom of one element different from an atom of another element? Atoms of different elements contain different numbers of protons. The number of protons is the element's atomic number. Therefore, different elements have different atomic numbers. For example, an oxygen atom has eight protons. Its atomic number is eight. If an atom has seven protons, it is a nitrogen atom, and its atomic number is seven.

Neutral atoms of different elements also have different numbers of electrons. In a neutral atom, the number of electrons equals the number of protons. Thus, the number of positive charges equals the number of negative charges. The figure below illustrates three neutral atoms. Each atom has the same number of electrons as protons.

Different Elements

6 electrons   7 electrons   8 electrons

6 protons   7 protons   8 protons

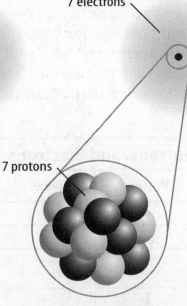

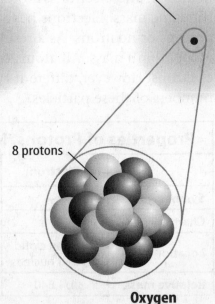

Carbon   **Nitrogen**   **Oxygen**

Neutrons and Isotopes

You have read that atoms of the same element have the same numbers of protons. However, atoms of the same element can have different numbers of neutrons. For example, all carbon atoms have six protons. However, some carbon atoms have six neutrons, some have seven neutrons, and some have eight neutrons. These three different types of carbon atoms, shown in the table below, are called isotopes. **Isotopes** *are atoms of the same element that have different numbers of neutrons*. Most elements have several isotopes.

Naturally Occurring Isotopes of Carbon			
Isotope	Carbon-12	Carbon-13	Carbon-14
Abundance	98.89%	<1.11%	<0.01%
Protons	6	6	6
Neutrons	+ 6	+ 7	+ 8
Mass Number	12	13	14

Protons, Neutrons, and Mass Number

The **mass number** *of an atom is the sum of the number of protons and neutrons in an atom.* This is shown in the following equation.

mass number = number of protons + number of neutrons

You can determine any one of these three quantities if you know the value of the other two quantities. For example, to determine the mass number of an atom, you must know the number of neutrons and the number of protons in the atom.

The mass numbers of the isotopes of carbon are shown in the table above. An isotope is often written with the element name followed by the mass number. Using this method, the isotopes of carbon are written as carbon-12, carbon-13, and carbon-14.

Average Atomic Mass

You might have noticed that the periodic table does not list mass numbers or the numbers of neutrons. This is because each element can have several isotopes. However, you might notice that a decimal number is listed with most elements. This decimal number is the average atomic mass of the element. *The* **average atomic mass** *of an element is the average mass of the element's isotopes, weighted according to the abundance of each isotope.*

Copyright © Glencoe/McGraw-Hill, a division of The McGraw-Hill Companies, Inc.

Interpreting Tables

4. Recognize Why do the isotopes of carbon in the table have different mass numbers?

_____

_____

_____

Math Skills ✖️➗

You can calculate the average atomic mass of an element if you know the percentage of each isotope in the element. Lithium (Li) contains 7.5% Li-6 and 92.5% Li-7. What is the average atomic mass of Li?

a. Divide each percentage by 100 to change to decimal form.

$$\frac{7.5\%}{100} = 0.075$$

$$\frac{92.5\%}{100} = 0.925$$

b. Multiply the mass of each isotope by its decimal percentage.

$6 \times 0.075 = 0.45$

$7 \times 0.925 = 6.475$

c. Add the values together to get the average atomic mass.

$0.45 + 6.475 = 6.93$

5. Use Percentages
Nitrogen (N) contains 99.63% N-14 and 0.37% N-15. What is the average atomic mass of nitrogen?

6. Identify What properties of carbon were averaged to arrive at the 12.01 average atomic mass?

_____

_____

_____

ACADEMIC VOCABULARY

spontaneous
(adjective) occurring without external force or cause

Reading Check

7. State Why can't scientists change lead into gold through a chemical reaction?

_____

_____

_____

Average Atomic Mass

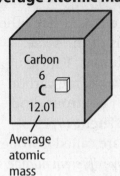

Average atomic mass

The figure on the left shows the block for carbon from the periodic table. It shows that the average atomic mass of carbon is 12.01. The table on the previous page shows that carbon has three isotopes. Why isn't the average atomic mass 13? After all, the average of the mass numbers 12, 13, and 14 is 13. The average atomic mass is weighted based on each isotope's abundance—how much of each isotope is present on Earth. Almost 99 percent of Earth's carbon is carbon-12, so the average atomic mass is close to 12.

Radioactivity

More than 1,000 years ago, people tried to change lead into gold by performing chemical reactions. However, they did not succeed. Why not? Today, scientists know that a chemical reaction does not change the number of protons in an atom's nucleus. If the number of protons does not change, the element does not change. But in the late 1800s, scientists discovered that some elements change into other elements spontaneously. How does this happen?

An Accidental Discovery

In 1896, a scientist named Henri Becquerel (1852–1908) studied minerals containing the element uranium. When these minerals were exposed to sunlight, they gave off a type of energy that could pass through paper. If Becquerel covered a photographic plate with black paper, this energy would pass through the paper and expose the film. An image of the mineral appeared on the plate. One day, Becquerel left the mineral in a drawer next to a wrapped, unexposed plate. Later, he unwrapped the plate and found that it contained an image of the mineral. The mineral spontaneously emitted energy, even in the dark! What was this energy?

Radioactivity

Becquerel shared his discovery with fellow scientists Pierre and Marie Curie. Marie Curie (1867–1934) called *elements that spontaneously emit radiation* **radioactive.** Becquerel and the Curies discovered that the radiation released by uranium was made of energy and particles. This radiation came from the nuclei of the uranium atoms. When this happens, the number of protons in one atom of uranium changes. When uranium releases radiation, it changes to a different element!

Types of Decay

Elements that are radioactive contain unstable nuclei. **Nuclear decay** *is a process that occurs when an unstable atomic nucleus changes into another more stable nucleus by emitting radiation.* Nuclear decay can produce three different types of radiation—alpha particles, beta particles, and gamma rays, as shown in the figure below. Alpha and beta decay change one element into another element. ✅

Types of Nuclear Decay

Alpha Decay

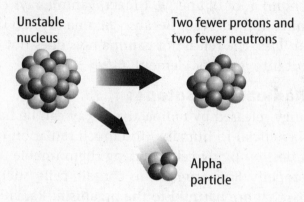

Unstable nucleus

Two fewer protons and two fewer neutrons

Alpha particle

Beta Decay

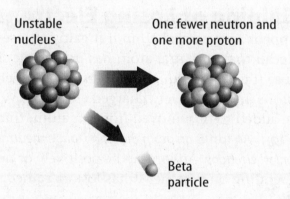

Unstable nucleus

One fewer neutron and one more proton

Beta particle

Gamma Decay

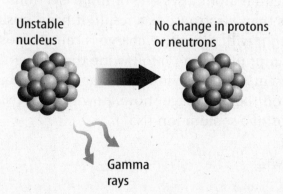

Unstable nucleus

No change in protons or neutrons

Gamma rays

✅ **Reading Check**

8. Name What does an atomic nucleus give off in the process of nuclear decay?

_____

_____

✅ **Visual Check**

9. Explain the change in atomic number for each type of decay.

_____

_____

_____

_____

_____

_____

_____

_____

Alpha Decay An alpha particle is made of two protons and two neutrons. When an atom releases an alpha particle, its atomic number decreases by two. Uranium-238 decays to thorium-234 through the process of alpha decay.

Beta Decay In beta decay, a neutron in an atom changes into a proton and a high-energy electron called a beta particle. The new proton becomes part of the nucleus. The beta particle is released. In beta decay, the atomic number of an atom increases by one because it has gained a proton.

Gamma Decay Gamma rays do not contain particles, but they do contain a lot of energy. In fact, gamma rays can pass through thin sheets of lead! Because gamma rays do not contain particles, the release of gamma rays does not change one element into another element. 🔑

Uses of Radioactive Isotopes

The energy released by radioactive decay can be harmful as well as beneficial to humans. Too much radiation can damage or destroy living cells, making them unable to function properly. Some organisms contain cells, such as cancer cells, that are harmful to the organism. Radiation therapy can be beneficial to humans by destroying these harmful cells.

Ions—Gaining or Losing Electrons

What happens to a neutral atom if it gains or loses electrons? Recall that a neutral atom has no overall charge. This is because it contains equal numbers of positively charged protons and negatively charged electrons. When electrons are added to or removed from an atom, that atom becomes an ion. *An **ion** is an atom that is no longer neutral because it has gained or lost electrons.* An ion can be positively or negatively charged, depending on whether it has lost or gained electrons. ✔

Positive Ions

When a neutral atom loses one or more electrons, it has more protons than electrons. As a result, it has a positive charge. An atom with a positive charge is called a positive ion. A superscript plus sign ($^+$) following the element's symbol represents a positive ion. For example, the top half of the figure on the next page shows how sodium (Na) becomes a positive sodium ion (Na^+).

🔑 **Key Concept Check**

10. Describe. What happens during radioactive decay?

_____

_____

_____

💡 **Think it Over**

11. Apply Why do people who work with radioactive materials wear special protective clothing?

_____

_____

_____

✔ **Reading Check**

12. Differentiate How is an ion different from a neutral atom?

_____

_____

_____

Formation of Positive and Negative Ions

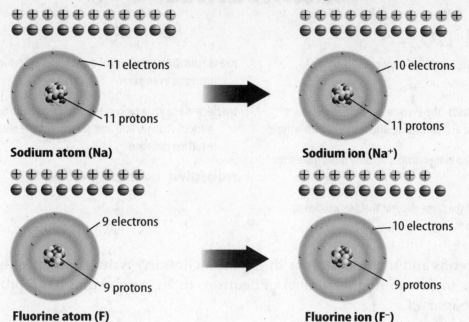

Sodium atom (Na) → Sodium ion (Na⁺)

Fluorine atom (F) → Fluorine ion (F⁻)

Negative Ions

When a neutral atom gains one or more electrons, it has more electrons than protons. As a result, the atom has a negative charge. An atom with a negative charge is called a negative ion. A superscript negative sign (⁻) following the element's symbol represents a negative ion. The bottom half of the figure above shows how fluorine (F) becomes a fluorine ion (F⁻).

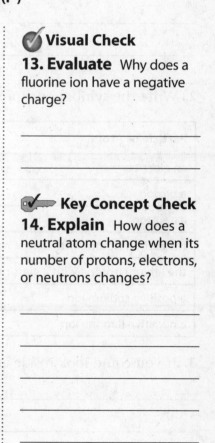

✔ **Visual Check**

13. Evaluate Why does a fluorine ion have a negative charge?

_____

_____

Key Concept Check

14. Explain How does a neutral atom change when its number of protons, electrons, or neutrons changes?

_____

_____

_____

_____

_____

Mini Glossary

atomic number: the number of protons in an atom of an element

average atomic mass: the average mass of the element's isotopes, weighted according to the abundance of each isotope

ion: an atom that is no longer neutral because it has gained or lost electrons

isotope: an atom of the same element that has a different number of neutrons

mass number: the sum of the number of protons and neutrons in an atom

nuclear decay: a process that occurs when an unstable atomic nucleus changes into another more stable nucleus by emitting radiation

radioactive: spontaneously emits radiation

1. Review the terms and their definitions in the Mini Glossary. Write a sentence that explains how to determine the number of neutrons in an isotope that has 6 protons and a mass number of 13.

_____

_____

2. Write the symbol or expression that represents each particle or atom in the table.

Particle or Atom	Symbol or Expression
an electron	e−
a proton	
a neutron	
the carbon isotope with a mass number of 12	carbon-12
the lithium isotope with a mass number of 6	
a positive sodium ion	
a negative fluorine ion	

3. If you could look inside an atom, what would you see?

_____

_____

What do you think NOW?

Reread the statements at the beginning of the lesson. Fill in the After column with an A if you agree with the statement or a D if you disagree. Did you change your mind?

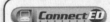

Log on to ConnectED.mcgraw-hill.com and access your textbook to find this lesson's resources.

END OF LESSON

Elements and Chemical Bonds

Electrons and Energy Levels

·············· **Before You Read** ··············

What do you think? Read the two statements below and decide whether you agree or disagree with them. Place an A in the Before column if you agree with the statement or a D if you disagree. After you've read this lesson, reread the statements to see if you have changed your mind.

Before	Statement	After
	1. Elements rarely exist in pure form. Instead, combinations of elements make up most of the matter around you.	
	2. Chemical bonds that form between atoms involve electrons.	

·············· **Read to Learn** ··············

The Periodic Table

The periodic table presents information about the elements in an organized way. A copy of the periodic table is on the inside back cover of this book. The table has more than 100 blocks. Each known element is a block. Each block describes basic properties of one element, such as its state of matter at room temperature, its atomic number, and its atomic mass. The atomic number is the number of protons in each atom of the element. The atomic mass is the average mass of all the different isotopes of an element.

Periods and Groups

You can learn about some properties of an element from its position on the periodic table. Elements are organized in periods (rows) and groups (columns). The periodic table lists elements in order of atomic number. The atomic number increases from left to right as you move across a period. Elements in each group have similar chemical properties and react with other elements in similar ways. ✓

Metals, Nonmetals, and Metalloids

There are three main regions of elements on the periodic table. The regions classify elements as metals, nonmetals, or metalloids.

Key Concepts

- How is an electron's energy related to its distance from the nucleus?
- Why do atoms gain, lose, or share electrons?

Study Coach

Ask Questions Read the headings in this lesson. Write questions about the information given under each heading. Take turns with a partner asking and answering the questions. Use the questions as a study guide.

✓ **Reading Check**

1. Explain How is the periodic table organized?

_____

_____

_____

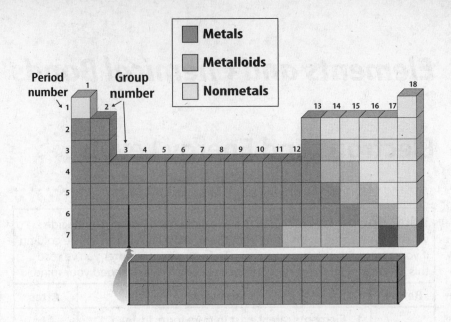

		Metals															
		Metalloids															
		Nonmetals															

Visual Check

2. Identify the metalloids in the figure and highlight them.

Reading Check

3. Locate Where are metals, nonmetals, and metalloids on the periodic table?

_____

_____

_____

_____

REVIEW VOCABULARY

compound
matter that is made up of two or more different kinds of atoms joined together by chemical bonds

The table above shows the three main regions of the periodic table. Except for hydrogen, elements on the left side are metals. Metals are good conductors of electricity and thermal energy. Metals can easily be hammered into sheets.

Metalloids form a narrow stair-step pattern between the metals and nonmetals. Metalloids have properties in common with both metals and nonmetals. They are often used as semiconductors in electronic devices.

Nonmetals are on the right side of the table. Nonmetals do not conduct electricity or thermal energy well. Most nonmetals are gases at room temperature. Those that are solids tend to be brittle. ✓

Atoms Bond

Elements rarely exist in pure form in nature. Instead, atoms of different elements chemically combine and form <u>compounds</u>. Compounds make up most of the matter around you, including living and nonliving things.

There are about 115 elements. These elements combine and form millions of compounds. Chemical bonds hold the compounds together. *A **chemical bond** is a force that holds two or more atoms together in a compound.*

Electron Number and Arrangement

Atoms contain protons, neutrons, and electrons, as shown at the top of the next page. A proton has a positive charge. A neutron has no charge. An electron has a negative charge. Each element's atomic number is the number of protons in each atom of that element. In a neutral (uncharged) atom, the number of protons equals the number of electrons.

Positions of Electrons

The exact position of the electrons in an atom cannot be determined. Electrons constantly move around the nucleus. However, each electron is usually in a certain area around the nucleus. Some are in areas close to the nucleus and some are in areas farther away.

Electrons and Energy

Different electrons in an atom have different amounts of energy. The areas in which electrons move around the nucleus are called energy levels. The figure on the right shows the energy levels of a fluorine atom. The amount of energy an electron has is related to its distance from the nucleus. Electrons closest to the nucleus are in the lowest energy level. These electrons have the least energy. Electrons farthest from the nucleus are in the highest energy level. These electrons have the most energy. Notice that the lowest energy level in the atom can hold only two electrons. The second energy level can hold up to eight electrons. 🔑

Electrons and Bonding Imagine two magnets. The closer they are to each other, the stronger the attraction of their opposite ends. Like magnets, the negatively charged electrons are attracted to the positively charged nucleus of an atom. Electrons close to the nucleus have a strong attraction to it. It is difficult for these electrons to interact with the electrons of atoms nearby. However, electrons far from the nucleus have a weak attraction to it. The nuclei of other atoms can easily attract these outer electrons. This attraction between the positive nucleus of one atom and the negative electrons of another atom creates a chemical bond.

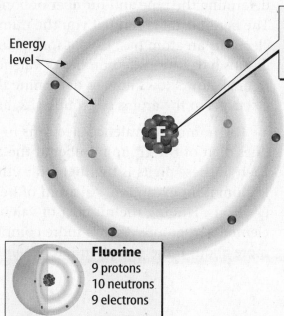

Lithium atom

The positively charged nucleus attracts the negatively charged electrons.

Fluorine
9 protons
10 neutrons
9 electrons

✔️ **Visual Check**

4. Highlight Use one color to highlight the protons in the lithium atom. Use a second color to highlight the neutrons. Use a third color to highlight the electrons.

✔️ **Visual Check**

5. Identify Circle the electrons that have the greatest amount of energy.

🔑 **Key Concept Check**

6. Analyze How is an electron's energy related to its position in an atom?

_____

_____

_____

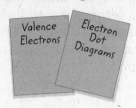
Valence Electrons

Electrons farthest from their nucleus are easily attracted to the nuclei of nearby atoms. These outermost electrons are the only electrons that form chemical bonds. Even atoms that have only a few electrons, such as hydrogen and lithium, can form chemical bonds. This is because these electrons are still the outermost electrons and are exposed to nuclei of other atoms. *A* **valence electron** *is an outermost electron of an atom that participates in chemical bonding.* Valence electrons have the most energy of all the electrons in an atom.

The number of valence electrons in an atom helps determine the type and number of bonds the atom can form. The periodic table can tell you the number of valence electrons an atom has. Except for helium, elements in certain groups have the same number of valence electrons. The figure below shows how to determine the number of valence electrons in the atoms of groups 1, 2, and 13–18.

The number of valence electrons in an atom equals the ones digit of the group number at the top of the column. Each of these digits is highlighted in the figure below. Helium is an exception. An atom of helium has two valence electrons. Finding the number of valence electrons for elements in groups 3–12 is more complicated. You will learn more about these groups in later chemistry courses.

Visual Check

7. Interpret How many valence electrons does an atom of phosphorous (P) have? (Circle the correct answer.)

a. 1

b. 5

c. 15

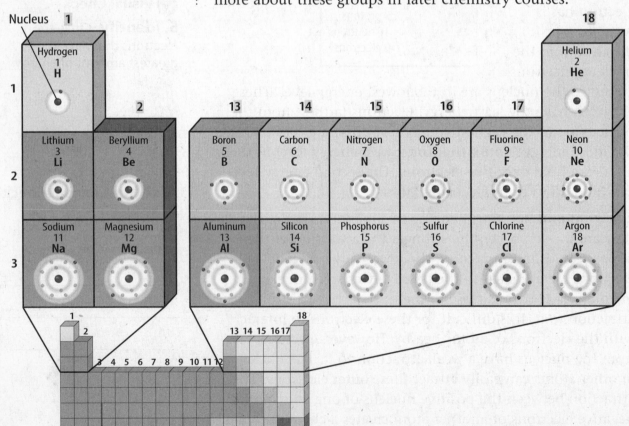

Electron Dot Diagrams

An electron dot diagram is a simple way to show an element's valence electrons. *An **electron dot diagram** is a model that represents valence electrons in an atom as dots around the element's chemical symbol.*

Electron dot diagrams can help you predict how an atom will bond with other atoms. Dots, representing valence electrons, are placed one-by-one on each side of an element's symbol. Then the dots are paired up until all the dots are used. The number of unpaired dots represents the number of bonds an atom can form. ✓

The figure below shows the steps for writing electron dot diagrams. Period 2 elements are shown. Remember that every element in a group has the same number of valence electrons. As a result, every element in a group has the same number of dots in its electron dot diagram.

Notice that a neon atom, Ne, has eight valence electrons, or four pairs of dots. It has no unpaired dots. Atoms with eight valence electrons are chemically stable. They do not react easily with other atoms. Atoms with 1–7 valence electrons are chemically unstable. These atoms easily bond with other atoms and form chemically stable compounds.

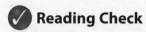
Reading Check

8. Explain Why are electron dot diagrams useful?

_____

_____

_____

Visual Check

9. Conclude How many electron pairs does an atom of Argon, Ar, have? (Circle the correct answer.)

a. 1

b. 3

c. 4

Steps for writing a dot diagram	Beryllium	Carbon	Nitrogen	Neon
1 Identify the element's group number on the periodic table.	2	14	15	18
2 Identify the number of valence electrons. • This equals the ones digit of the group number.	2	4	5	8
3 Draw the electron dot diagram. • Place one dot at a time on each side of the symbol (top, right, bottom, left). Pair up the dots until all are used.	Be·	·C·	·N·	:Ne:
4 Determine if the atom is stable. • An atom is stable if all dots on the electron dot diagram are paired.	Unstable	Unstable	Unstable	Stable
5 Determine how many bonds this atom can form. • Count the dots that are unpaired.	2	4	3	0

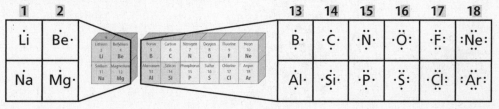

Noble Gases

The elements in Group 18 are called noble gases. With the exception of helium, noble gases have eight valence electrons and are chemically stable. Stable atoms do not easily react, or form bonds, with other atoms. The figure below shows the electron structures of two noble gases—neon and helium. Notice that all dots are paired.

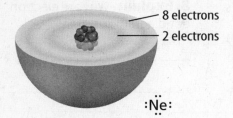
— 8 electrons
— 2 electrons

:Ne:

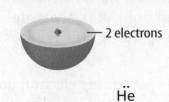
— 2 electrons

He

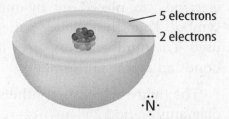
— 5 electrons
— 2 electrons

·N·

Neon has 10 electrons: 2 inner electrons and 8 valence electrons. A neon atom is chemically stable because it has 8 valence electrons. All dots in the dot diagram are paired.

Helium has 2 electrons. Because an atom's lowest energy level can hold only 2 electrons, the 2 dots in the dot diagram are paired. Helium is chemically stable.

Nitrogen has 7 electrons: 2 inner electrons and 5 valence electrons. Its dot diagram has 1 pair of dots and 3 unpaired dots. Nitrogen atoms become more stable by forming chemical bonds.

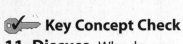 **Visual Check**

10. Analyze Can neon easily react, or form bonds, with other atoms? Explain.

_____

_____

_____

Key Concept Check

11. Discuss Why do atoms gain, lose, or share electrons?

_____

_____

_____

Stable and Unstable Atoms

Atoms with unpaired dots in their electron dot diagrams are reactive, or chemically unstable. For example, nitrogen, shown above at right, is reactive because it has three unpaired dots. Unstable atoms such as nitrogen become more stable by forming chemical bonds with other atoms.

When an atom forms a bond, it gains, loses, or shares valence electrons with other atoms. By forming bonds, atoms become more chemically stable. Recall that atoms are most stable with eight valence electrons. Therefore, atoms with less than eight valence electrons form bonds and become stable.

Mini Glossary

chemical bond: a force that holds two or more atoms together in a compound

electron dot diagram: a model that represents valence electrons in an atom as dots around the element's chemical symbol

valence electron: an outermost electron of an atom that participates in chemical bonding

1. Review the terms and their definitions in the Mini Glossary. Write a sentence describing how the number of valence electrons is used to make an electron dot diagram.

_____

_____

_____

2. Complete the table below about the three elements listed. Use a copy of the periodic table to help you.

Element	Metal, Nonmetal, or Metalloid	Group Number	Number of Valence Electrons
Sodium	metal	1	
Chlorine		17	7
Silicon	metalloid		4

3. How is the interaction of electrons with the nucleus of an atom similar to the interaction of two magnets?

_____

_____

_____

_____

What do you think NOW?

Reread the statements at the beginning of the lesson. Fill in the After column with an A if you agree with the statement or a D if you disagree. Did you change your mind?

 Connect ED

Log on to ConnectED.mcgraw-hill.com and access your textbook to find this lesson's resources.

END OF LESSON

Elements and Chemical Bonds

Compounds, Chemical Formulas, and Covalent Bonds

Key Concepts 🔑

- How do elements differ from the compounds they form?
- What are some common properties of a covalent compound?
- Why is water a polar compound?

Study Coach ▶

Make an Outline
Outline the information in this lesson. Use the headings as the main divisions of your outline. Include important details under each heading. Use your outline to review the lesson.

🔑 **Key Concept Check**
1. Differentiate How is a compound different from the elements that make it up?

_____

_____

_____

⋯⋯⋯⋯⋯⋯ Before You Read ⋯⋯⋯⋯⋯⋯

What do you think? Read the two statements below and decide whether you agree or disagree with them. Place an A in the Before column if you agree with the statement or a D if you disagree. After you've read this lesson, reread the statements to see if you have changed your mind.

Before	Statement	After
	3. The atoms in a water molecule are more chemically stable than they would be as individual atoms.	
	4. Many substances dissolve easily in water because opposite ends of a water molecule have opposite charges.	

⋯⋯⋯⋯⋯ Read to Learn ⋯⋯⋯⋯⋯
From Elements to Compounds

Have you ever baked cupcakes? The ingredients in the cupcakes—flour, baking soda, salt, sugar, eggs, vanilla, milk, and butter—all have unique physical and chemical properties. But when you mix the ingredients together and bake them, a new product results—cupcakes. The cupcakes have properties after being baked that are different from the original ingredients.

In some ways, compounds are like the cupcakes. Recall that a compound is a substance made up of two or more different elements. Just as cupcakes are different from their ingredients, compounds are different from their elements. An element is made of one type of atom. Compounds are chemical combinations of different types of atoms. Compounds and the elements that make them up often have different properties.

Chemical bonds join atoms together. Recall that a chemical bond is a force that holds atoms together in a compound. In this lesson, you will learn how atoms can form bonds by sharing valence electrons. You will also learn how to write and read a chemical formula. 🔑

Covalent Bonds—Electron Sharing

Recall that atoms can become more chemically stable by sharing valence electrons. When unstable, nonmetal atoms bond, they do this by sharing valence electrons. *A* **covalent bond** *is a chemical bond formed when two atoms share one or more pairs of valence electrons*. The atoms then form a stable covalent compound.

A Noble Gas Electron Arrangement

Hydrogen and oxygen can react to form water (H₂O), as shown in the figure below. Before the reaction, the atoms are chemically unstable. Each hydrogen atom is unstable with one valence electron. The oxygen atom is unstable with six valence electrons.

Recall that most atoms are chemically stable with eight valence electrons. This is the same electron arrangement as a noble gas. An atom with less than eight valence electrons becomes stable by forming chemical bonds until it has eight valence electrons. Therefore, an oxygen atom forms two bonds to become stable. A hydrogen atom is stable with two valence electrons. It forms one bond to become stable.

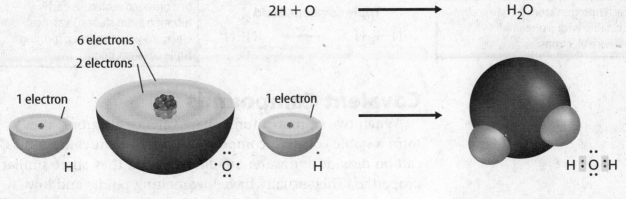

Shared Electrons

Look again at the figure above. How did the oxygen atom and the hydrogen atoms become chemically stable? They shared their unpaired valence electrons and formed a stable covalent compound. Each covalent bond has two valence electrons—one from the hydrogen atom and one from the oxygen atom. These electrons are shared in the bond. They count as valence electrons for both atoms.

Look at the dot diagram for water, on the right side of the figure. Each hydrogen atom now has two valence electrons. The oxygen atom bonded with two hydrogen atoms. As a result, oxygen now has eight valence electrons. All three atoms have the electron arrangement of a noble gas. The compound is stable.

FOLDABLES

Make three quarter-sheet note cards to organize information about single, double, and triple covalent bonds.

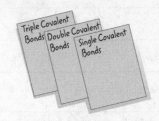

Visual Check

2. State How many valence electrons did the oxygen atom have before bonding?

3. Compare Is the bond stronger between atoms in hydrogen gas (H_2) or nitrogen gas (N_2)? Why?

_____

_____

_____

Double and Triple Covalent Bonds

Look at the figure below. In a single covalent bond, two atoms share one pair of valence electrons. In a double covalent bond, two atoms share two pairs of valence electrons. In a triple covalent bond, two atoms share three pairs of valence electrons. The more electrons that two atoms share, the stronger the covalent bond is between them. Double bonds are stronger than single bonds. Triple bonds are stronger than double bonds.

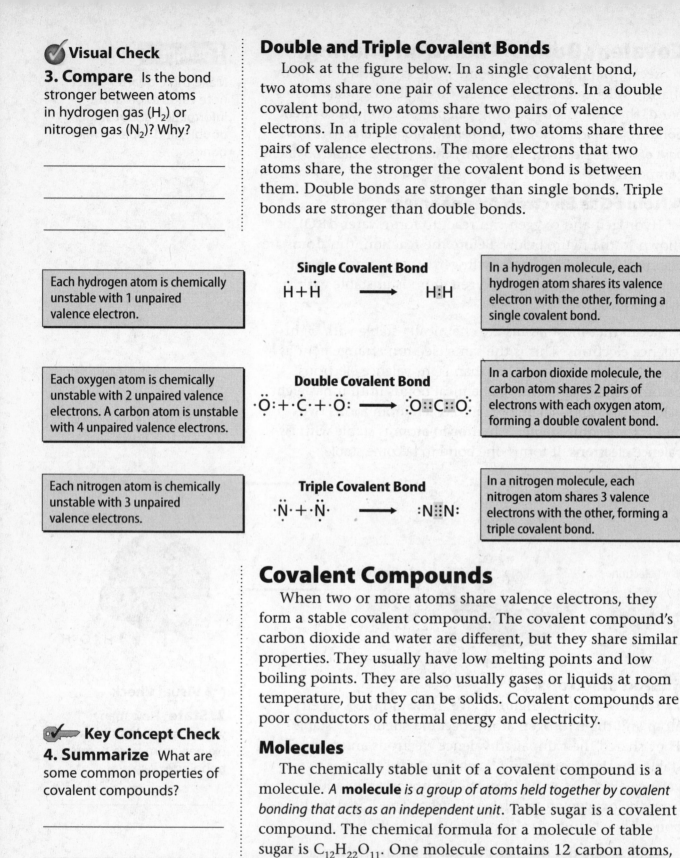

Each hydrogen atom is chemically unstable with 1 unpaired valence electron.

Single Covalent Bond

$\dot{H} + \dot{H} \longrightarrow H\!:\!H$

In a hydrogen molecule, each hydrogen atom shares its valence electron with the other, forming a single covalent bond.

Each oxygen atom is chemically unstable with 2 unpaired valence electrons. A carbon atom is unstable with 4 unpaired valence electrons.

Double Covalent Bond

$\cdot\ddot{O}\!:\,+\,\dot{\underset{.}{C}}\,+\,\cdot\ddot{O}\!: \longrightarrow \ddot{O}\!::\!C\!::\!\ddot{O}$

In a carbon dioxide molecule, the carbon atom shares 2 pairs of electrons with each oxygen atom, forming a double covalent bond.

Each nitrogen atom is chemically unstable with 3 unpaired valence electrons.

Triple Covalent Bond

$\cdot\ddot{N}\cdot\,+\,\cdot\ddot{N}\cdot \longrightarrow :N\!:\!:\!:\!N:$

In a nitrogen molecule, each nitrogen atom shares 3 valence electrons with the other, forming a triple covalent bond.

Covalent Compounds

When two or more atoms share valence electrons, they form a stable covalent compound. The covalent compound's carbon dioxide and water are different, but they share similar properties. They usually have low melting points and low boiling points. They are also usually gases or liquids at room temperature, but they can be solids. Covalent compounds are poor conductors of thermal energy and electricity.

🔑 **Key Concept Check**

4. Summarize What are some common properties of covalent compounds?

_____

_____

_____

_____

Molecules

The chemically stable unit of a covalent compound is a molecule. *A* **molecule** *is a group of atoms held together by covalent bonding that acts as an independent unit*. Table sugar is a covalent compound. The chemical formula for a molecule of table sugar is $C_{12}H_{22}O_{11}$. One molecule contains 12 carbon atoms, 22 hydrogen atoms, and 11 oxygen atoms. All these atoms are covalently bonded together. The only way to further break down the molecule would be to chemically separate the carbon, hydrogen, and oxygen atoms. Trillions of sugar molecules make up one grain of sugar. 🔑

Copyright © Glencoe/McGraw-Hill, a division of The McGraw-Hill Companies, Inc.

Water and Other Polar Molecules

In a covalent bond, one atom can attract the shared electrons more strongly than the other atom can. In a water molecule, shown in the model below on the left, the oxygen atom attracts the electrons more strongly than each hydrogen atom does. As a result, the shared electrons are pulled closer to the oxygen atom. Because electrons have a negative charge, the oxygen atom has a partial negative charge. The hydrogen atoms have a partial positive charge. As a result, a water molecule is polar. *A **polar molecule** is a molecule that has a partial positive end and a partial negative end because of unequal sharing of electrons.*

The charges on the ends of a polar molecule affect its properties. Sugar, for example, dissolves easily in water because both sugar and water are polar. The negative end of a water molecule pulls on the positive end of a sugar molecule. Also, the positive end of a water molecule pulls on the negative end of a sugar molecule. This causes the sugar molecules to separate from one another and mix with the water molecules. ☞

Copyright © Glencoe/McGraw-Hill, a division of The McGraw-Hill Companies, Inc.

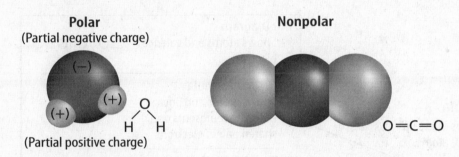

Polar
(Partial negative charge)

(−)

(+) (+)

(Partial positive charge)

Nonpolar

O=C=O

Nonpolar Molecules

A molecule is nonpolar if its atoms pull equally on the shared valence electrons. A hydrogen molecule, H_2, is a nonpolar molecule. Because the two hydrogen atoms are identical, their attraction for shared electrons is equal. A carbon dioxide molecule, CO_2, is shown in the model on the right above. This molecule is also nonpolar because the carbon atom and the oxygen atoms pull equally on the shared electrons.

A nonpolar compound will not easily dissolve in a polar compound. For example, oil is a nonpolar compound. It will not dissolve in water, which is a polar compound. However, "like dissolves like." Polar compounds dissolve in other polar compounds. Nonpolar compounds dissolve in other nonpolar compounds.

Elements and Chemical Bonds **143**

✓ **Key Concept Check**

5. State Why is water a polar compound?

_____

_____

_____

_____

✓ **Visual Check**

6. Identify Which atoms in a water molecule have a partial positive charge?

_____

_____

💡 **Think it Over**

7. Infer Imagine that you mix table salt with water, and the salt dissolves. Are the molecules of table salt polar or nonpolar? How do you know?

_____

_____

_____

_____

Chemical Formulas and Molecular Models

How do you know which elements make up a compound? *A **chemical formula** is a group of chemical symbols and numbers that represent the elements and the number of atoms of each element that make up a compound.* Just as a recipe lists the ingredients, a chemical formula lists the elements in a compound. For example, the chemical formula for carbon dioxide is CO_2. The formula uses chemical symbols to identify the elements in the compound. CO_2 is made up of carbon (C) and oxygen (O). The small number after a chemical symbol is a subscript. A subscript shows the number of atoms of that element in the compound. A symbol without a subscript means one atom. Carbon dioxide (CO_2) contains two atoms of oxygen bonded to one atom of carbon.

A chemical formula identifies the types of atoms in a compound or a molecule. However, a formula does not explain the shape or appearance of the molecule. Models can provide different information about a molecule. Each one can show the molecule in a different way. Common types of models for CO_2 are shown below. ✓

Copyright © Glencoe/McGraw-Hill, a division of The McGraw-Hill Companies, Inc.

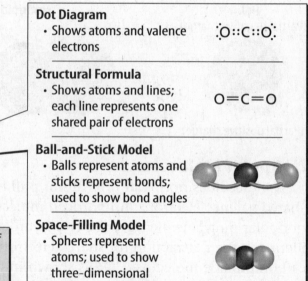

Dot Diagram
- Shows atoms and valence electrons

Structural Formula
- Shows atoms and lines; each line represents one shared pair of electrons

Ball-and-Stick Model
- Balls represent atoms and sticks represent bonds; used to show bond angles

Space-Filling Model
- Spheres represent atoms; used to show three-dimensional arrangement of atoms

Chemical Formula

A carbon dioxide molecule is made up of carbon (C) and oxygen (O) atoms.

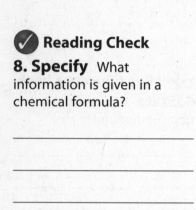

A symbol without a subscript indicates one atom. Each molecule of carbon dioxide has one carbon atom.

The 2 indicates two atoms of oxygen. Each molecule of carbon dioxide has two oxygen atoms.

Reading Check

8. Specify What information is given in a chemical formula?

_____

_____

_____

Visual Check

9. Interpret According to the structural formula, how many pairs of shared electrons are in a molecule of carbon dioxide?

_____

_____

Mini Glossary

chemical formula: a group of chemical symbols and numbers that represent the elements and the number of atoms of each element that make up a compound

covalent bond: a chemical bond formed when two atoms share one or more pairs of valence electrons

molecule: a group of atoms held together by covalent bonding that acts as an independent unit

polar molecule: a molecule that has a partial positive end and a partial negative end because of unequal sharing of electrons

1. Review the terms and their definitions in the Mini Glossary. Write a sentence explaining what the chemical formula of glucose, a simple sugar, means. Glucose's chemical formula is $C_6H_{12}O_6$.

_____

_____

2. Complete the table below to compare the different types of covalent bonds.

	Single Bond	**Double Bond**	**Triple Bond**
Electrons shared			three pairs of electrons
Example	hydrogen gas		

3. How did making an outline of the lesson help you organize information about compounds, covalent bonds, and chemical formulas?

_____

_____

_____

What do you think NOW?

Reread the statements at the beginning of the lesson. Fill in the After column with an A if you agree with the statement or a D if you disagree. Did you change your mind?

Log on to ConnectED.mcgraw-hill.com and access your textbook to find this lesson's resources.

END OF LESSON

Elements and Chemical Bonds

Ionic and Metallic Bonds

Key Concepts

- What is an ionic compound?
- How do metallic bonds differ from covalent and ionic bonds?

···········Before You Read···········

What do you think? Read the two statements below and decide whether you agree or disagree with them. Place an A in the Before column if you agree with the statement or a D if you disagree. After you've read this lesson, reread the statements to see if you have changed your mind.

Before	Statement	After
	5. Losing electrons can make some atoms more chemically stable.	
	6. Metals are good electrical conductors because they tend to hold onto their valence electrons very tightly.	

···········Read to Learn···········

Understanding Ions

As you read in Lesson 2, the atoms of two or more nonmetals form compounds by sharing valence electrons. However, when a metal and a nonmetal bond, they do not share electrons. Instead, one or more valence electrons transfers from the metal atom to the nonmetal atom. After electrons transfer, the atoms bond and form a chemically stable compound. Transferring valence electrons results in atoms with the same number of valence electrons as a noble gas.

When an atom loses or gains a valence electron, it becomes an ion. *An **ion** is an atom that is no longer electrically neutral because it has lost or gained valence electrons.* Because electrons have a negative charge, gaining or losing an electron changes the overall charge of the atom. An atom that loses valence electrons becomes an ion with a positive charge. This is because after an atom loses an electron, the atom has more protons than electrons. The atom is now an ion with a positive charge. An atom that gains valence electrons becomes an ion with a negative charge. This is because the number of protons is now less than the number of electrons. ✓

Study Coach

Write a Quiz As you read, write one question for each paragraph on one side of a sheet of paper. Write the answers on the back of the paper. Exchange quizzes with a partner and take each other's quiz.

✓ **Reading Check**

1. Explain Why does an atom that gains an electron become an ion with a negative charge?

_____

_____

_____

_____

Losing Valence Electrons

Sodium (Na) is a metal. Its atomic number is 11. This means each sodium atom has 11 protons and 11 electrons. Sodium is in group 1 on the periodic table. Therefore, sodium atoms have one valence electron and are chemically unstable.

Metal atoms, such as sodium, become more stable when they lose valence electrons and form a chemical bond with a nonmetal. The figure below describes the process of losing and gaining valance electrons. When a sodium atom loses one valence electron, the electrons in the next-lower energy level become the new valence electrons. The sodium atom then has eight valence electrons, the same as the noble gas neon (Ne). The sodium atom is chemically stable.

Gaining Valence Electrons

In Lesson 2, you learned that nonmetal atoms can share valence electrons with other nonmetal atoms. Nonmetal atoms can also gain valence electrons from metal atoms. Either way, they achieve the electron arrangement of a noble gas. The nonmetal chlorine (Cl) has an atomic number of 17. Chlorine atoms have seven valence electrons, as shown in the figure below. If a chlorine atom gains one valence electron, it will have eight valence electrons. It will then have the same electron arrangement as the stable noble gas argon (Ar).

When a sodium atom loses a valence electron, it becomes a positively charged ion. This is shown by a plus (+) sign. When a chlorine atom gains a valence electron, it becomes a negatively charged ion. This is shown by a negative (−) sign. ✔️

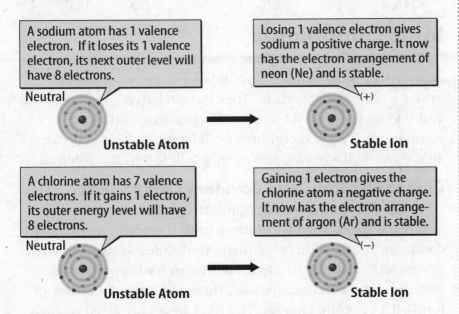

A sodium atom has 1 valence electron. If it loses its 1 valence electron, its next outer level will have 8 electrons.

Neutral

Unstable Atom

Losing 1 valence electron gives sodium a positive charge. It now has the electron arrangement of neon (Ne) and is stable.

(+)

Stable Ion

A chlorine atom has 7 valence electrons. If it gains 1 electron, its outer energy level will have 8 electrons.

Neutral

Unstable Atom

Gaining 1 electron gives the chlorine atom a negative charge. It now has the electron arrangement of argon (Ar) and is stable.

(−)

Stable Ion

Copyright © Glencoe/McGraw-Hill, a division of The McGraw-Hill Companies, Inc.

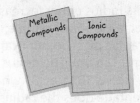

Make two quarter-sheet note cards to summarize information about ionic and metallic compounds.

Metallic Compounds

Ionic Compounds

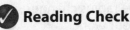 **Reading Check**

2. Predict Are atoms of a group 16 element more likely to gain or lose valence electrons?

_____

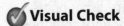 **Visual Check**

3. Identify What would an ion's charge be if the atom gained two electrons?

_____

An atom's radius is measured in picometers (pm), 1 trillion times smaller than a meter. When an atom becomes an ion, its radius either increases or decreases. For example, the radius of a sodium (Na) atom is 186 pm. The radius of a Na^+ ion is 102 pm. By what percentage does the radius change as the ion forms?

a. Subtract the ion's radius from the atom's radius.

102 pm − 186 pm = −84 pm

b. Divide the difference by the atom's radius.

−84 pm ÷ 186 pm = −0.45

c. Multiply the answer by 100 and add a % sign.

−0.45 × 100 = −45%

A negative value means a decrease in size. A positive value means an increase.

4. Calculate Percentage The radius of an oxygen (O) atom is 73 pm. The radius of an oxygen ion (O^{2-}) is 140 pm. By what percentage does the radius change? Show your work.

_____

_____

_____

Key Concept Check
5. Explain What holds ionic compounds together?

_____

_____

_____

Determining an Ion's Charge

Atoms are electrically neutral because they have the same number of protons (+) and electrons (−). Once an atom gains or loses electrons, it becomes a charged ion. For example, the atomic number for nitrogen (N) is 7. This means that each N atom has 7 protons and 7 electrons. It is electrically neutral. When forming an ionic bond, N atoms gain 3 electrons. The N ion then has 10 electrons. To determine the charge of the ion, subtract the number of electrons in the ion from the number of protons.

7 protons − 10 electrons = −3 charge

A nitrogen ion has a −3 charge. This is written as N^{3-}.

Ionic Bonds—Electron Transferring

Recall that metal atoms lose valence electrons and nonmetal atoms gain valence electrons. When forming a chemical bond, the nonmetal atoms gain the electrons lost by the metal atoms. In (NaCl), or table salt, a sodium atom loses one valence electron. The electron is transferred to a chlorine atom. The sodium atom is now a positively charged ion. The chlorine atom is now a negatively charged ion. These ions attract each other and form a stable ionic compound, as shown below. *The attraction between positively and negatively charged ions in an ionic compound is an* **ionic bond.**

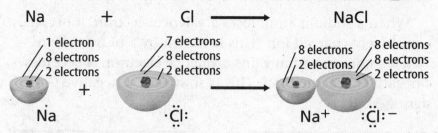

Ionic Compounds

The ions of ionic compounds are strongly attracted to each other. As a result, ionic compounds are usually solid and brittle at room temperature. They have relatively high melting and boiling points. Water that contains dissolved ionic compounds is a good conductor of electricity. This is because an electrical charge can pass from ion to ion in the solution.

Comparing Ionic and Covalent Compounds

Recall that in a covalent bond, two or more nonmetal atoms share electrons and form a unit, or molecule. Covalent compounds are made up of many molecules. However, when nonmetal ions bond to metal ions in an ionic compound, there are no molecules. Instead, there is a large collection of ions with opposite charges. The ions are all attracted to each other and are held together by ionic bonds.

Metallic Bonds—Electron Pooling

Recall that metal atoms typically lose valence electrons when forming compounds. Metal atoms form compounds with one another by combining, or pooling, their valence electrons, as shown in the table below. *A metallic bond is a bond formed when many metal atoms share their pooled valence electrons*. In aluminum (Al), atoms lose their valence electrons and become positive ions. The negatively charged valence electrons move from ion to ion. Valence electrons in metals do not bond to one atom. Instead, a "sea of electrons" surrounds the positive ions.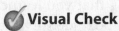

Properties of Metallic Compounds

Metals are good conductors of thermal energy and electricity. Because the valence electrons can move from ion to ion, they can easily <u>conduct</u> an electric charge. When a metal is hammered into a sheet or drawn into a wire, it does not break. The metal ions can slide past one another in the electron sea and move to new positions. Metals are shiny because the valence electrons at the surface interact with light. The table below compares the covalent, ionic, and metallic bonds that you studied in this chapter.

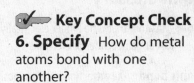 **Key Concept Check**

6. Specify How do metal atoms bond with one another?

_____

_____

ACADEMIC VOCABULARY
conduct
(verb) to serve as a medium through which something can flow

Visual Check

7. Identify Circle the bond that results in a compound that conducts thermal energy well.

Type of Bond	What is bonding?	Properties of Compounds
Covalent—share valence electrons Water	nonmetal atoms; nonmetal atoms	• gas, liquid, or solid • low melting and boiling points • often not able to dissolve in water • poor conductors of thermal energy and electricity • dull appearance
Ionic—transfer valence electrons Na⁺ Cl⁻ Salt	nonmetal ions; metal ions	• solid crystals • high melting and boiling points • dissolves in water • solids are poor conductors of thermal energy and electricity • ionic compounds in water solutions conduct electricity
Metallic—pool valence electrons Aluminum	metal ions; metal ions	• usually solid at room temperature • high melting and boiling points • do not dissolve in water • good conductors of thermal energy and electricity • shiny surface • can be hammered into sheets and pulled into wires

Mini Glossary

ion: an atom that is no longer electrically neutral because it has lost or gained valence electrons

ionic bond: the attraction between positively and negatively charged ions in an ionic compound

metallic bond: a bond formed when many metal atoms share their pooled valence electrons

1. Review the terms and their definitions in the Mini Glossary. Write two sentences that describe the difference between a positively charged ion and a negatively charged ion.

_____

_____

2. Fill in the organizer below with the following terms to compare the three types of bonding that you learned about in this chapter: *ionic, metallic,* and *covalent.* In each bottom box, identify the type of bond by writing *transfer valence electrons, pool valence electrons,* or *share valence electrons.*

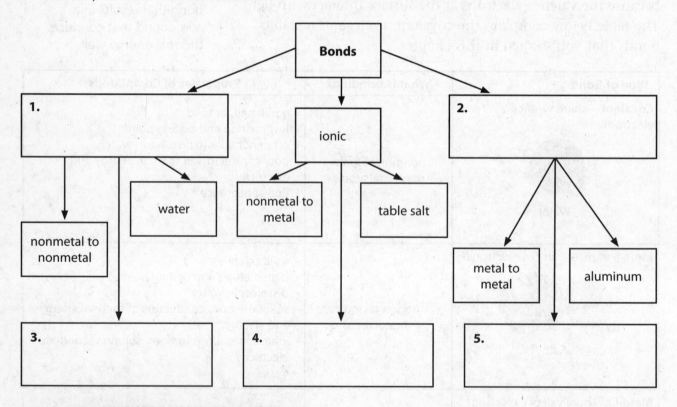

What do you think NOW?

Reread the statements at the beginning of the lesson. Fill in the After column with an A if you agree with the statement or a D if you disagree. Did you change your mind?

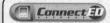 **Connect ED**

Log on to ConnectED.mcgraw-hill.com and access your textbook to find this lesson's resources.

 END OF LESSON

Chemical Reactions and Equations

Understanding Chemical Reactions

········ **Before You Read** ············

What do you think? Read the two statements below and decide whether you agree or disagree with them. Place an A in the Before column if you agree with the statement or a D if you disagree. After you've read this lesson, reread the statements to see if you have changed your mind.

Before	Statement	After
	1. If a substance bubbles, you know a chemical reaction is occurring.	
	2. During a chemical reaction, some atoms are destroyed and new atoms are made.	

············· **Read to Learn** ···············

Changes in Matter

When you put liquid water in a freezer, it changes to solid water, or ice. When you pour brownie batter into a pan and bake it, the liquid batter changes to a solid. In both cases, a liquid changes to a solid. Are these changes the same?

Physical Changes

Recall that matter can undergo two types of changes—chemical or physical. A physical change does not produce new substances. The substances that exist before and after the change are the same, although they might have different physical properties. This is what happens when liquid water changes to ice. Its physical properties change from a liquid to a solid. But the water, H_2O, does not change into a different substance. Water molecules are always made up of two hydrogen atoms bonded to one oxygen atom, regardless of whether the water is solid, liquid, or gas.

Chemical Changes

Recall that during a chemical change, one or more substances change into new substances. The starting substances and the substances produced have different physical and chemical properties.

Key Concepts 🔑

- What are some signs that a chemical reaction might have occurred?
- What happens to atoms during a chemical reaction?
- What happens to the total mass in a chemical reaction?

Mark the Text

Identify the Main Ideas
Highlight two or three phrases in each paragraph that summarize the main ideas. After you have finished the lesson, review the highlighted text.

✓ **Reading Check**

1. Contrast How is a chemical change different from a physical change?

_____

_____

_____

Physical and Chemical Properties During baking, brownie batter changes physically from a liquid to a solid. But a chemical change also occurs. Many substances in the batter change to new substances in the baked brownies. As a result, baked brownies have physical and chemical properties that are different from those of brownie batter. ✓

Chemical Reaction A chemical change also is called a chemical reaction. These terms mean the same thing. *A* **chemical reaction** *is a process in which atoms of one or more substances rearrange to form one or more new substances.*

Signs of a Chemical Reaction

How can you tell if a chemical reaction has taken place? You have read that the substances before and after a reaction have different properties. One way to detect a chemical reaction is to look for changes in properties. Changes in color, state of matter, and odor are signs that a chemical reaction might have occurred. Another sign of a chemical reaction is a change in energy. If substances get warmer or cooler or if they give off light or sound, a reaction probably has occurred. The table below describes some signs of a chemical reaction.

Change in Properties	
Change in color Bright copper changes to green when the copper reacts with certain gases in the air.	**Formation of bubbles** Bubbles of carbon dioxide form when baking soda is added to vinegar.
Change in odor When food burns or rots, a change in odor is a sign of a chemical change.	**Formation of a precipitate** A precipitate is a solid formed when two liquids react.
Changes in Energy	
Warming or cooling Thermal energy is either given off or absorbed during a chemical change.	**Release of light** A firefly gives off light as the result of a chemical change.

However, these signs are not proof of a chemical change. For example, bubbles appear when water boils. But bubbles also appear when baking soda and vinegar react and form carbon dioxide gas. How can you be sure that a chemical reaction has taken place? The only way to know is to study the chemical properties of the substances before and after the change. If they have different chemical properties, then the substances have undergone a chemical reaction. 🔑

What happens in a chemical reaction?

In a chemical reaction, one or more substances react and form one or more new substances. How do these new substances form?

Atoms Rearrange and Form New Substances

Recall that there are two types of substances—elements and compounds. Substances have a fixed arrangement of atoms. For example, a single drop of water has trillions of oxygen and hydrogen atoms. However, all of these atoms are arranged in the same way—two atoms of hydrogen are bonded to one atom of oxygen. If this arrangement changes, the substance is no longer water. Instead, a different substance forms with different physical and chemical properties. This is the kind of change that happens during a chemical reaction. Atoms of elements or compounds rearrange and form different elements or compounds.

Bonds Break and Bonds Form

Atoms rearrange when <u>chemical bonds</u> between atoms break and other chemical bonds form. All substances, including solids, are made of particles that move constantly. As particles move, they collide. If the particles collide with enough energy, the bonds between atoms can break. The atoms separate and rearrange, and new bonds can form.

The figure below shows the reaction that forms hydrogen and oxygen from water. Adding electric energy to water molecules can cause this reaction. The added energy causes bonds between the hydrogen and oxygen atoms to break. Then new bonds can form between pairs of hydrogen atoms and between pairs of oxygen atoms. The reaction creates no new atoms. Instead, it rearranges the existing atoms. 🔑

REVIEW VOCABULARY

chemical bond
an attraction between atoms when electrons are shared, transferred, or pooled

🔑 **Key Concept Check**

5. Describe What happens to atoms during a chemical reaction?

_____

_____

_____

✅ **Visual Check**

6. Explain How can you tell that this reaction created no new atoms?

_____

_____

_____

Chemical Bonds in a Reaction

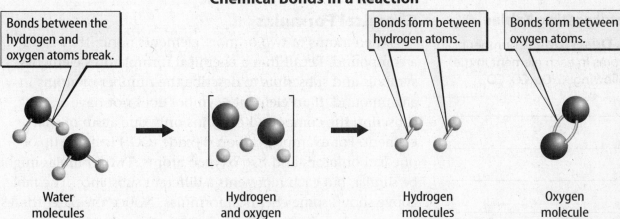

Bonds between the hydrogen and oxygen atoms break.

Bonds form between hydrogen atoms.

Bonds form between oxygen atoms.

Water molecules (H₂O)

Hydrogen and oxygen atoms

Hydrogen molecules (H₂)

Oxygen molecule (O₂)

Chemical Equations

In your science laboratory, you usually describe a chemical reaction in the form of a chemical equation. A **chemical equation** *is a description of a reaction using element symbols and chemical formulas.* Element symbols represent elements. Chemical formulas represent compounds.

Element Symbols

Element symbols appear in the periodic table. The symbol for carbon is C. Copper is Cu. Each element can exist as just one atom. Some elements exist in nature as diatomic molecules—two atoms bonded together. A diatomic element's formula includes the element's symbol and the subscript 2. Subscripts describe the number of atoms of an element in a compound. Oxygen (O_2) and hydrogen (H_2) are diatomic molecules. Element symbols are shown below. ✓

Copyright © Glencoe/McGraw-Hill, a division of The McGraw-Hill Companies, Inc.

Reading Check

7. Apply What does the subscript mean in the diatomic molecule O_2?

_____

_____

_____

Symbols and Formulas of Some Elements and Compounds

Substance	Formula	# of Atoms	Substance	Formula	# of Atoms
Carbon	C	C: 1	Carbon monoxide	CO	C: 1 O: 1
Copper	Cu	Cu: 1	Water	H_2O	H: 2 O: 1
Cobalt	Co	Co: 1	Hydrogen peroxide	H_2O_2	H: 2 O: 2
Oxygen	O_2	O: 2	Glucose	$C_6H_{12}O_6$	C: 6 H: 12 O: 6
Hydrogen	H_2	H: 2	Sodium chloride	NaCl	Na: 1 Cl: 1
Chlorine	Cl_2	Cl: 2	Magnesium hydroxide	$Mg(OH)_2$	Mg: 1 O: 2 H: 2
Carbon dioxide	CO_2	C: 1 O: 2			

Interpreting Tables

8. Describe the number of atoms in each element in the following: C, Co, CO, CO_2.

_____

_____

_____

_____

Chemical Formulas

When atoms of two or more elements bond, they form a compound. Recall that a chemical formula uses elements' symbols and subscripts to describe the number of atoms in a compound. If an element's symbol does not have a subscript, the compound contains only one atom of that element. For example, carbon dioxide (CO_2) is made up of one carbon atom and two oxygen atoms. Two formulas might be similar, but each represents a different substance. The table above shows some chemical formulas. Notice the parentheses in magnesium hydroxide. This means the subscript applies to both elements within the parentheses.

Writing Chemical Equations

A chemical equation includes the substances that react and the substances that form in a chemical reaction. *The starting substances in a chemical reaction are* **reactants.** *The substances produced by the chemical reaction are* **products.** The figure below shows how to write a chemical equation. Chemical formulas describe the reactants and the products. Write the reactants to the left of the arrow. Write the products to the right of the arrow. Separate two or more reactants or products with a plus sign. The structure for an equation is:

$$\text{reactant} + \text{reactant} \rightarrow \text{product} + \text{product}$$

Be sure to use correct chemical formulas for the reactants and the products. For example, suppose a chemical reaction produces carbon dioxide and water. The product carbon dioxide is CO_2, not CO. CO is the formula for carbon monoxide, which is not the same compound as CO_2. Water is H_2O, not H_2O_2, the formula for hydrogen peroxide.

✓ **Visual Check**

9. Identify Highlight the symbol that separates the reactants from the products in a chemical equation.

Parts of an Equation

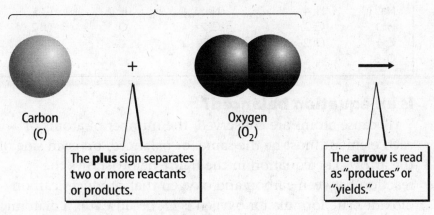

Reactants are written to the left of the arrow.

Products are written to the right of the arrow.

Carbon (C)

Oxygen (O_2)

Carbon dioxide (CO_2)

The **plus** sign separates two or more reactants or products.

The **arrow** is read as "produces" or "yields."

Conservation of Mass

Antoine Lavoisier (AN twan · luh VWAH see ay) (1743–1794), a French chemist, discovered something interesting about chemical reactions. In a series of experiments, Lavoisier measured the masses of substances before and after a chemical reaction inside a closed container. He found that the total mass of the reactants always equaled the total mass of the products. Lavoisier's results led to the law of conservation of mass. *The* **law of conservation of mass** *states that the total mass of the reactants before a chemical reaction is the same as the total mass of the products after the chemical reaction.*

🔑 **Key Concept Check**

10. Explain What happens to the total mass of the reactants in a chemical reaction?

_____

_____

_____

Atoms are conserved.

The discovery of atoms helped explain Lavoisier's observations. Mass is conserved in a reaction because atoms are conserved. During a chemical reaction, bonds break and new bonds form. However, a reaction does not destroy atoms, and it does not form new atoms. All atoms at the start of a chemical reaction are present at the end of the reaction.

Suppose you attach a balloon with baking soda inside to a flask of vinegar. You place the flask on a scale and record the mass. Then you mix the two substances. They react, and the balloon fills with gas. You find that the products after the reaction have the same mass as the reactants. Mass is conserved. The atoms also are conserved, as shown in the equation below.

Conservation of Mass

Mass is equal.

baking soda + vinegar $\longrightarrow$ sodium acetate + water + carbon dioxide

Atoms are equal.

$NaHCO_3$	$HC_2H_3O_2$	$NaC_2H_3O_2$	H_2O	CO_2
Na: 1	H: 4	Na: 1	H: 2	C: 1
H: 1	C: 2	C: 2	O: 1	O: 2
C: 1	O: 2	H: 3		
O: 3		O: 2		

Is an equation balanced?

Because atoms are conserved, the number of atoms of each element must be the same, or balanced, on each side of the arrow. The equation in the figure below shows the reaction between carbon and oxygen that produces carbon dioxide. The formula for oxygen is O_2 because it is a diatomic molecule. The formula for carbon dioxide is CO_2. There is one carbon atom on the left of the arrow and one on the right. Carbon is balanced. Two oxygen atoms are on each side of the arrow. Oxygen also is balanced. The atoms of all elements are balanced. So, the equation is balanced.

A Balanced Chemical Equation

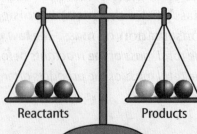

Reactants			Product		Balanced
C	+	O_2	$\longrightarrow$	CO_2	
1 carbon atom		2 oxygen atoms		1 carbon atom 2 oxygen atoms	Reactants Products

Interpreting Tables

11. Identify How many atoms of hydrogen are on each side of the equation in the table?

_____

_____

✔ **Reading Check**

12. Recognize How do you know that a chemical equation is balanced?

_____

_____

_____

✔ **Visual Check**

13. Explain Were the atoms conserved in this equation? How do you know?

_____

_____

An Unbalanced Chemical Equation

Reactants   **Product**   **Unbalanced**

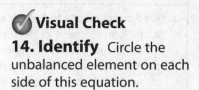

H_2 + O_2 $\longrightarrow$ H_2O

2 hydrogen atoms   2 oxygen atoms   2 hydrogen atoms
                                              1 oxygen atom

You might think a balanced equation happens automatically when you write the symbols and formulas for reactants and products. However, this usually is not the case.

For example, the reaction between hydrogen (H_2) and oxygen (O_2) that forms water (H_2O) is shown in the figure above. Count the number of hydrogen atoms on each side of the arrow. There are two hydrogen atoms in the product and two in the reactants. They are balanced.

Now count the number of oxygen atoms on each side of the arrow. Did you notice that there are two oxygen atoms in the reactants and only one in the product? Because they are not equal, this equation is not balanced. To accurately represent this reaction, you need to balance the equation.

Balancing Chemical Equations

Balancing a chemical equation is the process of counting the atoms in the reactants and the products and then adding coefficients to balance the atoms. *A* **coefficient** *is a number placed in front of an element symbol or chemical formula in an equation.* A coefficient tells the number of units of a substance in the reaction. For example, the coefficient *2* added to H_2O is written as $2H_2O$. This means that two molecules of water take part in the reaction. ✓

If one molecule of water contains two hydrogen atoms and one oxygen atom, how many H and O atoms are in two molecules of water ($2H_2O$)? Multiply each by 2.

$$2 \times 2 \text{ H atoms} = 4 \text{ H atoms}$$
$$2 \times 1 \text{ O atom} = 2 \text{ O atoms}$$

When no coefficient is present, only one unit of that substance takes part in the reaction.

Copyright © Glencoe/McGraw-Hill, a division of The McGraw-Hill Companies, Inc.

✓**Visual Check**

14. Identify Circle the unbalanced element on each side of this equation.

FOLDABLES

Make a four-tab book to study the steps of balancing equations.

✓**Reading Check**

15. Name the coefficient of $3O_2$. What does it tell you?

_____

_____

_____

Balancing a Chemical Equation

1. **Write the unbalanced equation.** Make sure that all chemical formulas are correct.	H_2 + O_2 $\longrightarrow$ H_2O reactants products
2. **Count atoms of each element in the reactants and in the products.** **a.** Note which, if any, elements have a balanced number of atoms on each side of the equation. Which atoms are not balanced? **b.** If all of the atoms are balanced, the equation is balanced.	H_2 + O_2 $\longrightarrow$ H_2O reactants products H = 2 H = 2 O = 2 O = 1
3. **Add coefficients to balance the atoms.** **a.** Pick an element in the equation that is not balanced, such as oxygen. Write a coefficient in front of a reactant or a product that will balance the atoms of that element. **b.** Recount the atoms of each element in the reactants and the products. Note which atoms are not balanced. Some atoms that were balanced before might no longer be balanced. **c.** Repeat step 3 until the atoms of each element are balanced.	H_2 + O_2 $\longrightarrow$ $2H_2O$ reactants products H = 2 H = 2 O = 2 O = 1 $2H_2$ + O_2 $\longrightarrow$ $2H_2O$ reactants products H = 2 H = 2 O = 2 O = 1
4. **Write the balanced chemical equation** including the coefficients.	$2H_2$ + O_2 $\longrightarrow$ $2H_2O$

Interpreting Tables

16. Identify In step 2 above, which element is not balanced?

_____

In the first equation of step 3, which element is not balanced?

_____

The table above shows the steps of balancing a chemical equation. Notice that adding the coefficient 2 in front of H_2O in the equation balances the oxygen atoms but unbalances the hydrogen atoms. Adding the coefficient 2 in front of the reactant H_2 brings the hydrogen atoms back into balance.

Mini Glossary

chemical equation: a description of a reaction using element symbols and chemical formulas

chemical reaction: a process in which atoms of one or more substances rearrange to form one or more new substances

coefficient: a number placed in front of an element symbol or chemical formula in an equation

law of conservation of mass: states that the total mass of the reactants before a chemical reaction is the same as the total mass of the products after the chemical reaction

product: a substance produced by a chemical reaction

reactant: a starting substance in a chemical reaction

1. Review the terms and their definitions in the Mini Glossary. Write a sentence that describes how a chemical equation and a chemical reaction are related.

_____

_____

2. Count the number of atoms of each element on both sides of the chemical equation below. Then determine whether the equation is balanced or unbalanced.

$$CH_4 + 2O_2 \quad \Longrightarrow \quad CO_2 + 2H_2O$$

C = _____   C = _____

H = _____   H = _____

O = _____   O = _____

balanced or unbalanced? _____

3. When water boils, bubbles form. Is this a chemical change or a physical change? Explain your answer.

_____

_____

What do you think  **NOW?**

Reread the statements at the beginning of the lesson. Fill in the After column with an A if you agree with the statement or a D if you disagree. Did you change your mind?

 Connect ED

Log on to ConnectED.mcgraw-hill.com and access your textbook to find this lesson's resources.

END OF LESSON

Chemical Reactions and Equations

Types of Chemical Reactions

Key Concepts 🔑

- How can you recognize the type of chemical reaction by the number or type of reactants and products?
- What are the different types of chemical reactions?

Mark the Text

Sticky Notes As you read, use sticky notes to mark information that you do not understand. Read the text carefully a second time. If you still need help, write a list of questions to ask your teacher.

FOLDABLES

Make a four-door book to organize your notes about the different types of chemical reactions.

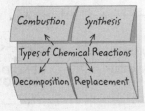

·············· **Before You Read** ··············

What do you think? Read the two statements below and decide whether you agree or disagree with them. Place an A in the Before column if you agree with the statement or a D if you disagree. After you've read this lesson, reread the statements to see if you have changed your mind.

Before	Statement	After
	3. Reactions always start with two or more substances that react with each other.	
	4. Water can be broken down into simpler substances.	

·············· **Read to Learn** ··············

Patterns in Reactions

If you have ever used hydrogen peroxide, you might have noticed that it comes in a dark bottle. This is because light causes hydrogen peroxide to change into other substances. You might have seen video of an old building or a sports facility being demolished by an explosion. How is the reaction between hydrogen and light similar to a building being demolished? In both, one reactant breaks down into two or more products.

The breakdown of one reactant into two or more products is one of four major types of chemical reaction. Each type of chemical reaction follows a unique pattern in the way atoms in reactants rearrange to form products. In this lesson, you will read how chemical reactions are classified by recognizing patterns in the way the atoms recombine.

Types of Chemical Reactions

Many different types of reactions can occur. It would be impossible to memorize them all. However, most chemical reactions fit into four major categories. Understanding these categories of reactions can help you predict how compounds will react and what products will form.

Synthesis

A **synthesis** (SIHN thuh sus) *is a type of chemical reaction in which two or more substances combine and form one compound.* In the synthesis reaction on the left below, magnesium (Mg) reacts with oxygen (O_2) in the air and forms magnesium oxide (MgO). You can recognize a synthesis reaction because two or more reactants form only one product.

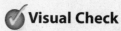 **Visual Check**

1. Identify How many products result from a synthesis reaction?

_____

Synthesis and Decomposition Reactions

Synthesis Reactions	Decomposition Reactions
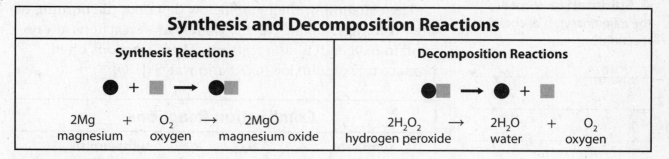	
$2Mg$ + O_2 → $2MgO$ magnesium oxygen magnesium oxide	$2H_2O_2$ → $2H_2O$ + O_2 hydrogen peroxide water oxygen

Decomposition

In a **decomposition** *reaction, one compound breaks down and forms two or more substances.* You can recognize a decomposition reaction because one reactant forms two or more products. For example, hydrogen peroxide (H_2O_2), shown on the right above, decomposes and forms water (H_2O) and oxygen gas (O_2). Notice that decomposition is the reverse of synthesis.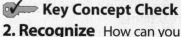

Replacement

In a replacement reaction, an atom or a group of atoms replaces part of a compound. Two types of replacement reactions can occur. The figure below describes both types.

In a **single-replacement** *reaction, one element replaces another element in a compound.* In this type of reaction, an element and a compound react and form a different element and a different compound. *In a* **double-replacement** *reaction, the negative ions in two compounds switch places, forming two new compounds.* In this type of reaction, two compounds react and form two new compounds.

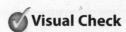 **Key Concept Check**

2. Recognize How can you tell the difference between synthesis and decomposition reactions?

_____

_____

_____

_____

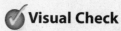 **Visual Check**

3. Interpret In the single-replacement equation, which elements switched places?

Replacement Reactions

Single Replacement	Double Replacement
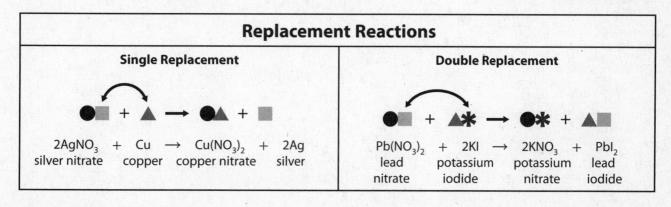	
$2AgNO_3$ + Cu → $Cu(NO_3)_2$ + $2Ag$ silver nitrate copper copper nitrate silver	$Pb(NO_3)_2$ + $2KI$ → $2KNO_3$ + PbI_2 lead nitrate potassium iodide potassium nitrate lead iodide

Copyright © Glencoe/McGraw-Hill, a division of The McGraw-Hill Companies, Inc.

Combustion

Combustion *is a chemical reaction in which a substance combines with oxygen and releases energy.* A combustion reaction usually releases energy as thermal energy and light energy. For example, burning is a common combustion reaction. The burning of fossil fuels produces the energy we use to cook food, power vehicles, and light cities.

The equation in the figure below describes the burning of the fossil fuel propane (C_3H_8). Combustion reactions always contain oxygen (O_2) as a reactant. These reactions often produce carbon dioxide (CO_2) and water (H_2O).

Key Concept Check

4. Summarize What are the different types of chemical reactions?

_____

_____

_____

Visual Check

5. Identify Which element is common to all combustion reactions?

_____

Combustion Reactions				
substance	+	O_2	→	substance(s)
C_3H_8 propane	+	$5O_2$ oxygen	→	$3CO_2$ carbon dioxide + $4H_2O$ water

Mini Glossary

combustion: a chemical reaction in which a substance combines with oxygen and releases energy

decomposition: a type of chemical reaction in which one compound breaks down and forms two or more substances

double replacement: a type of chemical reaction in which the negative ions in two compounds switch places, forming two new compounds

single replacement: a type of chemical reaction in which one element replaces another element in a compound

synthesis (SIHN thuh sus): a type of chemical reaction in which two or more substances combine and form one compound

1. Review the terms and their definitions in the Mini Glossary. Write a sentence that identifies the types of energy that combustion usually releases.

_____

_____

2. Identify each type of chemical reaction described in the diagram.

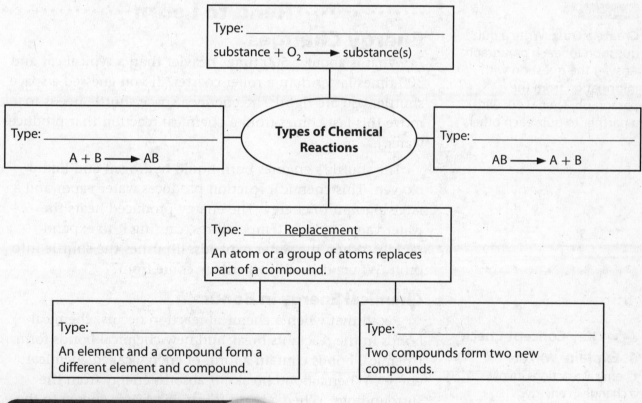

Type: _____

substance + O_2 ⟶ substance(s)

Types of Chemical Reactions

Type: _____

A + B ⟶ AB

Type: _____

AB ⟶ A + B

Type: _____ Replacement _____

An atom or a group of atoms replaces part of a compound.

Type: _____

An element and a compound form a different element and compound.

Type: _____

Two compounds form two new compounds.

What do you think NOW?

Reread the statements at the beginning of the lesson. Fill in the After column with an A if you agree with the statement or a D if you disagree. Did you change your mind?

Connect ED

Log on to ConnectED.mcgraw-hill.com and access your textbook to find this lesson's resources.

END OF LESSON

Chemical Reactions and Equations

Energy Changes and Chemical Reactions

Key Concepts 🔑

- Why do chemical reactions always involve a change in energy?
- What is the difference between an endothermic reaction and an exothermic reaction?
- What factors can affect the rate of a chemical reaction?

Study Coach ▶

Create a Quiz Write a quiz question for each paragraph. Answer the question with information from the paragraph. Then work with a partner to quiz each other.

🔑 **Key Concept Check**

1. Explain Why do chemical reactions involve a change in energy?

_____

_____

_____

_____

·············· **Before You Read** ··············

What do you think? Read the two statements below and decide whether you agree or disagree with them. Place an A in the Before column if you agree with the statement or a D if you disagree. After you've read this lesson, reread the statements to see if you have changed your mind.

Before	Statement	After
	5. Reactions that release energy require energy to get started.	
	6. Energy can be created in a chemical reaction.	

·············· **Read to Learn** ··············

Energy Changes

What is about 1,500 times heavier than a typical car and 300 times faster than a roller coaster? If you guessed a space shuttle, you are right! The energy a space shuttle needs to move this fast comes from a chemical reaction that produces water.

The shuttle's engines burn liquid hydrogen and liquid oxygen. This chemical reaction produces water vapor and a large amount of energy. The energy produced heats the water vapor to high temperatures, causing it to expand rapidly. When the water expands, it pushes the shuttle into orbit. Where does all this energy come from?

Chemical Energy in Bonds

Recall that when a chemical reaction occurs, chemical bonds in the reactants break and new chemical bonds form. Chemical bonds contain a form of energy called chemical energy. When a bond breaks, it absorbs energy from the surroundings. When a bond forms, it releases energy to the surroundings. Some chemical reactions release more energy than they absorb. Some chemical reactions absorb more energy than they release. You can feel this energy change as a change in the temperature of the surroundings. Keep in mind that in all chemical reactions, energy is conserved. 🔑

Endothermic Reactions—Energy Absorbed

On a very warm day, have you ever heard someone say that the sidewalk was hot enough to fry an egg? To fry, the egg must absorb energy.

Chemical reactions that absorb thermal energy are **endothermic** *reactions.* For an endothermic reaction to continue, energy must be constantly added.

reactants + thermal energy → products

In an endothermic reaction, more energy is required to break the bonds of the reactants than is released when the products form. Therefore, the overall reaction absorbs energy.

The reaction in the figure below is an endothermic reaction. Notice that the products hold more energy than the reactants. The chemical bonds absorbed energy during the reaction.

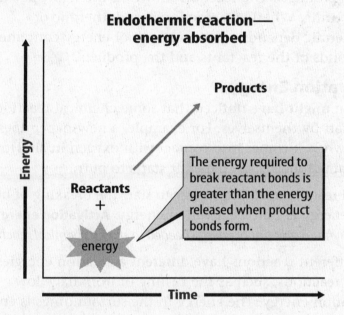

Endothermic reaction—energy absorbed

Products

Energy

Reactants
+
energy

The energy required to break reactant bonds is greater than the energy released when product bonds form.

Time

Exothermic Reactions—Energy Released

Most chemical reactions release energy as opposed to absorbing it. *An* **exothermic** *reaction is a chemical reaction that releases thermal energy.*

reactants → products + thermal energy

In an exothermic reaction, more energy is released when the products form than is required to break the bonds in the reactants. Therefore, the overall reaction releases energy.

Copyright © Glencoe/McGraw-Hill, a division of The McGraw-Hill Companies, Inc.

FOLDABLES

Make a three-tab Venn book to compare and contrast energy in chemical reactions.

Exothermic Reaction

Both

Endothermic Reaction

✓ Visual Check

2. Interpret Why does the arrow point upward?

_____

_____

_____

💡 Think it Over

3. Analyze The engines of a space shuttle burn liquid hydrogen and liquid oxygen. Is this an endothermic or an exothermic reaction? How do you know?

_____

_____

_____

_____

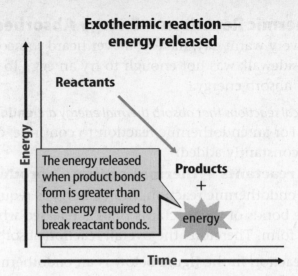

Copyright © Glencoe/McGraw-Hill, a division of The McGraw-Hill Companies, Inc.

Visual Check

4. Interpret Why does the arrow point downward?

_____

_____

_____

Key Concept Check

5. Contrast What is the difference between an endothermic reaction and an exothermic reaction?

_____

_____

_____

_____

Reading Check

6. Explain Why doesn't the book on your desk burst into flames?

_____

_____

_____

_____

An exothermic reaction releases energy. As a result, the bonds of the reactants contain more energy than the bonds of the products. The reaction shown in the figure above is exothermic. Whether a reaction is endothermic or exothermic depends on the amount of energy contained in the bonds of the reactants and the products.

Activation Energy

You might have noticed that some chemical reactions do not start by themselves. For example, a newspaper does not burn when it comes into contact with oxygen in air. However, if a flame touches the paper, it starts to burn.

All reactions require energy to start the breaking of bonds. This energy is called activation energy. **Activation energy** is *the minimum amount of energy needed to start a chemical reaction.*

Different reactions have different activation energies. Some reactions, such as the rusting of iron, have low activation energy. The energy in the surroundings is enough to start these reactions.

If a reaction has high activation energy, more energy is needed to start the reaction. For example, wood requires the thermal energy of a flame to start burning. Once the reaction starts, it releases enough energy to keep the reaction going. The figure at the top of the next page shows the role that activation energy plays in endothermic and exothermic reactions.

Activation Energy

Endothermic

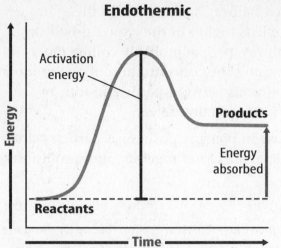

Exothermic

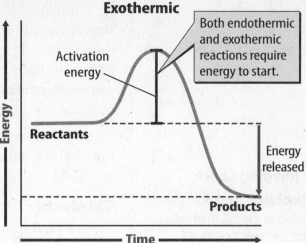

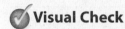
Both endothermic and exothermic reactions require energy to start.

Reaction Rates

Some chemical reactions, such as the rusting of a bicycle wheel, happen slowly. Other chemical reactions, such as the explosion of fireworks, happen in less than a second. The rate of a reaction is the speed at which the reaction occurs. What controls how fast a chemical reaction occurs?

Recall that particles must collide before they can react. Chemical reactions occur faster if particles collide more often or if the particles move faster when they collide. Several factors affect how often particles collide and how fast particles move.

Surface Area

Surface area is the amount of the exposed outer area of a solid. Increased surface area increases reaction rate. This is because more particles on the surface of a solid come into contact with the particles of another substance. For example, if you place a piece of chalk in vinegar, the chalk reacts slowly with the acid. This is because the acid contacts only the particles on the surface of the chalk. But, if you grind the chalk into powder, more chalk particles contact the acid, and the reaction occurs faster.

Temperature

Imagine a crowded hallway. If everyone in the hallway were running, they would probably collide with each other more often and with more energy than if everyone were walking. This is also true when particles move faster. At higher temperatures, the average speed of particles is greater. This speeds reactions in two ways. First, particles collide more often. Second, collisions with more energy are more likely to break chemical bonds.

Visual Check
7. Analyze How can a reaction absorb energy to start but still be exothermic?

Math Skills

The area of a side of a 1-cm cube is 1 cm × 1 cm or 1 cm². The cube has 6 equal sides. Its total surface area is 6 × 1 cm² or 6 cm². How much surface area is gained by cutting the cube in half?

a. Two surfaces are made, each with an area:

$$1 \text{ cm} \times 1 \text{ cm} = 1 \text{ cm}^2$$

b. Multiply by the number of new surfaces.

$$1 \text{ cm}^2 \times 2 = 2 \text{ cm}^2$$

The surface area is increased by 2 cm².

8. Use Geometry How much surface area is gained when a 2 cm cube is cut in half?

Concentration and Pressure

Think of a crowded hallway again. Because the concentration of people is higher in the crowded hallway than in an empty hallway, people probably collide more often. Similarly, increasing the concentration of one or more reactants increases collisions between particles. More collisions result in a faster reaction rate.

In gases, an increase in pressure pushes gas particles closer together. When particles are closer together, more collisions occur.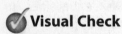

Catalysts

A **catalyst** *is a substance that increases reaction rate by lowering the activation energy of a reaction.* One way catalysts speed reactions is by causing reactant particles to contact each other more often. Look at the figure below. Notice that the activation energy of the reaction is lower with a catalyst than it is without a catalyst. The reaction doesn't change the catalyst, and the catalyst doesn't change the reactants or products. A catalyst doesn't increase the amount of reactant used or amount of product made. A catalyst only makes a reaction happen faster. Catalysts are not reactants in a reaction.

Effect of a Catalyst on Activation Energy

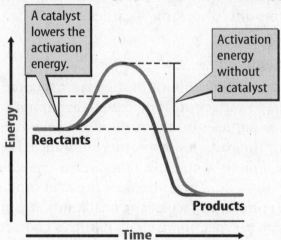

A catalyst lowers the activation energy.

Activation energy without a catalyst

Energy

Reactants

Products

Time

Your body is filled with catalysts called enzymes. *An* **enzyme** *is a catalyst that speeds up chemical reactions in living cells.* For example, the enzyme protease (PROH tee ays) breaks the protein molecules in the food you eat into smaller molecules that your intestine can absorb. Without enzymes, these reactions would occur too slowly for life to exist.

Reading Check

9. Explain Why does an increase in pressure on a gas speed the reaction rate?

_____

_____

_____

Visual Check

10. Interpret How does a catalyst increase the reaction rate?

_____

_____

Think it Over

11. Apply Protease breaks protein molecules into smaller molecules. This speeds the reaction rate by increasing ___. (Circle your answer.)

a. temperature

b. pressure

c. surface area

Inhibitors

Recall that an enzyme is a molecule that speeds reactions in organisms. However, some organisms, such as bacteria, are harmful to humans. Some medicines contain molecules that attach to enzymes in bacteria. These medicines prevent enzymes in bacteria or viruses from working. If the enzymes in bacteria can't work, the bacteria die and can no longer infect a human. The active ingredients in these medicines are called inhibitors. *An **inhibitor** is a substance that slows, or even stops, a chemical reaction.* Inhibitors can slow or stop the reactions caused by enzymes.

Inhibitors are also important in the food industry. Preservatives in food are substances that inhibit, or slow, food spoilage. 🗝

🗝 Key Concept Check

12. Identify What factors can affect the rate of a chemical reaction?

_____

_____

_____

_____

_____

_____

Copyright © Glencoe/McGraw-Hill, a division of The McGraw-Hill Companies, Inc.

Mini Glossary

activation energy: the minimum amount of energy needed to start a chemical reaction

catalyst: a substance that increases reaction rate by lowering the activation energy of a reaction

endothermic: a chemical reaction that absorbs thermal energy

enzyme: a catalyst that speeds up chemical reactions in living cells

exothermic: a chemical reaction that releases thermal energy

inhibitor: a substance that slows, or even stops, a chemical reaction

1. Review the terms and their definitions in the Mini Glossary. Write a sentence that explains how an enzyme and a catalyst are related.

_____

_____

2. The diagram below summarizes how four factors affect the speed of chemical reactions.

Complete the diagram by writing "increase" or "decrease" next to each factor.

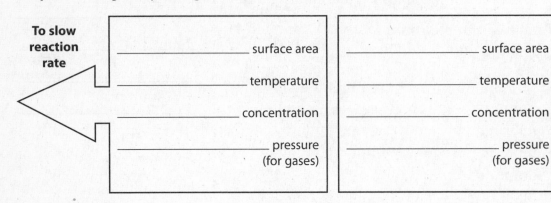

To slow reaction rate

_____ surface area

_____ temperature

_____ concentration

_____ pressure (for gases)

To speed up reaction rate

_____ surface area

_____ temperature

_____ concentration

_____ pressure (for gases)

3. In the space below, write a question from your partner's quiz that helped you learn an important concept. Then answer the question.

_____

_____

What do you think NOW?

Reread the statements at the beginning of the lesson. Fill in the After column with an A if you agree with the statement or a D if you disagree. Did you change your mind?

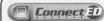
Log on to ConnectED.mcgraw-hill.com and access your textbook to find this lesson's resources.

END OF LESSON

Mixtures, Solubility, and Acid/Base Solutions

Substances and Mixtures

Copyright © Glencoe/McGraw-Hill, a division of The McGraw-Hill Companies, Inc.

·············· **Before You Read** ··············

Before	Statement	After
	What do you think? Read the two statements below and decide whether you agree or disagree with them. Place an A in the Before column if you agree with the statement or a D if you disagree. After you've read this lesson, reread the statements to see if you have changed your mind.	
	1. You can identify a mixture by looking at it without magnification.	
	2. A solution is another name for a homogeneous mixture.	

·················· **Read to Learn** ··················

Matter: Substances and Mixtures

Many different types of matter exist around you. On your way to school, you might see metal, plastic, rocks, concrete, plants, fabric, water, and skin. You might notice that you can group some types of matter together into one category. For example, keys, coins, and paper clips are made of metal. Grouping matter into categories helps you understand how some things are similar to each other, but different from other things. Nearly all types of matter can be sorted into just two major categories—substances and mixtures.

What is a substance?

A **substance** *is matter that is always made up of the same combination of atoms.* There are two types of substances—elements and compounds. Recall that an element is matter made of only one type of atom, such as carbon (C), oxygen (O_2), and chlorine (Cl_2). A compound is matter made of atoms of two or more elements that are chemically bonded together. For example, water (H_2O) always contains two atoms of hydrogen bonded to one atom of oxygen. Table salt (NaCl) always contains one atom of sodium bonded to one atom of chlorine. The composition of elements and compounds doesn't change. Therefore, all elements and compounds are substances. ✓

Key Concepts
- How do substances and mixtures differ?
- How do solutions compare and contrast with heterogeneous mixtures?
- In what three ways do compounds differ from mixtures?

Study Coach

Make a Table with two columns to contrast substances and mixtures. Label one column *Substances* and the other column *Mixtures*. Complete the table as you read this lesson.

✓ **Reading Check**

1. Analyze Is table salt an element or a compound? Explain.

_____

_____

_____

_____

Combine a two-tab and a four-tab book to organize your notes on matter.

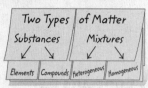

Key Concept Check

2. Contrast How do substances and mixtures differ?

_____

_____

_____

_____

Key Concept Check

3. Distinguish How can you determine whether a mixture is homogeneous or heterogeneous?

_____

_____

_____

_____

Visual Check

4. Interpret Can a mixture be made only of elements? Explain.

_____

_____

_____

_____

What is a mixture?

A **mixture** is two or more substances that are physically blended but are not chemically bonded together. Therefore, the relative amounts of each substance in a mixture can vary. Granite, a type of rock, is a mixture. Granite contains bits that are white, black, and other colors. Two pieces of granite will have different amounts of each color bit. The composition of rocks varies. Air is also a mixture. Air contains about 78 percent nitrogen, 21 percent oxygen, and 1 percent other substances. But this composition varies. Air in a scuba tank can have more than 21 percent oxygen and less of the other substances.

Rocks and air are examples of the two different types of mixtures—heterogeneous (he tuh roh JEE nee us) and homogeneous (hoh muh JEE nee us). It's not always easy to identify a mixture. A rock looks like a mixture, but air does not.

Heterogeneous Mixtures _A **heterogeneous mixture** is a mixture in which substances are not evenly mixed._ For example, granite is a heterogeneous mixture. The substances that make up granite are unevenly mixed. When you look at a piece of granite, you can easily see the different parts.

Often, you can see the different substances and parts of a heterogeneous mixture with unaided eyes. Sometimes you can see them only with a microscope. For example, blood looks evenly mixed—its color and texture are the same throughout. But suppose you compare two samples of blood under a microscope. They will look different. The samples will contain different amounts of the components.

Solutions—Homogeneous Mixtures Many mixtures look evenly mixed, even when you view them with a powerful microscope. These mixtures are homogeneous. _A **homogeneous mixture** is a mixture in which two or more substances are evenly mixed on the atomic level but not bonded together._ The individual atoms or compounds of each substance are mixed. The mixture looks the same throughout under a microscope because individual atoms and compounds are too small to see. Air is a homogeneous mixture. If you view air under a microscope, you can't see the individual substances that make it up.

Another name for a homogeneous mixture is **solution.** As you read, remember that the terms _homogeneous mixture_ and _solution_ mean the same thing. Review the figure on the next page. It summarizes the characteristics of substances and mixtures.

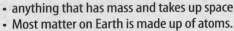

Matter
- anything that has mass and takes up space
- Most matter on Earth is made up of atoms.

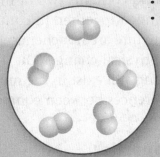

Substances
- matter with a composition that is always the same
- two types of substances: elements and compounds

Elements
- consists of just one type of atom
- organized on the periodic table
- Elements can exist as single atoms or as a diatomic molecule—two atoms bonded together.

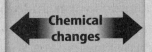
Chemical changes

Compounds
- made of two or more types of atoms bonded together
- can't be separated by physical methods
- properties are different from the properties of the elements that make them up
- two types: ionic and covalent

Separating mixtures
- filtering
- boiling
- using a magnet

Physical changes

Combining substances
- mixing
- dissolving

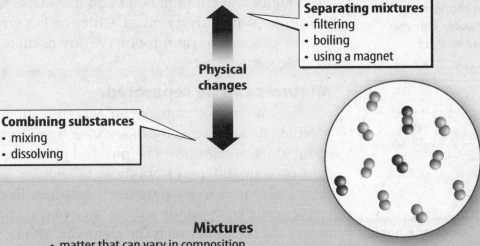

Mixtures
- matter that can vary in composition
- made of two or more substances mixed but not bonded together
- can be separated into substances by physical methods
- Two types of mixtures: heterogeneous and homogeneous

Heterogeneous mixtures
- two or more substances unevenly mixed
- uneven mixing is visible with unaided eyes or a microscope

Homogeneous mixtures (solutions)
- two or more substances evenly mixed
- Homogeneous mixtures appear uniform under a microscope.

How do compounds and mixtures differ?

You have read that a compound contains two or more elements that are chemically bonded together. In contrast, the substances that make up a mixture are not chemically bonded. Therefore, mixing is a physical change. The substances that exist before mixing still exist in the mixture. This leads to two important differences between compounds and mixtures.

Substances keep their properties.

Substances that make up a mixture are not changed chemically. Therefore, you can observe some of their properties in the mixture. Sugar water is a mixture of two compounds—sugar and water. After the sugar is mixed in, you can't see the sugar in the water, but you can still taste its property of sweetness. You can also observe some properties of the water, such as its liquid state.

In contrast, the properties of a compound can be different from the properties of the elements that make it up. Sodium and chlorine bond and form table salt. Sodium is a soft, opaque, silvery metal. Chlorine is a greenish, poisonous gas. You cannot observe any of these properties in table salt.

Mixtures can be separated.

The substances that make up a mixture are not bonded together. As a result, you can separate them from each other using physical methods. The physical properties of one substance are different from those of another. You can use these differences to separate the substances. In contrast, you can separate compounds only by a chemical change that breaks the bonds between the elements.

Copyright © Glencoe/McGraw-Hill, a division of The McGraw-Hill Companies, Inc.

Think it Over

5. Apply Seawater is a mixture of mainly salts and water. What properties of these substances can you observe in seawater?

_____

_____

_____

_____

Key Concept Check

6. Identify In what three ways do compounds differ from mixtures?

_____

_____

_____

_____

_____

Mini Glossary

heterogeneous (he tuh roh JEE nee us) mixture: a mixture in which substances are not evenly mixed

homogeneous (hoh muh JEE nee us) mixture: a mixture in which two or more substances are evenly mixed on the atomic level but not bonded together

mixture: two or more substances that are physically blended but are not chemically bonded together

solution: a homogeneous mixture

substance: matter that is always made up of the same combination of atoms

1. Review the terms and their definitions in the Mini Glossary. Write a sentence that explains how a homogeneous mixture and a solution are related.

_____

_____

2. Complete the concept web by writing the types of matter in the correct boxes.

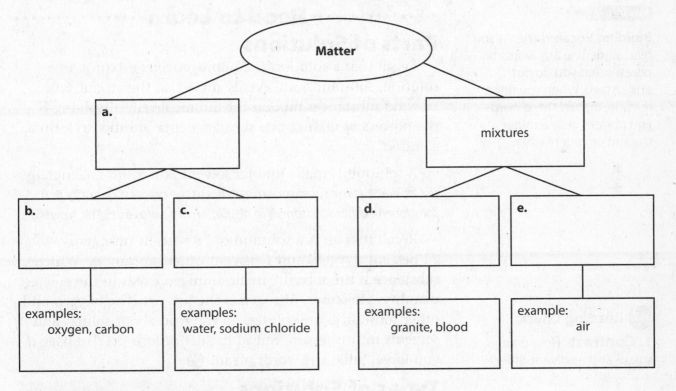

Matter

a.

mixtures

b.

c.

d.

e.

examples:
    oxygen, carbon

examples:
    water, sodium chloride

examples:
    granite, blood

example:
    air

What do you think NOW?

Reread the statements at the beginning of the lesson. Fill in the After column with an A if you agree with the statement or a D if you disagree. Did you change your mind?

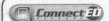

Log on to ConnectED.mcgraw-hill.com and access your textbook to find this lesson's resources.

END OF LESSON

Mixtures, Solubility, and Acid/Base Solutions

Properties of Solutions

Copyright © Glencoe/McGraw-Hill, a division of The McGraw-Hill Companies, Inc.

Key Concepts 🔑
- Why do some substances dissolve in water and others do not?
- How do concentration and solubility differ?
- How can the solubility of a solute be changed?

Mark the Text ▶

Building Vocabulary As you read, underline any words or phrases that you do not understand. When you finish reading, discuss these words and phrases with another student or your teacher.

✔ **Reading Check**

1. Contrast How do a solute and a solvent differ?

_____

_____

_____

_____

············· **Before You Read** ·············

What do you think? Read the two statements below and decide whether you agree or disagree with them. Place an A in the Before column if you agree with the statement or a D if you disagree. After you've read this lesson, reread the statements to see if you have changed your mind.

Before	Statement	After
	3. Solutions can be solids, liquids, or gases.	
	4. A teaspoon of soup is less concentrated than a cup of the same soup.	

············· **Read to Learn** ·············

Parts of Solutions

Recall that a solution is a homogeneous mixture. In a solution, substances are evenly mixed on the atomic level. How do substances mix on the atomic level? Dissolving is the process of mixing one substance into another to form a solution.

A solution is made up of a solvent and solutes. Generally, *the* **solvent** *is the substance that exists in the greatest quantity in a solution. All other substances in a solution are* **solutes** (SAHL yewts).

Recall that air is a solution of 78 percent nitrogen, 21 percent oxygen, and 1 percent other substances. Which substance is the solvent? In air, nitrogen exists in the greatest quantity. Therefore, nitrogen is the solvent. The oxygen and other substances are solutes. As you read about solutes and solvents in this lesson, reread the definitions on this page if you forget what the words mean. ✔

Types of Solutions

You might think of a solution as a liquid. But solutions can exist in all three states of matter—solid, liquid, or gas.

The solvent exists in the greatest quantity. Therefore, the state of the solvent determines the state of the solution. Read the table on the next page. It contrasts solid, liquid, and gaseous solutions.

Types of Solutions

State of Solution	Solvent Is:	Solute Can Be:
Solid	solid	**gas or solid (called alloys)** A saxophone is a solid solution of solid copper and solid zinc.
Liquid	liquid	**solid, liquid, and/or gas** Soda is a liquid solution of liquid water, gaseous carbon dioxide, and solid sugar and other flavorings.
Gas	gas	**gas** A gaseous mixture of gaseous argon and gaseous mercury produces the light you see in many brightly colored signs.

Water as a Solvent

Water exists naturally in all three states—solid, liquid, and gas. In nature, water almost always exists as a solution. Why does nearly all water on Earth contain dissolved solutes? The answer has to do with the structure of the water molecule.

The Polarity of Water

A water molecule, such as the one illustrated in the figure below, is a covalent compound. Recall that atoms are held together with covalent bonds when sharing electrons. In a water molecule, one oxygen atom shares electrons with two hydrogen atoms.

These electrons are not shared equally. The electrons in the oxygen-hydrogen bonds are closer to the oxygen atom than they are to the hydrogen atoms. This unequal sharing of electrons gives the end with the oxygen atom a slightly negative charge and the end with the hydrogen atoms a slightly positive charge.

Polarity of a Water Molecule

The electrons spend more time near the oxygen atom. This makes the end with the oxygen atom slightly negative (−).

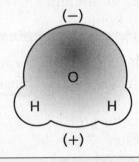

The end with the hydrogen atoms is slightly positive (+).

Because of the unequal sharing of electrons, a water molecule is said to be polar. *A* **polar molecule** *is a molecule with a slightly negative end and a slightly positive end.* Nonpolar molecules have an even distribution of charge. Solutes and solvents can be polar or nonpolar. ✔

Interpreting Tables

2. Name When the solvent is a solid and the solute is a gas, what will the state of the solution be?

_____

Visual Check

3. Analyze Shared electrons in a water molecule are closer to the oxygen atom than to the hydrogen atoms. Why does this create a slightly negative charge on the oxygen end of the molecule?

_____

_____

Reading Check

4. Identify What makes a molecule polar?

_____

_____

_____

Like Dissolves Like

Water is often called the universal solvent because it dissolves many substances. But water can't dissolve everything. Why does water dissolve some substances but not others? Water is a polar solvent. Polar solvents dissolve polar solutes easily. Nonpolar solvents dissolve nonpolar solutes easily. In other words: "Like dissolves like." Because water is a polar solvent, it dissolves most polar and ionic solutes. 🔑

Polar Solvents and Polar Molecules

Because water molecules are polar, water dissolves groups of other polar molecules. The figure to the right shows rubbing alcohol in solution with water. Notice that molecules of rubbing alcohol also are polar. When alcohol and water mix, the positive ends of the water molecules attract the negative ends of the alcohol molecules. Similarly, the negative ends of the water molecules attract the positive ends of the alcohol molecules. In this way, alcohol molecules dissolve in the solvent.

Polar Molecules in Solution

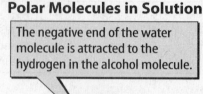

The negative end of the water molecule is attracted to the hydrogen in the alcohol molecule.

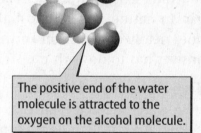

The positive end of the water molecule is attracted to the oxygen on the alcohol molecule.

Polar Solvents and Ionic Compounds

Many ionic compounds are also soluble in water. Recall that ionic compounds are made of alternating positive and negative ions. Sodium chloride (NaCl) is an ionic compound. It is made of sodium ions (Na^+) and chloride ions (Cl^-).

When sodium chloride dissolves, these ions are pulled apart. Look at the figure to the right. The negative ends of the water molecules attract the positive sodium ions. The positive ends of the water molecules attract the negative chloride ions.

Ionic Compound in Solution

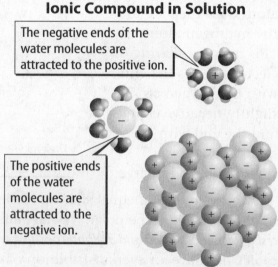

The negative ends of the water molecules are attracted to the positive ion.

The positive ends of the water molecules are attracted to the negative ion.

Copyright © Glencoe/McGraw-Hill, a division of The McGraw-Hill Companies, Inc.

---

5. Explain Why do some substances dissolve in water and others do not?

_____

_____

_____

_____

✓ Visual Check

6. Compare How is an alcohol molecule similar to a water molecule?

_____

_____

_____

FOLDABLES

Make a four-tab shutterfold to collect information about which solvents dissolve which solutes.

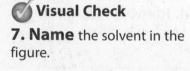

Polar solvents dissolve:	Nonpolar solvents dissolve:
Like Dissolves Like	
Polar solvents do not dissolve:	Nonpolar solvents do not dissolve:

✓ Visual Check

7. Name the solvent in the figure.

_____

_____

Concentration—How much is dissolved?

Have you ever tasted a soup and wished it had more salt in it? Your taste buds were evaluating the concentration of salt in the soup. **Concentration** *is the amount of a particular solute in a given amount of solution.* In the soup, salt is a solute. Saltier soup has a higher concentration of salt. Soup with less salt has a lower concentration of salt. Suppose two cups each contain an equal volume of soup. You stir a teaspoon of salt into one cup of soup. You stir a half-teaspoon of salt into the other cup of soup. The soup in the first cup has a higher concentration of the solute salt than the soup in the second cup.

Concentrated and Dilute Solutions

One way to describe the soup in the first cup is to say that it is saltier. The salt is more concentrated. The less-salty soup is more dilute. The terms *concentrated* and *dilute* are one way to describe how much solute is dissolved in a solution. However, these terms don't state the exact amount of solute dissolved. One person might think that a solution is concentrated. Another person might think that the same solution is dilute. Soup that tastes too salty to you might be perfect for someone else.

Describing Concentration Using Quantity

A more precise way to describe concentration is to state the quantity of solute in a given quantity of solution. When a solution is made of a solid dissolved in a liquid, such as salt in water, concentration is the mass of solute in a given volume of solution. Mass usually is stated in grams, and volume usually is stated in liters. For example, concentration can be stated as grams of solute per 1 L of solution. However, concentration can be stated using any units of mass or volume.

Calculating Concentration—Mass per Volume

You can calculate concentration using this equation:

$$\text{Concentration } (C) = \frac{\text{mass of solute } (m)}{\text{volume of solution } (V)}$$

To calculate concentration, you must know the mass of solute and the volume of solution that contains this mass. Then divide the mass of solute by the volume of solution.

Copyright © Glencoe/McGraw-Hill, a division of The McGraw-Hill Companies, Inc.

Math Skills ×÷

Suppose you want to calculate the concentration of salt in a 0.4 L can of soup. The label says it contains 1.6 g of salt. What is its concentration in g/L? In other words, how much salt would 1 L of soup have?

a. This is what you know:

mass: 1.6 g volume: 0.4 L

b. You need to find:

concentration: C

c. Use this formula:

$$C = \frac{m}{V}$$

d. Substitute and divide:

$$C = \frac{1.6}{0.4} = 4$$

e. Determine the units:

units of concentration =

$$\frac{\text{units of mass}}{\text{units of volume}} = \frac{g}{L} = g/L$$

Answer: The concentration is 4 g/L. As you might expect, 0.4 L of soup contains less salt (1.6 g) than 1 L of soup (4 g). However, the concentration of both amounts of soup is the same: 4 g/L.

8. Solve for Concentration

1. What is the concentration of 5 g of sugar in 0.2 L of solution?

_____

2. How many grams of salt are in 5 L of a solution with a concentration of 3 g/L?

_____

Concentration—Percent by Volume

Not all solutions are a solid dissolved in a liquid. A solution can contain only liquids or gases. The concentration of the solution is then stated as the volume of solute in a given volume of solution. In this case, the units of volume must be the same—usually mL or L. Because the units match, you can state the concentration as a percentage. ✓

To calculate percent by volume, first divide the volume of solute by the total volume of solution. Then multiply the quotient by 100. For example, if a container of orange drink contains 3 mL of acetic acid in a 1,000-mL container, the concentration is 0.3 percent.

$$3 \text{ mL} \div 1,000 = 0.003 \times 100 = 0.3\%$$

Solubility—How much can dissolve?

What happens if you put a lot of sugar into a glass of iced tea? Not all of the sugar dissolves. You stir and stir, but some sugar still remains at the bottom of the glass. That is because there is a limit to how much solute (sugar) can dissolve in a solvent (water). **Solubility** (sahl yuh BIH luh tee) *is the maximum amount of solute that can dissolve in a given amount of solvent at a given temperature and pressure.* If a substance has a high solubility, more of it can dissolve in a given solvent. ⚷✓

Saturated and Unsaturated Solutions

If you add water to a dry sponge, the sponge absorbs the water. If you keep adding water, the sponge becomes saturated. It can't hold any more water. This is <u>analogous</u> (uh NA luh gus), or similar, to what happens when you try to stir a lot of sugar into iced tea. Some sugar dissolves, but the excess sugar does not dissolve. The solution has become saturated.

A **saturated solution** *is a solution that contains the maximum amount of solute the solution can hold at a given temperature and pressure.* An **unsaturated solution** *is a solution that can still dissolve more solute at a given temperature and pressure.*

Factors that Affect How Much Can Dissolve

Can you change a solvent so that a larger or smaller amount of a particular solute can dissolve in it? Yes. Recall the definition of solubility—the maximum amount of solute that can dissolve in a given amount of solvent at a given temperature and pressure. Changing either the temperature or the pressure of the solvent changes how much solute can dissolve in a solvent.

✓ Reading Check

9. State What is the concentration of a solution?

_____

_____

_____

⚷ Key Concept Check

10. Contrast How do concentration and solubility differ?

_____

_____

_____

_____

_____

ACADEMIC VOCABULARY

analogous
(adjective) showing a likeness in some ways between things that are otherwise different

Effect of Temperature Have you noticed that more sugar dissolves in hot tea than in iced tea? The solubility of sugar in water increases as the temperature of the water increases. This is true for many solid solutes, as shown in the figure below. But notice that some solids are less soluble in warmer liquids than in cooler ones. The difference depends on the chemical structure of the solid. ✓

How does temperature affect the solubility of a gas in a liquid? Soda, or soft drinks, contains carbon dioxide gas dissolved in liquid water. The bubbles in soda are undissolved carbon dioxide. More carbon dioxide bubbles out when you open a warm can of soda than when you open a cold can. This is because the solubility of a gas in a liquid decreases when the temperature of the solution increases.

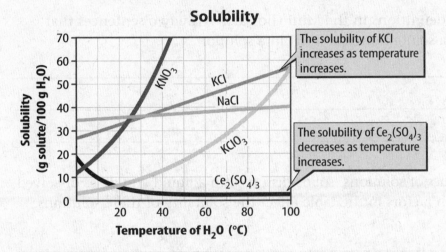

Solubility

The solubility of KCl increases as temperature increases.

The solubility of $Ce_2(SO_4)_3$ decreases as temperature increases.

Effect of Pressure What keeps carbon dioxide dissolved in an unopened can of soda? In a can, the carbon dioxide in the space above the liquid soda is under pressure. This causes the gas to move to an area of lower pressure—the solvent. The gas moves into the solvent and forms a solution. When you open the can, this pressure is released, and the carbon dioxide gas leaves the solution. Pressure does not affect the solubility of a solid solute in a liquid. 🔑

How Fast a Solute Dissolves

If solute and solvent particles come into contact more often, the solute dissolves faster. You can increase the contact between solvent and solute particles by stirring the solution. Or, you can crush the solute into smaller particles. Heating a solution will also increase this contact. Each of these methods will make a solute dissolve faster. However, stirring the solution or crushing the solute will not make more solute dissolve.

Reading Check

11. Summarize How does the temperature of water affect the solubility of sugar?

_____

_____

_____

Visual Check

12. Interpret How many grams of KNO_3 will dissolve in 100 g of water at 10°C?

_____

Key Concept Check

13. Generalize How can the solubility of a solute be changed?

_____

_____

_____

Mini Glossary

concentration: the amount of a particular solute in a given amount of solution

polar molecule: a molecule with a slightly negative end and a slightly positive end

saturated solution: a solution that contains the maximum amount of solute the solution can hold at a given temperature and pressure

solubility (sahl yuh BIH luh tee): the maximum amount of solute that can dissolve in a given amount of solvent at a given temperature and pressure

solute: any substance in a solution that is not the solvent

solvent: the substance that exists in the greatest quantity in a solution

unsaturated solution: a solution that can still dissolve more solute at a given temperature and pressure

1. Review the terms and their definitions in the Mini Glossary. Write two sentences that explain how to distinguish a solute from a solvent in a solution.

_____

_____

_____

2. The table below lists two types of solutions: solids dissolved in a liquid and gases dissolved in a liquid. Determine which factors in the table affect the solubility of these solutions. Write *yes* or *no* in each cell.

Do these factors . . .	affect the solubility of . . .	
	solids in a liquid?	**gases in a liquid?**
Temperature change		
Pressure change		
Stirring the solution		

3. If more solvent is added to a solution, what happens to the concentration of the solution?

_____

What do you think NOW?

Reread the statements at the beginning of the lesson. Fill in the After column with an A if you agree with the statement or a D if you disagree. Did you change your mind?

Log on to ConnectED.mcgraw-hill.com and access your textbook to find this lesson's resources.

END OF LESSON

Mixtures, Solubility, and Acid/Base Solutions

Acid and Base Solutions

·············· **Before You Read** ··············

What do you think? Read the two statements below and decide whether you agree or disagree with them. Place an A in the Before column if you agree with the statement or a D if you disagree. After you've read this lesson, reread the statements to see if you have changed your mind.

Before	Statement	After
	5. Acids are found in many foods.	
	6. You can determine the exact pH of a solution by using pH paper.	

Copyright © Glencoe/McGraw-Hill, a division of The McGraw-Hill Companies, Inc.

·············· **Read to Learn** ··············

What are acids and bases?

Would someone ever drink an acid? At first thought, you might answer no. After all, when people think of acids, they often think of the acids such as those in batteries or in acid rain. However, acids exist in other items, including milk, vinegar, fruits, and green leafy vegetables.

Along with the word *acid*, you might have heard the word *base*. Like acids, you can also find bases in your home. Detergent, antacids, and baking soda are examples of items that contain bases. But acids and bases are found in more than just household goods. As you will learn in this lesson, acids and bases are necessities for our daily life.

Acids

Have you ever tasted the sourness of a lemon or a grapefruit? The acid in the fruit creates this sour taste. *An* **acid** *is a substance that produces a hydronium ion (H_3O^+) when dissolved in water.*

Nearly all acid molecules contain one or more hydrogen atoms (H). When an acid mixes with water, this hydrogen atom separates from the acid. It quickly combines with a water molecule, resulting in a hydronium ion. *A* **hydronium ion,** H_3O^+, *is a positively charged ion formed when an acid dissolves in water.*

Key Concepts

- What happens when acids and bases dissolve in water?
- How does the concentration of hydronium ions affect pH?
- What methods can be used to measure pH?

> **Study Coach**
>
> **Preview Headings** Before you read the lesson, preview all the headings. Make a chart and write a question for each heading beginning with *What* or *How*. As you read, write the answers to your questions.

> **FOLDABLES**
>
> Draw and label a pH scale in a shutterfold and shade it with colored pencils. Use the scale to compare acid and base solutions.

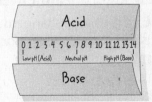

Acids in Water

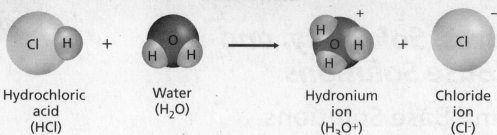

Hydrochloric acid (HCl) + Water (H₂O) → Hydronium ion (H₃O⁺) + Chloride ion (Cl⁻)

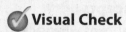 **Visual Check**

1. Identify Circle the atom in hydrochloric acid that joins water in solution.

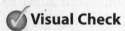 **Visual Check**

2. Compare How is dissolving an acid, shown above, similar to dissolving ammonia, shown below?

_____

_____

The figure above shows what happens when an acid molecule combines with water. In this example, a hydrogen atom separates from a molecule of hydrochloric acid and combines with a water molecule. The result is a hydronium ion and a chloride ion. The hydronium ion has a positive charge. The chloride ion has a negative charge.

Bases

A **base** *is a substance that produces hydroxide ions (OH⁻) when dissolved in water.* When a hydroxide compound such as sodium hydroxide (NaOH) mixes with water, hydroxide ions separate from the base and form hydroxide ions (OH⁻) in water. The top part of the figure below illustrates this process.

Bases in Water

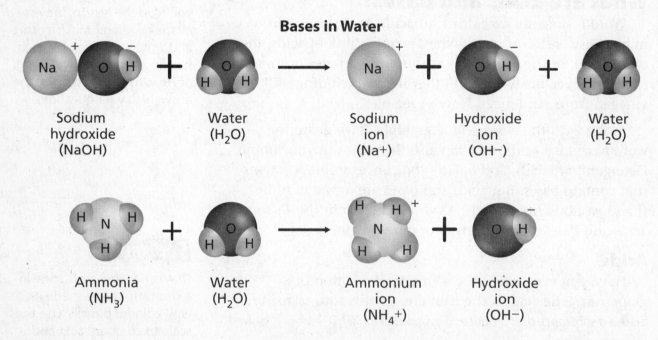

Sodium hydroxide (NaOH) + Water (H₂O) → Sodium ion (Na⁺) + Hydroxide ion (OH⁻) + Water (H₂O)

Ammonia (NH₃) + Water (H₂O) → Ammonium ion (NH₄⁺) + Hydroxide ion (OH⁻)

🔑 **Key Concept Check**

3. Explain what happens when acids and bases dissolve in water.

_____

_____

Some bases, such as ammonia (NH₃), do not contain hydroxide ions. These bases produce hydroxide ions by taking hydrogen atoms away from water, leaving hydroxide ions (OH⁻). The bottom part of the figure above illustrates this process. 🔑

Properties and Uses of Acids and Bases

	Acids	Bases
Ions produced	Acids produce H_3O^+ in water.	Bases produce OH^- ions in water.
Examples	• hydrochloric acid, HCl • acetic acid, CH_3COOH • citric acid, $H_3C_6H_5O_7$ • lactic acid, $C_3H_6O_3$	• sodium hydroxide, NaOH • ammonia, NH_3 • sodium carbonate, Na_2CO_3 • calcium hydroxide, $Ca(OH)_2$
Some properties	• Acids provide the sour taste in food (never taste acids in the laboratory). • Most can damage skin and eyes. • Acids react with some metals to produce hydrogen gas. • H_3O^+ ions can conduct electricity in water. • Acids react with bases to form neutral solutions.	• Bases provide the bitter taste in food (never taste bases in the laboratory). • Most can damage skin and eyes. • Bases are slippery when mixed with water. • OH^- ions can conduct electricity in water. • Bases react with acids to form neutral solutions.
Some uses	• Acids are responsible for natural and artificial flavoring in foods, such as fruits. • Milk contains lactic acid. • Acid in your saliva and stomach breaks down food. • Blueberries, strawberries, and many vegetable crops grow better in acidic soil. • Acids are used to make products such as fertilizers, detergents, and plastics.	• Bases are found in natural and artificial flavorings in food, such as cocoa beans. • Antacids neutralize stomach acid, relieving heartburn. • Cleaners such as shampoo, dish detergent, and window cleaner contain bases. • Many flowers grow better in basic soil. • Bases are used to make products such as rayon and paper.

The table above shows some properties and uses of acids and bases. As you can see, acids and bases are part of everyday life.

What is pH?

Have you ever seen someone test the water in a swimming pool? That person was probably testing the pH of the water. Swimming pool water should have a pH around 7.4. If the pH of the water is higher or lower than 7.4, the water might become cloudy, burn swimmers' eyes, or contain too many bacteria. What does a pH of 7.4 mean?

Hydronium Ions

The **pH** *is an inverse measure of the concentration of hydronium ions (H_3O^+) in a solution.* What does *inverse* mean? It means that as one thing increases, another thing decreases. In this case, as the concentration of hydronium ions increases, pH decreases. A solution with a lower pH is more acidic. As the concentration of hydronium ions decreases, the pH increases. A solution with a higher pH is more basic. 🗝️

Interpreting Tables

4. Identify List one use of an acid and one use of a base that are part of your everyday life.

_____

_____

_____

🗝️ **Key Concept Check**

5. Describe How does the concentration of hydronium ions affect pH?

_____

_____

_____

_____

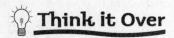
6. Classify How would you classify a soft drink that has a higher concentration of hydronium than hydroxide ions? (Circle the correct answer.)

a. acid

b. neutral

c. base

7. Interpret Is a tomato more or less acidic than detergent? What is the difference in acidity?

_____

_____

_____

Balance of Hydronium and Hydroxide Ions

All acid and base solutions contain both hydronium and hydroxide ions. In a neutral solution, such as water, the concentrations of hydronium and hydroxide ions are equal. What distinguishes an acid from a base is which of the two ions is present in the greater concentration. Acids have a greater concentration of hydronium ions (H_3O^+) than hydroxide ions (OH^-). Bases have a greater concentration of hydroxide ions than hydronium ions. Brackets around a chemical formula mean *concentration*.

Acids	$[H_3O^+] > [OH^-]$
Neutral	$[H_3O^+] = [OH^-]$
Bases	$[H_3O^+] < [OH^-]$

The pH Scale

The pH scale, shown in the figure below, indicates how acidic or basic a solution is. Notice that the scale contains values that range from below 0 to above 14. Acids have a pH below 7. Bases have a pH above 7. Solutions that are neutral have a pH of 7—they are neither acidic nor basic.

What do the numbers on the pH scale mean? How is the concentration of hydronium ions different in solutions with a different pH? A change in one pH unit represents a tenfold change in the acidity or basicity of a solution. For example, suppose one solution has a pH of 1 and a second solution has a pH of 2. The first solution is not twice as acidic as the second solution; it is ten times more acidic.

The difference in acidity or basicity between two solutions is represented by 10^n, where n is the difference between the two pH values. For example, how much more acidic is a solution with a pH of 1 than a solution with a pH of 3? First, calculate the difference, n, between the two pH values: $n = 3 - 1 = 2$. Then use the formula, 10^n, to calculate the difference in acidity: $10^2 = 100$. A solution with a pH of 1 is 100 times more acidic than a solution with a pH of 3.

The pH Scale

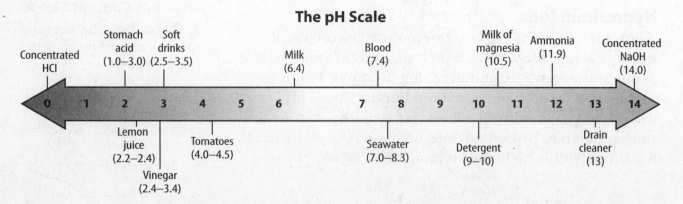

Copyright © Glencoe/McGraw-Hill, a division of The McGraw-Hill Companies, Inc.

How is pH measured?

How is the pH of a solution, such as swimming pool water, measured? Water test kits contain chemicals called indicators. When a person adds the chemicals to a water sample, the chemicals change color. The color identifies the solution as an acid or a base.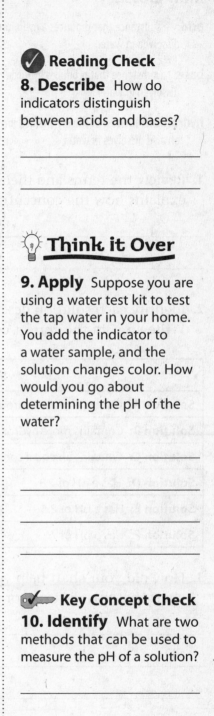

pH Indicators

Indicators can measure the approximate pH of a solution. *An* **indicator** *is a compound that changes color at different pH values when it reacts with acidic or basic solutions.* The pH of a solution is measured by adding a drop or two of the indicator to the solution. When the solution changes color, the person doing the test matches the color to a set of standard colors. Each standard color corresponds to a certain pH value.

There are many different indicators. Each indicator changes color over a specific range of pH values. For example, bromthymol blue is an indicator that changes from yellow to green to blue between pH 6 and pH 7.6.

pH Testing Strips

You can also use pH testing strips to measure pH. The strips contain an indicator. An indicator changes to a variety of colors over a range of pH values. To use pH strips, dip the strip into the solution. Then match the resulting color to the list of standard colors that represent specific pH values.

pH Meters

Although pH strips are quick and easy, they provide only an approximate pH value. A more accurate way to measure pH is to use a pH meter. A pH meter is an electronic instrument with an electrode that senses the hydronium ion concentration in solution.

Copyright © Glencoe/McGraw-Hill, a division of The McGraw-Hill Companies, Inc.

Reading Check

8. Describe How do indicators distinguish between acids and bases?

_____

_____

Think it Over

9. Apply Suppose you are using a water test kit to test the tap water in your home. You add the indicator to a water sample, and the solution changes color. How would you go about determining the pH of the water?

_____

_____

_____

_____

Key Concept Check

10. Identify What are two methods that can be used to measure the pH of a solution?

_____

_____

_____

Mini Glossary

acid: a substance that produces a hydronium ion (H_3O^+) when dissolved in water

base: a substance that produces hydroxide ions (OH^-) when dissolved in water

hydronium ion: a positively charged ion (H_3O^+) formed when an acid dissolves in water

indicator: a compound that changes color at different pH values when it reacts with acidic or basic solutions

pH: an inverse measure of the concentration of hydronium ions (H_3O^+) in a solution

1. Review the terms and their definitions in the Mini Glossary. Write a sentence that explains how the concentration of hydronium ions is related to pH.

_____

_____

2. Identify each solution described in the table as an acid, a base, or a neutral solution. Write an *X* in the correct column.

Solutions	acid	base	neutral
Solution A: Contains more hydronium ions than hydroxide ions			
Solution B: Contains more hydroxide ions than hydronium ions			
Solution C: Contains an equal concentration of hydronium and hydroxide ions			
Solution D: Has a pH of 2.3			
Solution E: Has a pH of 8.5			
Solution F: Has a pH of 7			

3. How did your chart help you learn about acids and bases? Choose a question from your chart that was difficult for you. Write the question and its answer below.

_____

_____

_____

What do you think NOW?

Reread the statements at the beginning of the lesson. Fill in the After column with an A if you agree with the statement or a D if you disagree. Did you change your mind?

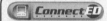 Connect ED

Log on to ConnectED.mcgraw-hill.com and access your textbook to find this lesson's resources.

 END OF LESSON

The Solar System

The Structure of the Solar System

·············· **Before You Read** ··············

What do you think? Read the two statements below and decide whether you agree or disagree with them. Place an A in the Before column if you agree with the statement or a D if you disagree. After you've read this lesson, reread the statements to see if you have changed your mind.

Before	Statement	After
	1. Astronomers measure distances between space objects using astronomical units.	
	2. Gravitational force keeps planets in orbit around the Sun.	

·············· **Read to Learn** ··············

What is the solar system?

The Sun and the group of objects that move around it make up the solar system. A few of the tiny specks of light that you can see in the night sky are planets in the solar system, but most are stars. It is hard to tell the difference. Stars are not part of the solar system. They are much farther away than any objects in our solar system.

Objects in the Solar System

The invention of the telescope in the 1600s led to the discovery of several planets and many other space objects. Before then, observers had seen only five planets—Mercury, Venus, Mars, Jupiter, and Saturn.

The Sun

The Sun is a <u>star</u> and is the largest object in the solar system. Its diameter is about 1.4 million km. The Sun is made mostly of hydrogen gas. The Sun's mass makes up about 99 percent of the mass of the entire solar system.

Inside the Sun, a process called nuclear fusion produces a huge amount of energy. The Sun emits some of this energy as light. This light shines on all of the planets every day. The Sun's gravitational force causes the planets and other objects in the solar system to move around, or orbit, the Sun.

Key Concepts

- How are the inner planets different from the outer planets?
- What is an astronomical unit and why is it used?
- What is the shape of a planet's orbit?

▶ Study Coach

K-W-L Fold a sheet of paper into three columns. In the first column, write what you know about the structure of the solar system. In the second column, write what you want to know. In the third column, write what you have learned after reading this lesson.

SCIENCE USE V. COMMON USE ···
star
Science Use an object in space made of gases in which nuclear fusion reactions occur that emit energy

Common Use a shape that usually has five or six points around a common center

·············

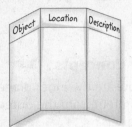

FOLDABLES

Make a tri-fold book to summarize information about the types of objects that make up the solar system.

Objects That Orbit the Sun

Planets, dwarf planets, asteroids, and comets orbit the Sun. These objects do not emit light. Instead, they reflect the Sun's light.

Planets The figure below shows the eight planets in the solar system. An object is a planet only if it orbits the Sun and has a nearly spherical shape. The object must also have a mass that is much larger than the total mass of all other objects in orbits nearby.

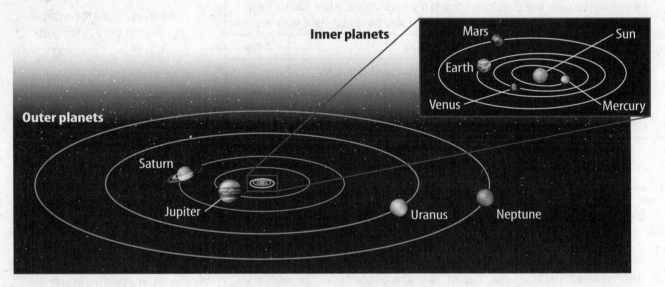

Visual Check

1. Interpret On the diagram, circle the inner planet that orbits the Sun closest to Earth.

Key Concept Check

2. Describe how the inner planets differ from the outer planets.

_____

_____

_____

Inner Planets and Outer Planets The inner planets are the four planets closest to the Sun—Mercury, Venus, Earth, and Mars. The inner planets are made of solid, rocky materials. The outer planets are the four planets farthest from the Sun—Jupiter, Saturn, Uranus (YOOR uh nus), and Neptune. The outer planets are made of ice and gases.

Dwarf Planets A dwarf planet is a spherical object that orbits the Sun and is not a moon of another planet. Unlike a planet, a dwarf planet does not have more mass than all the objects in nearby orbits. The figure on the next page locates the dwarf planets Ceres (SIHR eez), Eris (IHR is), Pluto, and Makemake (MAH kay MAH kay). Dwarf planets are made of rock and ice and are much smaller than Earth.

Asteroids *Millions of small, rocky objects called* **asteroids** *orbit the Sun in the asteroid belt between the orbits of Mars and Jupiter.* Unlike planets, asteroids usually are not spherical.

Comets *A* **comet** *is made of gas, dust, and ice and moves around the Sun in an oval-shaped orbit.* Comets come from the outer parts of the solar system. Possibly, 1 trillion comets orbit the Sun.

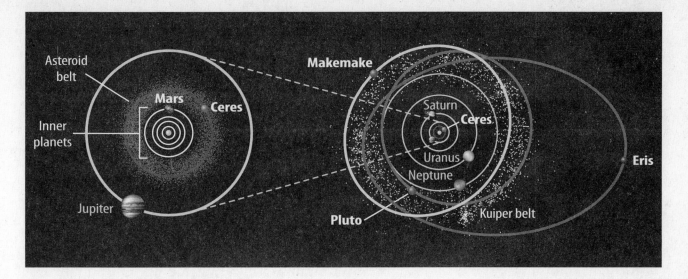

The Astronomical Unit

Astronomers need a unit larger than the kilometer to measure the great distances between objects in the solar system. They use astronomical units. *An **astronomical unit** (AU) is the average distance from Earth to the Sun—about 150 million km.* For example, the average distance of Mercury from the Sun is 0.39 AU, or about 58 million km. 🔑

The Motion of the Planets

The Sun's gravitational force pulls each planet toward the Sun. This force keeps each planet moving along a curved path around the Sun.

Revolution and Rotation

Objects in the solar system move in two ways. The objects orbit, or revolve, around the Sun. *The time it takes an object to travel once around the Sun is its **period of revolution.*** Earth's period of revolution is one year. The objects also spin, or rotate, as they orbit the Sun. *The time it takes for an object to complete one rotation is its **period of rotation.*** Earth has a period of rotation of one day.

Planetary Orbits and Speeds

A planet's orbit is an ellipse—a stretched-out circle. Two special points, each called a focus, are inside the ellipse. The focus points, or foci, control the shape of the ellipse. The Sun is at one focus, and the other focus is empty space.

A planet's distance from the Sun changes as the planet moves along its elliptical orbit. A planet's speed also changes as it orbits. The closer the planet is to the Sun, the faster it moves. This also means that planets farther from the Sun have longer periods of revolution. For example, Jupiter takes 12 times longer than Earth to revolve around the Sun. 🔑

Visual Check
3. Specify Which dwarf planet is farthest from the Sun?

_____

🔑 **Key Concept Check**
4. Identify Define what an astronomical unit is and explain why it is used.

_____

_____

_____

_____

_____

_____

🔑 **Key Concept Check**
5. Describe the shape of a planet's orbit.

_____

_____

_____

Mini Glossary

asteroid: one of millions of small, rocky objects that orbit the Sun in the asteroid belt between the orbits of Mars and Jupiter

astronomical unit (AU): the average distance from Earth to the Sun—about 150 million km

comet: a mixture of gas, dust, and ice that moves around the Sun in an oval-shaped orbit

period of revolution: the time it takes an object to travel once around the Sun

period of rotation: the time it takes an object to complete one rotation

1. Review the terms and their definitions in the Mini Glossary. Write a sentence that explains the difference between an asteroid and a comet.

_____

_____

2. Fill in the blanks below to identify the inner planets in the order of their distance from the Sun.

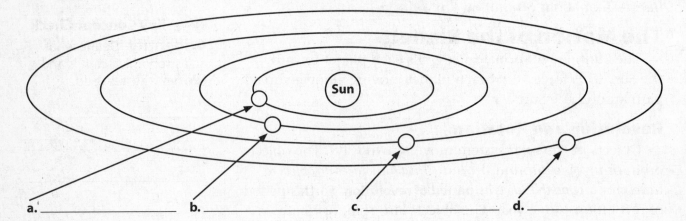

a._____   b._____   c._____   d._____

3. Based on your K-W-L chart, summarize the new facts you learned from this lesson.

_____

_____

_____

What do you think NOW?

Reread the statements at the beginning of the lesson. Fill in the After column with an A if you agree with the statement or a D if you disagree. Did you change your mind?

 **Connect ED**

Log on to ConnectED.mcgraw-hill.com and access your textbook to find this lesson's resources.

END OF LESSON

The Solar System

The Inner Planets

················· **Before You Read** ·············

What do you think? Read the two statements below and decide whether you agree or disagree with them. Place an A in the Before column if you agree with the statement or a D if you disagree. After you've read this lesson, reread the statements to see if you have changed your mind.

Before	Statement	After
	3. Earth is the only inner planet that has a moon.	
	4. Venus is the hottest planet in the solar system.	

················· **Read to Learn** ···············

Planets Made of Rock

The inner planets—Mercury, Venus, Earth, and Mars—are the four planets closest to the Sun. *Earth and the other inner planets are also called* **terrestrial planets.** All inner planets are made of rock and metals and have a solid outer layer. All inner planets also have a similar structure. All have a core, mantle, and crust. As the figure below shows, Mercury, Venus, and Earth have a solid inner core and a liquid outer core. The core of Mars is liquid with no solid part. The inner planets differ in their sizes, atmospheres, and surfaces.

Structure of Mercury, Venus, and Earth

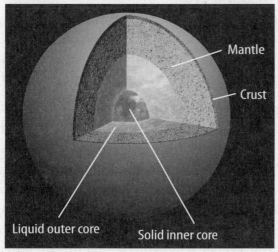

Mantle

Crust

Liquid outer core   Solid inner core

Key Concepts

• How are the inner planets similar?

• Why is Venus hotter than Mercury?

• What kind of atmospheres do the inner planets have?

▸ **Mark the Text**

Identify the Main Ideas Highlight the important ideas in each paragraph. Review these ideas as you study the lesson.

✓ **Visual Check**

1. Distinguish Number each structure in the figure from 1 to 4 in order from the outermost layer to the innermost layer.

Make a four-door book to organize your notes on the inner planets.

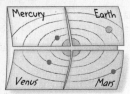

Mercury

Mercury is the smallest planet and the planet closest to the Sun. Mercury has no moon and no atmosphere. The strength of a planet's gravity depends on its mass. Mercury's small mass creates weak gravity that cannot hold on to an atmosphere. With no atmosphere there is no wind that moves energy across the planet's surface. This results in temperatures as high as 450°C on the side of Mercury facing the Sun and as cold as −170°C on the side facing away from the Sun.

Mercury's Surface

Impact craters cover the surface of Mercury. In addition, Mercury's surface has smooth plains of solidified lava from long-ago eruptions. Long, high cliffs are on its surface as well. Mercury also has many craters formed from the impact of objects. Without an atmosphere, almost no erosion occurs. As a result, surface features last for billions of years.

Mercury's Structure

The structures of the inner planets are similar. Like all inner planets, Mercury has a core made of iron and nickel. A mantle of silicon and oxygen surrounds the core. The crust is a thin, rocky layer above the mantle. Mercury has a large core, possibly formed by a collision with a large object during the planet's formation. 🔑

Venus

Venus is the second planet from the Sun. It has no moon. It is about the same size as Earth. Venus spins so slowly that its period of rotation is longer than its period of revolution. Unlike most planets, Venus rotates from east to west.

Venus's Atmosphere

The atmosphere of Venus is about 97 percent carbon dioxide. Its atmospheric pressure is nearly 90 times greater than Earth's. Venus has almost no water, yet a thick layer of clouds covers the planet. The clouds are made of acid.

The Greenhouse Effect on Venus

Venus is the hottest planet in the solar system. Its temperature averages 460°C. The greenhouse effect causes the high temperatures. *The greenhouse effect occurs when a planet's atmosphere traps solar energy and causes the surface temperature to increase.* The carbon dioxide in Venus's atmosphere traps energy and heats up the planet. The planet would be 450°C cooler without the greenhouse effect. 🔑

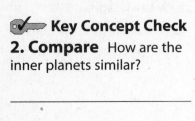

🔑 Key Concept Check

2. Compare How are the inner planets similar?

_____

_____

_____

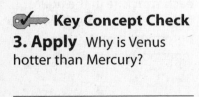

🔑 Key Concept Check

3. Apply Why is Venus hotter than Mercury?

_____

_____

_____

Venus's Structure and Surface

Solidified lava covers more than 80 percent of Venus's surface. The lava possibly came from volcanic eruptions.

Earth

Earth is the third planet from the Sun. Unlike Mercury and Venus, Earth has a moon.

Earth's Atmosphere

Gases and a small amount of water vapor make up most of Earth's atmosphere. The gases and vapor produce a greenhouse effect that increases Earth's average surface temperature. This effect and Earth's distance from the Sun warm Earth. As a result, large bodies of liquid water can exist on Earth. Earth's atmosphere also absorbs much of the Sun's radiation. This protects the surface below. Earth's protective atmosphere, liquid water, and moderate temperatures support a wide variety of life. ✓

Earth's Structure

Earth has a solid inner core surrounded by a liquid outer core. The outer core is surrounded by a mantle. Earth's crust is above the mantle. The crust is broken into large plates that slide past, away from, or into each other. It is made mostly of oxygen and silicon. Natural forces constantly destroy Earth's crust and create new crust.

Mars

Mars is the fourth planet from the Sun. It is about half the size of Earth. Unlike the other inner planets, Mars has no solid inner core. Its core is liquid. Mars has two moons. Space probes have visited Mars; however, they found no liquid water or life on the planet. ✓

Mars's Atmosphere

The atmosphere of Mars is much less dense than that of Earth. It is about 95 percent carbon dioxide. Temperatures range from about −125°C at the poles to about 20°C at the equator. Winds on Mars produce great dust storms. ✓

Mars's Surface

Iron oxide in the soil gives Mars its reddish color. The Martian canyon Valles Marineris is as long as the United States. The Martian volcano Olympus Mons is the largest known mountain in the solar system. Mars has polar ice caps made of frozen carbon dioxide and ice. Craters cover the southern hemisphere of Mars. The northern hemisphere is smoother and appears to be covered by lava flows.

✓ **Reading Check**

4. Explain Why is there life on Earth?

_____

_____

_____

✓ **Reading Check**

5. Compare Earth's inner core with the inner core of Mars.

_____

_____

_____

🔑 **Key Concept Check**

6. Describe the atmosphere of each inner planet.

_____

_____

_____

_____

_____

Copyright © Glencoe/McGraw-Hill, a division of The McGraw-Hill Companies, Inc.

Mini Glossary

greenhouse effect: occurs when a planet's atmosphere traps solar energy and causes the surface temperature to increase

terrestrial planet: Earth or another inner planet

1. Review the terms and their definitions in the Mini Glossary. Write a sentence that explains how the greenhouse effect benefits life on Earth.

_____

_____

2. Fill in the chart below to match the inner planets with their features.

Inner planets: Earth, Venus, Mars, and Mercury

a. _____	b. _____	c. _____	d. _____
Features: • Closest to the Sun • Smallest of all planets • No moon • Weak gravity • No atmosphere • Wide temperature range • Impact craters cover the surface	**Features:** • Second-closest to the Sun • No moon • High atmospheric pressure • Covered by acid clouds • Hottest planet in the solar system • Surface is mostly solidified lava	**Features:** • Third-closest to the Sun • One moon • Atmosphere protects surface from much of the Sun's radiation • Large bodies of liquid water • Moderate temperature range • Crust made up of large sliding plates	**Features:** • Fourth-closest to the Sun • Two moons • Great dust storms • Reddish color • Valles Marineris: a huge canyon • Olympus Mons: largest mountain in the solar system • Polar ice caps

3. Explain why Earth has an atmosphere but Mercury does not.

_____

_____

What do you think NOW?

Reread the statements at the beginning of the lesson. Fill in the After column with an A if you agree with the statement or a D if you disagree. Did you change your mind?

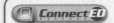 **ConnectED**

Log on to ConnectED.mcgraw-hill.com and access your textbook to find this lesson's resources.

 END OF LESSON

The Solar System

The Outer Planets

·············· Before You Read ··············

What do you think? Read the two statements below and decide whether you agree or disagree with them. Place an A in the Before column if you agree with the statement or a D if you disagree. After you've read this lesson, reread the statements to see if you have changed your mind.

Before	Statement	After
	5. The outer planets are also called the gas giants.	
	6. The atmospheres of Saturn and Jupiter are mainly water vapor.	

·············· Read to Learn ··············

The Gas Giants

The outer planets are Jupiter, Saturn, Uranus, and Neptune. The figure below shows the sizes of the outer planets compared to each other and to Earth. As you can see, their sizes are much greater than Earth's size. The outer planets are mostly made of hydrogen and helium. These elements are usually gases on Earth. Gases change to liquids at high pressure. This property of gases affects the outer planets. The outer planets are called the gas giants.

The huge size of each outer planet creates strong gravitational forces. These gravitational forces put so much pressure on the atmosphere of each planet that the gases change to liquids. As a result, the outer planets are mostly liquid inside. An outer planet has a thick gas and liquid layer covering a small solid core.

Key Concepts

- How are the outer planets similar?
- What are the outer planets made of?

Study Coach

Create a Quiz Write a quiz question for each paragraph. Answer the question with information from the paragraph. Refer to these questions and answers as you review the chapter.

Visual Check

1. Compare Which outer planet is the largest?

_____

Outer Planets

Jupiter, Fifth Planet from Sun

Saturn, Sixth Planet from Sun

Uranus, Seventh Planet from Sun

Neptune, Eighth Planet from Sun

Earth

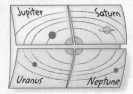
☞ Key Concept Check

2. Describe what makes up each of Jupiter's three distinct layers.

_____

_____

_____

Math Skills ×÷+

A ratio is a quotient—it is one quantity divided by another. Ratios can be used to compare distances.

For example, Jupiter is 5.20 AU from the Sun. Neptune is 30.05 AU from the Sun. Divide the larger distance by the smaller distance:

$$\frac{30.05 \text{ AU}}{5.20 \text{ AU}} = 5.78$$

Neptune is 5.78 times farther from the Sun than Jupiter is.

3. Use Ratios The figure on the right shows that Saturn is 9.58 AU from the Sun. Jupiter is 5.20 AU from the Sun. How many times farther from the Sun is Saturn than Jupiter?

Jupiter

Jupiter is the largest planet in the solar system. Its diameter is more than 11 times the diameter of Earth. Its mass is more than twice the mass of all the other planets combined. Jupiter takes almost 12 Earth years to complete one orbit of the Sun. Yet it spins faster than any other planet. Its period of rotation is less than 10 hours. Jupiter has a ring system.

Jupiter's Atmosphere

The atmosphere on Jupiter is about 90 percent hydrogen and 10 percent helium. The thickness of the atmosphere is about 1,000 km. The atmosphere holds several layers of dense, colorful clouds. Jupiter's fast rotation stretches the clouds into swirling bands of color. The Great Red Spot on the planet's surface is a storm of swirling gases.

Jupiter's Structure

Jupiter is about 80 percent hydrogen and 20 percent helium. Swirling gas covers a thick layer of liquid hydrogen over a solid core. The pressure at 1,000 km below the outer edge of the cloud layer is great. There, the hydrogen gas turns into a liquid. The core is probably rock and iron. ☞

The Moons of Jupiter

Jupiter has at least 63 moons, more than any other planet. In 1610, Galileo Galilei first spotted Jupiter's four largest moons. As a result, _the four largest moons of Jupiter—Io, Europa, Ganymede, and Callisto—are known as the_ **Galilean moons.** Collisions between Jupiter's moons and meteorites likely created the particles that make up the planet's faint rings.

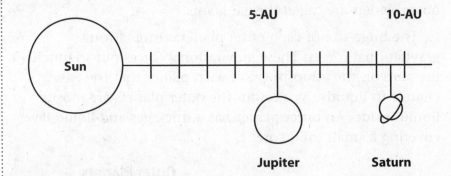

Saturn

The figure above shows the distances from the Sun of the fifth and sixth planets. Saturn is the sixth planet from the Sun. Like Jupiter, Saturn rotates rapidly and has bands of clouds. Saturn is about 90 percent hydrogen and 10 percent helium. It is the least-dense planet.

Saturn's Structure

Saturn is made up mostly of hydrogen and helium. Like Jupiter, Saturn's structure has an outer layer of gas, a thick layer of liquid hydrogen, and a solid core. Saturn's seven bands of rings are the largest in the solar system. The main ring system is more than 70,000 km wide. However, the ring system is likely less than 30 m thick. Ice particles mainly make up the rings. The particles range in size from specks to chunks as large as a house. 🗝️

Key Concept Check

4. Describe what makes up Saturn and its ring system.

_____

_____

_____

Saturn's Moons

Saturn has at least 60 moons. Its five largest moons are Titan, Rhea, Dione, Iapetus, and Tethys. Most of Saturn's moons are chunks of ice that are less than 10 km in diameter. Titan, the largest moon, is the only moon in the solar system with a dense atmosphere. In 2005, the *Huygens* (HOY guns) space probe landed on Titan.

Uranus

Uranus is the seventh planet from the Sun. It has narrow, dark rings. The diameter of Uranus is about four times that of Earth. The *Voyager 2* space probe explored the planet when it flew by Uranus in 1986. The deep atmosphere of Uranus is mostly hydrogen and helium. There is also a small amount of methane. Beneath the atmosphere is a thick, slushy layer of water, ammonia, and other materials. Uranus might also have a solid, rocky core. 🗝️

Key Concept Check

5. Identify the substances that make up the atmosphere and the thick, slushy layer on Uranus.

_____

_____

_____

Uranus's Axis and Moons

The sideways tilt of Uranus's axis of rotation is different from those of the other planets. Uranus's axis is tilted so that the planet moves around the Sun like a rolling ball.

Uranus has at least 27 moons. The two largest moons, Titania and Oberon, are much smaller than Earth's moon. Titania has an icy, cracked surface. At one time, an ocean might have covered Titania's surface.

Neptune

Like Uranus, Neptune has an atmosphere of mostly hydrogen and helium, with a little methane. Like Uranus, Neptune's interior is frozen water and ammonia with a core of rock and iron.

Neptune has at least 13 moons and faint dark rings. Triton (TRI tun) is Neptune's largest moon. Triton is made of rock, with an icy outer layer. Triton's surface is frozen nitrogen. Geysers on the surface erupt nitrogen gas. 🗝️

Key Concept Check

6. Compare How do the atmosphere and interior of Neptune compare with those of Uranus?

_____

_____

_____

Mini Glossary

Galilean moons: the four largest moons of Jupiter—Io, Europa, Ganymede, and Callisto

1. Review the term and its definition in the Mini Glossary. Write a sentence that explains how this group of moons got its name.

_____

_____

2. Fill in the diagram below to match the name of each outer planet with its key feature. Outer Planets: Jupiter, Saturn, Neptune, and Uranus

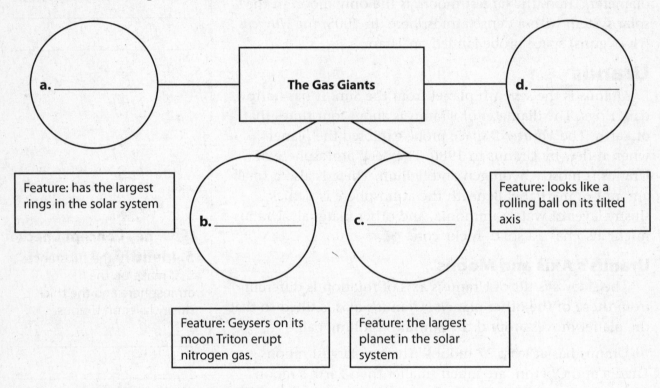

a. _____

The Gas Giants

d. _____

Feature: has the largest rings in the solar system

Feature: looks like a rolling ball on its tilted axis

b. _____

c. _____

Feature: Geysers on its moon Triton erupt nitrogen gas.

Feature: the largest planet in the solar system

3. How are the outer planets similar?

_____

_____

What do you think NOW?

Reread the statements at the beginning of the lesson. Fill in the After column with an A if you agree with the statement or a D if you disagree. Did you change your mind?

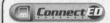

Connect ED

Log on to ConnectED.mcgraw-hill.com and access your textbook to find this lesson's resources.

END OF LESSON

The Solar System

Dwarf Planets and Other Objects

What do you think? Read the two statements below and decide whether you agree or disagree with them. Place an A in the Before column if you agree with the statement or a D if you disagree. After you've read this lesson, reread the statements to see if you have changed your mind.

Before	Statement	After
	7. Asteroids and comets are mainly rock and ice.	
	8. A meteoroid is a meteor that strikes Earth.	

················**Read to Learn**················

Dwarf Planets

The International Astronomical Union (IAU) defines a dwarf planet as an object that orbits a star. When a dwarf planet formed, there was enough mass and gravity for it to form a sphere. A dwarf planet has objects similar in mass orbiting nearby or crossing its orbital path. Astronomers classify Pluto, Ceres, Eris, Makemake, and Haumea (how MAY ah) as dwarf planets. Pluto was once considered to be a planet, but now it has the status of a dwarf planet.

All dwarf planets are smaller than Earth's moon. The figure below locates Ceres, Pluto, and Eris. These dwarf planets each have a rocky core surrounded by a thick layer of ice.

Key Concepts

- What is a dwarf planet?
- What are the characteristics of comets and asteroids?
- How does an impact crater form?

Mark the Text

Define Words Skim the lesson and underline words that you do not know. Then read the lesson to see if you can define those words. If you cannot, look up the word and write its definition in the margin to use as you study.

Visual Check

1. Interpret Which dwarf planet orbits closest to Earth?

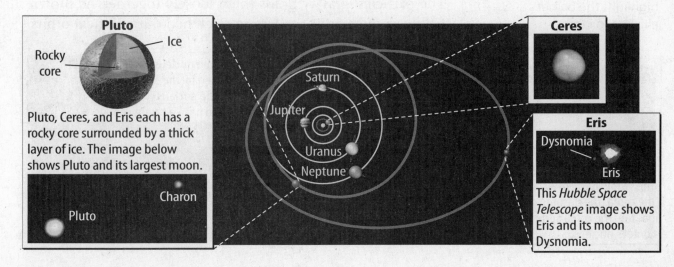

Pluto

Rocky core — Ice

Pluto, Ceres, and Eris each has a rocky core surrounded by a thick layer of ice. The image below shows Pluto and its largest moon.

Charon

Pluto

Saturn
Jupiter
Uranus
Neptune

Ceres

Eris

Dysnomia — Eris

This *Hubble Space Telescope* image shows Eris and its moon Dysnomia.

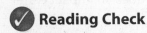
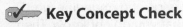 **Reading Check**

2. State Which dwarf planet is the largest? Which dwarf planet is the smallest?

_____

_____

 **Key Concept Check**

3. Specify Where do the orbits of most asteroids occur?

_____

_____

_____

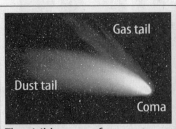 **Visual Check**

4. Identify In the figure, highlight the comet's elliptical orbit.

Ceres

Ceres is the smallest dwarf planet. It orbits the Sun in the asteroid belt. It might have a rocky core. A thin, dusty crust covers a layer of water ice that surrounds the core.

Pluto

Pluto is about two-thirds the size of the Moon. It is so far from the Sun that its period of revolution is about 248 Earth years. The surface of Pluto is so cold that it is covered with frozen nitrogen. Its average temperature is −230°C. Pluto has three known moons: Charon, Hydra, and Nix. Charon is Pluto's largest moon.

Eris

Eris is the largest dwarf planet. It was discovered in 2003. Eris takes about 557 Earth years to complete one orbit around the Sun. Dysnomia (dis NOH mee uh) is the only known moon of Eris.

Makemake and Haumea

Makemake and Haumea, named dwarf planets in 2008, orbit in the Kuiper (KI puhr) belt region of the solar system. Makemake is one of the largest objects in the Kuiper belt.

Asteroids

Recall that asteroids are chunks of rock and ice that never clumped together to form a planet. Most asteroids orbit the Sun in the asteroid belt. The asteroid belt is between the orbits of Mars and Jupiter. There are hundreds of thousands of asteroids. Pallas is the largest asteroid. Some astronomers suggest that asteroids are very old objects left over from the formation of the solar system.

Comets

Comets are mixtures of particles of rock, ice, and dust. The particles' gravity holds them loosely together. As shown below, comets orbit the Sun in stretched-out elliptical orbits.

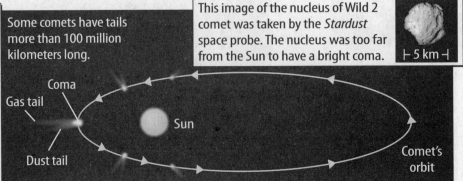

The visible parts of a comet are the coma, the dust tail, and the gas tail. The coma surrounds the comet's nucleus.

Gas tail
Dust tail
Coma

Some comets have tails more than 100 million kilometers long.

This image of the nucleus of Wild 2 comet was taken by the *Stardust* space probe. The nucleus was too far from the Sun to have a bright coma.

⊢ 5 km ⊣

Coma
Gas tail
Dust tail
Sun
Comet's orbit

The Structure of Comets

The solid, inner part of a comet is its nucleus. As a comet moves closer to the Sun, it gets hotter. Higher temperatures change the ice in the comet into a gas. Energy from the Sun pushes some of the gas and dust away from the comet's nucleus and makes it glow. This produces the comet's bright tail and glowing nucleus, called a coma. The coma surrounds the comet's nucleus. When energy from the Sun strikes the gas and dust in the comet's nucleus, it can create a two-part tail—a dust tail and a gas tail. The gas tail always points away from the Sun. 🗝

Short-Period and Long-Period Comets

A short-period comet takes less than 200 Earth years to orbit the Sun. Most short-period comets come from the Kuiper belt. The Kuiper belt extends from about the orbit of Neptune to about 50 AU from the Sun.

A long-period comet takes more than 200 Earth years to orbit the Sun. Long-period comets come from an area at the outer edge of the solar system called the Oort cloud. The Oort cloud surrounds the solar system and extends about 100,000 AU from the Sun. Some long-period comets take millions of years to orbit the Sun.

Meteoroids

Millions of particles called meteoroids enter Earth's atmosphere every day. *A **meteoroid** is a small, rocky particle that moves through space.* Most meteoroids are only about as big as a grain of sand. As a meteoroid passes through Earth's atmosphere, it creates friction. The friction makes the meteoroid and the air around it hot enough to glow. *A **meteor** is a streak of light in Earth's atmosphere made by a glowing meteoroid.* Most meteoroids burn up in Earth's atmosphere. Some are large enough that they reach Earth's surface before they burn up completely. When this happens, the meteoroid is then called a meteorite. *A **meteorite** is a meteoroid that strikes a planet or a moon.*

When a large meteorite strikes a moon or planet, it often forms a bowl-shaped impact crater. *An **impact crater** is a round depression formed on the surface of a planet, moon, or other space object by the impact of a meteorite.* Earth's surface has more than 170 impact craters. 🗝

🗝 **Key Concept Check**

5. Describe the characteristics of a comet.

_____

_____

_____

💡 **Think it Over**

6. Apply If you observed a long-period comet, would you ever be able to observe it again? Explain.

_____

_____

_____

🗝 **Key Concept Check**

7. Summarize What causes an impact crater to form?

_____

_____

Copyright © Glencoe/McGraw-Hill, a division of The McGraw-Hill Companies, Inc.

······· After You Read ·······

Mini Glossary

impact crater: a round depression formed on the surface of a planet, moon, or other space object by the impact of a meteorite

meteor: a streak of light in Earth's atmosphere made by a glowing meteoroid

meteorite: a meteoroid that strikes a planet or a moon

meteoroid: a small, rocky particle that moves through space

1. Review the terms and their definitions in the Mini Glossary. Write two sentences that explain how the term *meteoroid* relates to the term *meteor*.

_____

_____

_____

2. Write the name of each dwarf planet next to its description in the chart. You will need to write some names more than once.

Description	Dwarf Planet
a. Frozen nitrogen covers its surface.	
b. Largest dwarf planet	
c. Smallest dwarf planet	
d. Orbits in the Kuiper belt (two names)	
e. One of the largest objects in the Kuiper belt	
f. Orbits in the asteroid belt	
g. A thin, dusty crust covers its ice layer.	

3. Describe the characteristics of a dwarf planet.

_____

_____

What do you think NOW?

Reread the statements at the beginning of the lesson. Fill in the After column with an A if you agree with the statement or a D if you disagree. Did you change your mind?

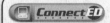

Log on to ConnectED.mcgraw-hill.com and access your textbook to find this lesson's resources.

Stars and Galaxies

The View from Earth

········· **Before You Read** ··············

What do you think? Read the two statements below and decide whether you agree or disagree with them. Place an A in the Before column if you agree with the statement or a D if you disagree. After you've read this lesson, reread the statements to see if you have changed your mind.

Before	Statement	After
	1. The night sky is divided into constellations.	
	2. A light-year is a measurement of time.	

··············· **Read to Learn** ·················

Looking at the Night Sky

Have you ever looked up at the sky on a clear, dark night and seen more stars than you could possibly count? If you have, you are lucky. Few people see a night sky dense with stars. Lights from towns and cities make the night sky too bright for people to see the faint stars.

If you look at a clear night sky for a long time, the stars seem to move. But what you are really seeing is Earth's movement. Earth spins, or rotates, once every 24 hours. Day turns to night and then back to day as Earth rotates. Because Earth rotates from west to east, objects in the sky rise in the east and set in the west.

Earth spins on its axis. Earth's axis is an imaginary line from the North Pole to the South Pole. The star Polaris is almost directly above the North Pole. As Earth spins, stars near Polaris appear to travel in a circle around Polaris. These stars never set when viewed from the northern hemisphere. They are always present in the night sky.

Naked-Eye Astronomy

You don't need expensive equipment to view the sky. *Naked-eye astronomy* means "gazing at the sky with just your eyes, without binoculars or a telescope." Long before the telescope was invented, people viewed the sky in this way.

Key Concepts

- How do astronomers divide the night sky?

- What can astronomers learn about stars from their light?

- How do scientists measure the distance and the brightness of objects in the sky?

◀ Study Coach

Create a Quiz Write five questions as you read this lesson. Exchange quizzes with a partner. After taking the quizzes, discuss your answers.

FOLDABLES

Make a horizontal two-tab book to organize your notes on astronomy.

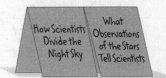

Constellations

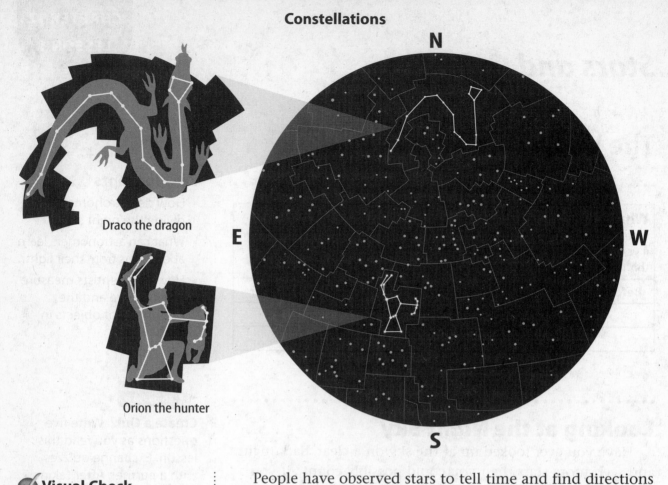

Draco the dragon

Orion the hunter

Visual Check

1. Analyze Why does east appear on the left and west appear on the right on the sky map? (Hint: Hold the map over your head, as you would view the sky. Position the map so that you are looking north.)

_____

_____

_____

Key Concept Check

2. Describe How do astronomers divide the night sky?

_____

_____

People have observed stars to tell time and find directions since ancient times. They learned about planets, seasons, and astronomical events merely by watching the sky. As you practice naked-eye astronomy, remember never to look directly at the Sun. Ultraviolet radiation from the Sun could damage your eyes.

Constellations

As people in ancient cultures gazed at the night sky, they saw patterns. The patterns resembled people, animals, or objects, such as the hunter and the dragon shown in the figure above. The Greek astronomer Ptolemy (TAH luh mee) identified dozens of star patterns nearly 2,000 years ago. Today, these patterns and others like them are known as ancient constellations.

Present-day astronomers use many ancient constellations to divide the sky into 88 regions. The sky map in the figure above shows some of these regions, which are also called constellations. Dividing the sky helps scientists communicate to others what area of sky they are studying.

Telescopes

Telescopes were invented in the early 1600s. They can collect much more light than the human eye can detect. Visible light is just one part of the electromagnetic spectrum.

Look at the figure below. The electromagnetic spectrum is a continuous range of wavelengths. Longer wavelengths have low energy. Shorter wavelengths have high energy. Different objects in space emit different ranges of wavelengths. The range of wavelengths that a star emits is the star's spectrum (plural, spectra).

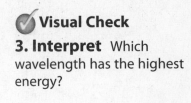

Visual Check

3. Interpret Which wavelength has the highest energy?

_____

_____

Electromagnetic Spectrum

Low energy
Long wavelength

High energy
Short wavelength

Wavelength

| Radio | Microwave | Infrared | Visible | Ultraviolet | X-ray | Gamma ray |

Radio waves can be used to study cold, dark regions of space.

Infrared waves can be used to study star-forming regions.

Ultraviolet waves can be used to study young stars.

X-rays and gamma rays can be used to study high-energy gas jets.

Spectroscopes

Scientists study the spectra of stars using an instrument called a spectroscope. *A* **spectroscope** *spreads light into different wavelengths.* Using spectroscopes, astronomers can study stars' characteristics, including temperatures, compositions, and energies. For example, newly formed stars emit mostly radio and infrared waves, which have low energy. Exploding stars emit mostly high-energy ultraviolet waves and X-rays.

Measuring Distances

Extend your arm, and hold up your thumb. Close one eye, and look at your thumb. Now open that eye, and close the other eye. Did your thumb seem to jump? This is an example of parallax. Parallax is the apparent change in an object's position caused by looking at it from two different points.

Astronomers use angles created by parallax to measure how far objects are from Earth. Astronomers do not use the eyes as the two points of view. Instead, they use two points in Earth's orbit around the Sun.

Key Concept Check

4. Assess What can astronomers learn from a star's spectrum?

_____

_____

_____

Reading Check

5. Explain What is parallax?

_____

_____

_____

Copyright © Glencoe/McGraw-Hill, a division of The McGraw-Hill Companies, Inc.

Distances Within the Solar System

The universe is too large to measure easily in meters or kilometers. Therefore, astronomers use other units of measurement. For distances within the solar system, they use astronomical units (AU). *An* **astronomical unit** *is the average distance between Earth and the Sun*, about 150 million km.

Astronomical units make it easy to compare distances between objects in the solar system and the distance between Earth and the Sun. The figure below shows that Jupiter is 5.2 AU from the Sun. This means that Jupiter is 5.2 times farther from the Sun than Earth is from the Sun. The most distant planet, Neptune, is 30 AU from the Sun.

Astronomical Units

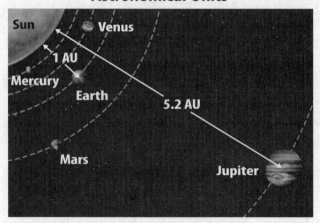

Distances Beyond the Solar System

Astronomers measure distances to objects beyond the solar system using a larger distance unit—the light-year. Despite its name, a light-year measures distance, not time. *A* **light-year** *is the distance light travels in 1 year*. Light travels at a rate of about 300,000 km/s. That means 1 light-year is about 10 trillion km! Proxima Centauri, the nearest star to the Sun, is 4.2 light-years away.

Looking Back in Time

Because it takes time for light to travel, you see a star not as it is today but as it was when light left it. At 4.2 light-years away, Proxima Centauri appears as it was 4.2 years ago. The farther away an object is, the longer it takes for its light to reach Earth.

Visual Check

6. Apply About how many million kilometers is Jupiter from the Sun?

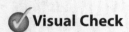

Proportions can be used to calculate distances to astronomical objects. Light can travel nearly 10 trillion km in 1 year (y). How many years would it take light to reach Earth from a star that is 100 trillion km away?

a. Set up a proportion.

$$\frac{10 \text{ trillion km}}{1 \text{ y}} = \frac{100 \text{ million km}}{x \text{ y}}$$

b. Cross multiply.

10 trillion km $\times$ (x)y =
100 trillion km $\times$ 1 y

c. Solve for x by dividing both sides by 10 trillion km.

$$x = \frac{100 \text{ trillion km}}{10 \text{ trillion km}} = 10 \text{ y}$$

7. Use Proportions How many years would it take light to reach Earth from a star 60 trillion km away?

Measuring Brightness

Some stars look dim and some look bright. Astronomers measure the brightness of stars in two ways: by how bright they appear from Earth and by how bright they actually are.

Apparent Magnitude

Scientists measure how bright stars appear using a scale developed by the ancient Greek astronomer Hipparchus (hi PAR kus). Hipparchus assigned a number to every star he saw based on the star's brightness. Astronomers today call these numbers magnitudes. *The **apparent magnitude** of an object is a measure of how bright it appears from Earth.*

As shown in the figure below, the fainter a star appears, the greater its <u>apparent</u> magnitude. Note that some objects have negative apparent magnitudes. That is because Hipparchus assigned a value of 1 to the brightest stars, but he did not assign values to the Sun, the Moon, or Venus. Astronomers later assigned negative numbers to the Sun, the Moon, Venus, and a few bright stars.

Apparent Magnitude

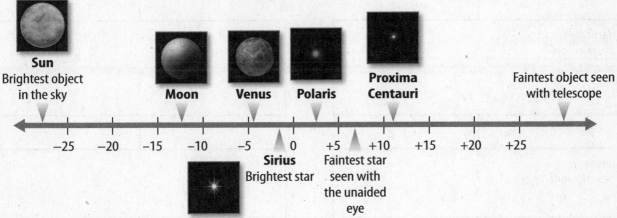

Sun — Brightest object in the sky
Moon
Venus
Polaris
Proxima Centauri
Faintest object seen with telescope

−25 −20 −15 −10 −5 0 +5 +10 +15 +20 +25

Sirius — Brightest star
Faintest star seen with the unaided eye

Absolute Magnitude

A star can appear bright or dim depending on its distance from Earth. But a star also has an actual, or absolute, magnitude. **Luminosity** (lew muh NAH sih tee) *is the true brightness of an object.* The luminosity of a star is measured on an absolute magnitude scale. A star's luminosity depends on the star's temperature and size, not on its distance from Earth. A star's luminosity, apparent magnitude, and distance from Earth are related. If scientists know two of these factors, they can determine the third using mathematical formulas.

ACADEMIC VOCABULARY

apparent
(*adjective*) appearing to the eye or mind

✓ **Visual Check**

8. Interpret What is the apparent magnitude of Sirius?

_____

_____

🔑 **Key Concept Check**

9. Summarize How do scientists measure the brightness of stars?

_____

_____

_____

_____

······· After You Read·······

Mini Glossary

apparent magnitude: a measure of how bright an object appears from Earth

astronomical unit (AU): the average distance between Earth and the Sun

light-year: the distance light travels in 1 year

luminosity (lew muh NAH sih tee): the true brightness of an object

spectroscope: an instrument that spreads light into different wavelengths

1. Review the terms and their definitions in the Mini Glossary. Write a sentence that explains why scientists use light-years rather than astronomical units to measure distances beyond the solar system.

_____

_____

2. In the diagram below, name and describe measurement units used in astronomy.

Units of Distance

name: description:

name: description:

Units of Brightness

name: description:

name: description:

3. Why do objects in the sky appear to rise in the east and set in the west?

_____

_____

What do you think NOW?

Reread the statements at the beginning of the lesson. Fill in the After column with an A if you agree with the statement or a D if you disagree. Did you change your mind?

Log on to ConnectED.mcgraw-hill.com and access your textbook to find this lesson's resources.

END OF LESSON

Stars and Galaxies

The Sun and Other Stars

What do you think? Read the two statements below and decide whether you agree or disagree with them. Place an A in the Before column if you agree with the statement or a D if you disagree. After you've read this lesson, reread the statements to see if you have changed your mind.

Before	Statement	After
	3. Stars shine because there are nuclear reactions in their cores.	
	4. Sunspots appear dark because they are cooler than nearby areas.	

······· **Read to Learn** ·······

How Stars Shine

The hotter something is, the more quickly its atoms move. As atoms move, they collide. If a gas is hot enough and its atoms move quickly enough, the nuclei of some of the atoms stick together. **Nuclear fusion** *is a process that occurs when the nuclei of several atoms combine into one larger nucleus.*

Nuclear fusion releases a great amount of energy. This energy powers stars. *A* **star** *is a large ball of gas held together by gravity with a core so hot that nuclear fusion occurs.* A star's core can reach hundreds of millions of degrees Celsius. When energy leaves a star's core, it travels throughout the star and radiates into space. As a result, the star shines.

Composition and Structure of Stars

The Sun is the closest star to Earth. Because it is so close, scientists easily can observe it. They can send probes to the Sun. They can study its spectrum using spectroscopes on Earth-based telescopes. Spectra of the Sun and other stars provide information about the composition of stars. The Sun and most stars are made almost entirely of hydrogen and helium gas. A star's composition changes slowly over time as hydrogen in its core fuses into more complex nuclei.

Key Concepts

- How do stars shine?
- How are stars layered?
- How does the Sun change over short periods of time?
- How do scientists classify stars?

Study Coach

Building Vocabulary Work with another student to write a question about each vocabulary term in this lesson. Answer the questions and compare your answers. Reread the text to clarify the meaning of the terms.

Key Concept Check

1. Explain How do stars shine?

_____

_____

_____

_____

Interior of Stars

When first formed, all stars fuse hydrogen into helium in their cores. Helium is denser than hydrogen, so it sinks to the inner part of the core after it forms. A typical star has three interior layers. The layers are shown in the figure at the bottom of this page. The core is the center layer. *The* **radiative zone** *is a shell of cooler hydrogen above a star's core.* Hydrogen in this layer is dense. Light energy bounces from atom to atom as it gradually makes its way upward, out of the radiative zone. Above the radiative zone is the **convection zone,** *where hot gas moves up toward the surface and cooler gas moves deeper into the interior.* Light energy moves quickly upward in the convection zone.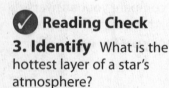

Atmosphere of Stars

Beyond the convection zone are the three outer layers of a star. These layers make up a star's atmosphere. *The* **photosphere** *is the apparent surface of a star.* In the Sun, it is the dense, bright part you can see, where light energy radiates into space. From Earth, the Sun's photosphere looks smooth. But like the rest of the Sun, it is made of gas.

Above the photosphere are the two outer layers of a star's atmosphere. *The* **chromosphere** *is the orange-red layer above the photosphere,* as shown in the figure below. *The* **corona** *is the wide, outermost layer of a star's atmosphere.* The temperature of the corona is higher than that of the photosphere or the chromosphere. It is irregular in shape and can extend for several million kilometers.

Key Concept Check

2. Name What are the interior layers of a star?

_____

_____

_____

Reading Check

3. Identify What is the hottest layer of a star's atmosphere?

_____

_____

Visual Check

4. Locate Where is the photosphere located in relation to the Sun's other layers?

_____

_____

_____

_____

Layers of the Sun

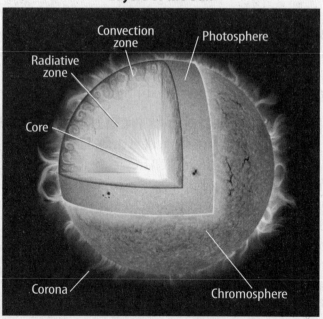

Convection zone

Photosphere

Radiative zone

Core

Corona

Chromosphere

The Sun's Changing Features

The interior features of the Sun are stable over millions of years. But the Sun's atmosphere can change over years, months, or even minutes. Some of these features are described below.

Sunspots Regions of strong magnetic activity are called sunspots. Sunspots are cooler than the rest of the photosphere and appear as dark splotches on the Sun. They seem to move across the Sun as the Sun rotates. The number of sunspots changes over time. They follow a cycle, peaking in number every 11 years. An average sunspot is about the size of Earth.

Prominences and Flares Prominences are clouds of gas that make loops and jets that extend into the corona. They sometimes last for weeks. Flares are sudden increases in brightness that often occur near sunspots or prominences. They are violent eruptions that last from minutes to hours. Both prominences and flares begin at or just above the photosphere.

Coronal Mass Ejections (CMEs) Huge bubbles of gas ejected from the corona are coronal mass ejections (CMEs). They are much larger than flares and occur over the course of several hours. Material from a CME can reach Earth, sometimes interfering with radio and satellite communications.

The Solar Wind Charged particles that stream continually away from the Sun create the solar wind. The solar wind passes Earth and extends to the edge of the solar system. The northern lights, or auroras, are curtains of light. Particles from the solar wind or a CME create auroras when they interact with Earth's magnetic field.

Groups of Stars

There are no other stars near the Sun. The star closest to the Sun is 4.2 light-years away. Many stars are single stars, such as the Sun. But most stars exist in star systems bound by gravity.

Binary Star Systems The most common star system is a binary system. In a binary system, two stars orbit each other. By studying the orbits of binary stars, astronomers can determine the stars' masses.

FOLDABLES

Make a vertical four-tab book to organize your notes about the changing features of the Sun.

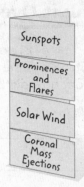

Sunspots

Prominences and Flares

Solar Wind

Coronal Mass Ejections

Key Concept Check

5. Name Which parts of the Sun change over short periods of time?

_____

_____

_____

_____

Think it Over

6. Relate Look up the word *binary* in a dictionary. How does the definition relate to a binary star system?

_____

_____

_____

_____

Many stars exist in large groups called clusters. There are two types of star clusters—open clusters and globular clusters. Open clusters contain fewer than 1,000 stars. Globular clusters can contain hundreds of thousands of stars.

All the stars in a cluster formed at about the same time and are the same distance from Earth. If astronomers know the distance to or the age of one star in a cluster, they know the distance to or the age of every star in the cluster.

Classifying Stars

How do you classify a star? Which properties are important? Scientists classify stars according to their spectra. Recall that a star's spectrum is the light the star emits spread out by wavelength. Stars have different spectra and different colors depending on their surface temperatures.

Temperature, Color, and Mass

Have you ever seen coals in a fire? Red coals are the coolest. Blue-white coals are the hottest. Stars are similar. Blue-white stars are hotter than red stars. The temperatures of orange, yellow, and white stars are between the hottest blue-white stars and cooler red stars.

There are some exceptions, but color in most stars is related to mass. Blue-white stars tend to have the greatest mass, followed by white stars, yellow stars, orange stars, and red stars. The most massive stars are normally the hottest. The smallest stars tend to be cooler and red. ✓

The Sun is a yellow star. It is tiny compared to blue-white stars. However, scientists suspect that most stars—as many as 90 percent—are smaller than the Sun. These stars are called red dwarfs.

Hertzsprung-Russell Diagram

When scientists plot the temperatures of stars against their luminosities, the result is a graph like that shown on the next page. *The **Hertzsprung-Russell diagram** (or H-R diagram) is a graph that plots luminosity v. temperature of stars.* The *y*-axis of the H-R diagram displays increasing luminosity. The *x*-axis displays decreasing temperature.

The H-R diagram is an important tool for categorizing stars. Astronomers also use it to determine distances of some stars. If a star has the same temperature as a star on the H-R diagram, astronomers often can determine its luminosity. As you read earlier, if astronomers know a star's luminosity, they can calculate its distance from Earth. 🔑

💡 **Think it Over**

7. Apply Star *A* is 12 billion years old. Star *B* is in the same cluster. About how old is Star *B?* Why?

_____

_____

_____

_____

✓ **Reading Check**

8. Relate How does star color relate to mass?

_____

_____

_____

_____

🔑 **Key Concept Check**

9. Describe What is the Hertzsprung-Russell diagram?

_____

_____

_____

_____

_____

_____

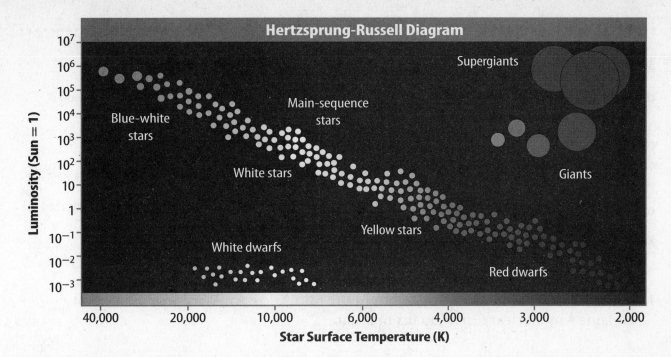

Hertzsprung-Russell Diagram

Supergiants

Main-sequence stars

Blue-white stars

Giants

White stars

Yellow stars

White dwarfs

Red dwarfs

Luminosity (Sun = 1)

Star Surface Temperature (K)

The Main Sequence

Most stars exist along the main sequence in the H-R diagram. The main sequence stars form a line that stretches from the upper left corner to the lower right corner of the graph. The mass of a main-sequence star determines both its temperature and its luminosity. High-mass stars have more gravity pulling inward than low-mass stars do. As a result, the cores of high-mass stars have higher temperatures and produce more energy through fusion.

As shown in the figure above, some groups of stars on the H-R diagram do not fit on the main sequence. Stars at the top right are cool yet luminous. This is because they are unusually large, not because they produce more energy. Massive stars are giants. The most massive are supergiants. The white dwarfs at the bottom of the H-R diagram are hot yet dim. This is because they are unusually small. You will read more about these stars in Lesson 3.

✅ **Visual Check**

10. Interpret Where is the Sun on this diagram? Draw a circle to indicate the approximate location.

✅ **Reading Check**

11. Explain Supergiants are cool yet luminous because _____. (Circle the correct answer.)

a. they produce more energy

b. they are farther from Earth

c. they are unusually large

Mini Glossary

chromosphere: the orange-red layer of a star above the photosphere

convection zone: a layer of a star where hot gas moves up toward the surface and cooler gas moves deeper into the interior

corona: the wide, outermost layer of a star's atmosphere

Hertzsprung-Russell diagram: a graph that plots luminosity v. temperature of stars

nuclear fusion: a process that occurs when the nuclei of several atoms combine into one larger nucleus

photosphere: the apparent surface of a star

radiative zone: a shell of cooler hydrogen above a star's core

star: a large ball of gas held together by gravity with a core so hot that nuclear fusion occurs

1. Review the terms and their definitions in the Mini Glossary. Write a sentence that explains why nuclear fusion occurs in a star.

_____

_____

2. Label each star below as *yellow*, *red*, *white*, *blue-white*, or *orange*, based on mass. Then label each end of the temperature arrow as *higher* or *lower*.

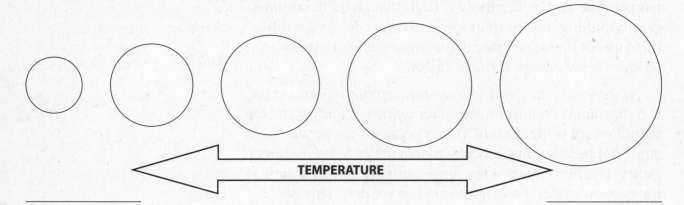

TEMPERATURE

_____   _____

3. How did writing and answering questions about vocabulary terms in this lesson help you learn about the Sun and other stars?

_____

_____

What do you think NOW?

Reread the statements at the beginning of the lesson. Fill in the After column with an A if you agree with the statement or a D if you disagree. Did you change your mind?

Connect ED

Log on to ConnectED.mcgraw-hill.com and access your textbook to find this lesson's resources.

END OF LESSON

Stars and Galaxies

Evolution of Stars

·············· **Before You Read** ··············

| **What do you think?** Read the two statements below and decide whether you agree or disagree with them. Place an A in the Before column if you agree with the statement or a D if you disagree. After you've read this lesson, reread the statements to see if you have changed your mind. |

Before	Statement	After
	5. The more matter a star contains, the longer it is able to shine.	
	6. Gravity plays an important role in the formation of stars.	

Key Concepts

- How do stars form?
- How does a star's mass affect its evolution?
- How is star matter recycled in space?

·············· **Read to Learn** ··············

Life Cycle of a Star

Stars have life cycles that can be compared to the life cycles of living things. They are "born," and after millions or billions of years, they "die." Stars die in different ways, depending on their masses. But all stars—from white dwarfs to supergiants—form in the same way.

Nebulae and Protostars

Stars form deep inside clouds of gas and dust. *A cloud of gas and dust is a* **nebula** (plural, nebulae). Star-forming nebulae are cold, dense, and dark. Gravity causes the densest parts to collapse, forming regions called protostars. Protostars continue to contract. As they contract, they pull in surrounding gas. Eventually, their cores are hot and dense enough for nuclear fusion to begin. As they contract, protostars produce enormous amounts of thermal energy.

Birth of a Star

Over many thousands of years, the energy produced by protostars heats the gas and dust around the protostars. Eventually, the gas and dust blow away, and the protostars become visible as stars. Some of this material might later become planets or other objects that orbit the star. During the star-formation process, nebulae glow brightly.

> **Study Coach**
>
> **Ask Questions** As you read, write your questions about stars on a sheet of paper. Answer your questions as you read the lesson a second time. Discuss any questions that you can't answer with your teacher.

Key Concept Check

1. Summarize How do stars form?

_____

_____

_____

_____

_____

Visual Check

2. Name what forms in only the most massive stars.

Main-Sequence Stars

Recall the main sequence of the Hertzsprung-Russell diagram. Stars spend most of their lives on the main sequence. A star becomes a main-sequence star as soon as it begins to fuse hydrogen into helium. It remains on the main sequence for as long as it continues to fuse hydrogen into helium. Lower-mass stars such as the Sun stay on the main sequence for billions of years. High-mass stars stay on the main sequence for only a few million years. Even though massive stars have more hydrogen than lower-mass stars, they process it at a much faster rate.

When a star's hydrogen supply is nearly gone, the star leaves the main sequence. It begins the next stage of its life cycle, as shown in the figure below. Not all stars go through all phases shown in the figure below. Lower-mass stars, such as the Sun, do not have enough mass to become supergiants.

A Massive Star's Life Cycle

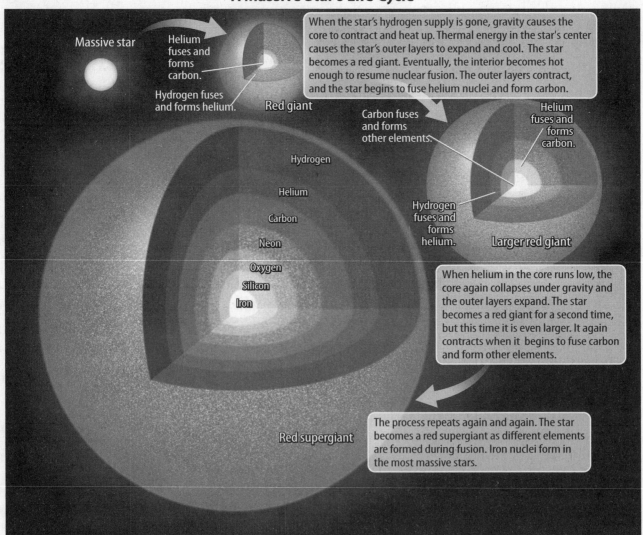

Massive star

Helium fuses and forms carbon.

Hydrogen fuses and forms helium.

Red giant

When the star's hydrogen supply is gone, gravity causes the core to contract and heat up. Thermal energy in the star's center causes the star's outer layers to expand and cool. The star becomes a red giant. Eventually, the interior becomes hot enough to resume nuclear fusion. The outer layers contract, and the star begins to fuse helium nuclei and form carbon.

Carbon fuses and forms other elements.

Helium fuses and forms carbon.

Hydrogen fuses and forms helium.

Larger red giant

Hydrogen
Helium
Carbon
Neon
Oxygen
Silicon
Iron

When helium in the core runs low, the core again collapses under gravity and the outer layers expand. The star becomes a red giant for a second time, but this time it is even larger. It again contracts when it begins to fuse carbon and form other elements.

Red supergiant

The process repeats again and again. The star becomes a red supergiant as different elements are formed during fusion. Iron nuclei form in the most massive stars.

End of a Star

All stars form in the same way. But stars die in different ways, depending on their masses. Massive stars collapse and explode. Lower-mass stars die more slowly.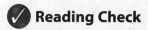

White Dwarfs

Lower-mass stars, such as the Sun, do not have enough mass to fuse elements beyond helium. They do not get hot enough. After helium in their cores is gone, the stars cast off their gases, exposing their cores. The core becomes a **white dwarf,** *a hot, dense, slowly cooling sphere of carbon.*

The Sun as a Red Giant What will happen to Earth and the solar system when the Sun runs out of fuel? When the Sun runs out of hydrogen, in about 5 billion years, it will become a red giant. Once helium fusion begins, the Sun will contract. When the helium is gone, the Sun will expand again, probably absorbing Mercury, Venus, and Earth, and pushing Mars and Jupiter outward.

The Sun as a White Dwarf Eventually, the Sun will become a white dwarf, as shown in the figure below. Imagine the mass of the Sun squeezed a million times until it is the size of Earth. That's the size of a white dwarf. ✓

Scientists hypothesize that all stars with masses less than 8–10 times that of the Sun will eventually become white dwarfs. With a white dwarf at the center, the solar system will be a cold, dark place.

The Sun as a White Dwarf

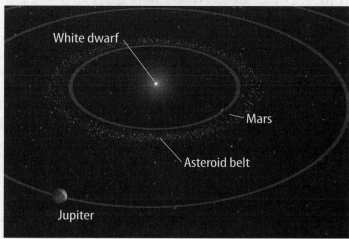

White dwarf

Mars

Asteroid belt

Jupiter

✓ Reading Check

3. Point Out What determines the way a star will die?

_____

_____

✓ Reading Check

4. Summarize What will happen to Earth when the Sun runs out of fuel?

_____

_____

_____

_____

_____

✓ Visual Check

5. Locate Circle the planet closest to the white dwarf.

Stars with more than 10 times the mass of the Sun do not become white dwarfs. Instead, they explode. A **supernova** (plural, supernovae) *is an enormous explosion that destroys a star.*

In the most massive stars, a supernova occurs when iron forms in the star's core. Iron is stable and does not fuse. After a star forms iron, it loses its internal energy source. Without its energy source, the core collapses quickly under the force of gravity. The collapse of the core releases so much energy that the star explodes. When it explodes, a star can become 1 billion times brighter and form elements even heavier than iron. ✓

Neutron Stars

Have you ever eaten cotton candy? A bag of cotton candy is spun from just a few spoonfuls of sugar. Cotton candy is mostly air. Similarly, atoms are mostly empty space. During a supernova, the collapse is so violent that it eliminates the normal spaces inside atoms, and a <u>neutron</u> star forms.

A **neutron star** *is a dense core of neutrons that remains after a supernova.* Neutron stars are only about 20 km wide. Their cores are so dense that a teaspoonful would weigh more than 1 billion tons.

Black Holes

For the most massive stars, atomic forces holding neutrons together are not strong enough to overcome so much mass in such a small volume. Gravity is too strong, and the matter crushes into a black hole. *A* **black hole** *is an object whose gravity is so great that no light can escape.* ✓

A black hole does not suck matter in like a vacuum cleaner. But a black hole's gravity is very strong because all of its mass is concentrated in a single point. Because astronomers cannot see a black hole, they only can infer its existence. For example, if they detect a star circling around something but they cannot see what that something is, they suspect it is a black hole.

Recycling Matter

At the end of a star's life cycle, much of its gas escapes into space. This gas is recycled. It becomes the building blocks of future generations of stars and planets.

✓ Reading Check

6. Explain Why does a massive star lose its internal energy source when iron forms in its core?

_____

_____

_____

REVIEW VOCABULARY

neutron
a neutral particle in the nucleus of an atom

🗝 Key Concept Check

7. Explain How does a star's mass determine if it will become a white dwarf, a neutron star, or a black hole?

_____

_____

_____

_____

_____

Planetary Nebulae

You read that lower-mass stars, such as the Sun, become white dwarfs. When a star becomes a white dwarf, it casts off hydrogen and helium gases in its outer layers. The expanding, cast-off matter of a white dwarf is a planetary nebula. Most of the star's carbon remains locked in the white dwarf. But the gases in the planetary nebula can be used to form new stars. ✓

Planetary nebulae have nothing to do with planets. They are called "planetary" because early astronomers thought they were regions where planets were forming.

Supernova Remnants

During a supernova, a massive star comes apart. This sends a shock wave into space. The expanding cloud of dust and gas is called a supernova remnant. Like a snowplow pushing snow, a supernova remnant pushes on the gas and dust it encounters.

In a supernova, a star releases the elements that formed inside it during nuclear fusion. Almost all of the elements in the universe other than hydrogen and helium were created by nuclear reactions inside the cores of massive stars and released in supernovae. This includes the oxygen in air, the silicon in rocks, and the carbon in you.

Gravity causes recycled gases and other matter to clump together in nebulae and form new stars and planets. As you will read in the next lesson, gravity also causes stars to clump together into even larger structures called galaxies. 🗝✓

✓ Reading Check

8. Relate How are a white dwarf and a planetary nebula related?

_____

_____

_____

_____

🗝 Key Concept Check

9. Describe How do stars recycle matter?

_____

_____

_____

_____

Mini Glossary

black hole: an object whose gravity is so great that no light can escape

nebula: a cloud of gas and dust

neutron star: a dense core of neutrons left from a supernova

supernova: an enormous explosion that destroys a star

white dwarf: a hot, dense, slowly cooling sphere of carbon

1. Review the terms and their definitions in the Mini Glossary. Write a sentence that describes how a supernova and a neutron star are related.

_____

_____

2. Complete the life cycle of a massive star by writing the following in the correct sequence in the circles of the diagram: larger red giant, protostar, red giant, red supergiant, supernova remnants.

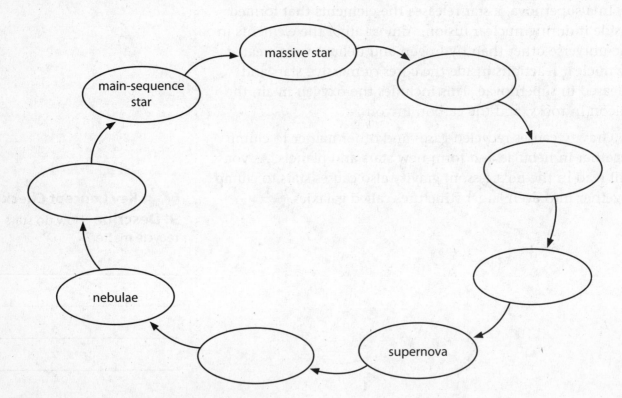

What do you think NOW?

Reread the statements at the beginning of the lesson. Fill in the After column with an A if you agree with the statement or a D if you disagree. Did you change your mind?

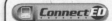

Log on to ConnectED.mcgraw-hill.com and access your textbook to find this lesson's resources.

END OF LESSON

Stars and Galaxies

Galaxies and the Universe

·············· **Before You Read** ··············

What do you think? Read the two statements below and decide whether you agree or disagree with them. Place an A in the Before column if you agree with the statement or a D if you disagree. After you've read this lesson, reread the statements to see if you have changed your mind.

Before	Statement	After
	7. Most of the mass in the universe is in stars.	
	8. The Big Bang theory is an explanation of the beginning of the universe.	

·············· **Read to Learn** ··············

Galaxies

Most stars exist in galaxies. **Galaxies** *are huge collections of stars.* The universe has hundreds of billions of galaxies. Each galaxy can contain hundreds of billions of stars.

Dark Matter

Gravity holds stars together. Gravity also holds galaxies together. Astronomers study how galaxies rotate and how gravity affects their movement. They have found that most matter in galaxies is invisible. *Matter that emits no light at any wavelength is* **dark matter.**

Scientists hypothesize that more than 90 percent of the universe's mass is dark matter. Scientists do not know what material this invisible dark matter contains.

Types of Galaxies

There are three major types of galaxies. Galaxies can be spiral, elliptical, or irregular.

Spiral Galaxies The stars, gas, and dust in a spiral galaxy exist in spiral arms that begin at a central disk. Some spiral arms are long and symmetrical. Others are short and stubby. Spiral galaxies are thicker near the center, a region called the central bulge. A spherical halo of globular clusters and older, redder stars surrounds the disk.

Key Concepts 🔑

- What are the major types of galaxies?
- What is the Milky Way, and how is it related to the solar system?
- What is the Big Bang theory?

◀ **Mark the Text**

Identify the Main Ideas To help you learn about galaxies and the universe, highlight each heading in one color. Then highlight the details that support and explain it in a different color. Refer to this highlighted text as you study the lesson.

🔑 **Key Concept Check**

1. Name What are the major types of galaxies?

_____

_____

_____

Make a horizontal single-tab matchbook to describe the contents of the Milky Way.

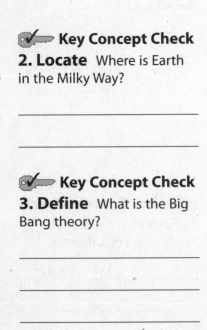

Elliptical Galaxies Unlike spiral galaxies, elliptical galaxies do not have an internal structure. Some are spheres, like basketballs. Others are shaped like footballs. Elliptical galaxies have higher percentages of old, red stars than spiral galaxies do. They contain little or no gas and dust. Scientists hypothesize that many elliptical galaxies form as the gravity of two or more spiral galaxies comes together.

Irregular Galaxies Galaxies that are oddly shaped are called irregular galaxies. Many form from the gravitational pull of neighboring galaxies. Irregular galaxies contain many young stars and have areas of intense star formation.

Groups of Galaxies

Galaxies are not distributed evenly in the universe. Gravity holds them together in groups called clusters. Some clusters of galaxies are enormous. The Virgo Cluster contains about 2,000 galaxies.

Most clusters exist in even larger structures called superclusters. Between superclusters are voids, which are regions of nearly empty space. Scientists hypothesize that the large-scale structure of the universe resembles a sponge.

The Milky Way

The Milky Way is a spiral galaxy that contains gas, dust, and almost 200 billion stars. The Sun, Earth, and the rest of our solar system is inside the disk of the Milky Way. The Milky Way is part of the Local Group, a cluster of about 30 galaxies.

The Andromeda Galaxy is the largest galaxy in the Local Group. Scientists expect the Milky Way will begin to merge with the Andromeda Galaxy in about 3 billion years. Because stars are far apart in galaxies, few stars are likely to collide during this event.

The Big Bang Theory

Is there a beginning to time? According to the **Big Bang theory,** *the universe began from one point billions of years ago and has been expanding ever since.*

Origin and Expansion of the Universe

Most scientists agree that the universe is 13–14 billion years old. The universe was so dense and hot at first that even atoms didn't exist. After a few hundred thousand years, the universe cooled enough for atoms to form. Eventually, stars formed, and gravity pulled them into galaxies.

Key Concept Check
2. Locate Where is Earth in the Milky Way?

_____

_____

Key Concept Check
3. Define What is the Big Bang theory?

_____

_____

_____

Copyright © Glencoe/McGraw-Hill, a division of The McGraw-Hill Companies, Inc.

Stretching Space As the universe expands, space stretches and galaxies move away from one another. The same thing happens in a loaf of unbaked raisin bread. As the dough rises, the raisins move apart. Scientists observe how space stretches by measuring the speed at which galaxies move away from Earth. As the galaxies move away, their wavelengths lengthen and stretch out. How does light stretch?

Doppler Shift

You have probably heard the siren of a speeding police car. As the figure below illustrates, when the car moves toward you, the sound waves compress. As the car moves away, the sound waves spread out. Similarly, when visible light travels toward you, its wavelength compresses. When light travels away from you, its wavelength stretches out. It shifts to the red end of the electromagnetic spectrum. *The shift to a different wavelength is called the* **Doppler shift.** Because the universe is expanding, light from galaxies is red-shifted. The more distant a galaxy is, the faster it moves away from Earth and the more it is red-shifted.

Doppler Shift

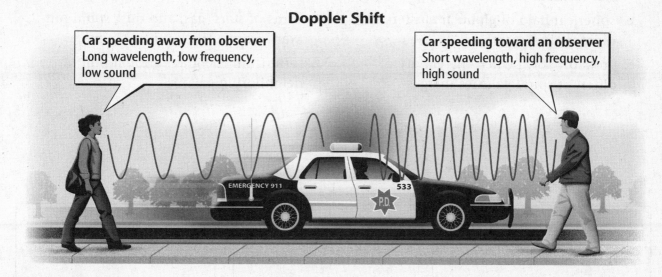

Car speeding away from observer
Long wavelength, low frequency, low sound

Car speeding toward an observer
Short wavelength, high frequency, high sound

EMERGENCY 911   P.D   533

Dark Energy

Will the universe expand forever? Or will gravity cause the universe to contract? Scientists have observed that galaxies are moving away from Earth faster over time. To explain this, they suggest that a force called dark energy is pushing the galaxies apart.

Dark energy, like dark matter, is an active area of research. There is still much to learn about the universe and all that it contains.

☑️ **Visual Check**

4. Contrast Describe how the sound waves in front of the car are different from the sound waves behind the car.

_____

_____

_____

Mini Glossary

Big Bang theory: the theory that the universe began from one point billions of years ago and has been expanding ever since

dark matter: matter that emits no light at any wavelength

Doppler shift: the shift to a different wavelength

galaxy: a huge collection of stars

1. Review the terms and their definitions in the Mini Glossary. Write a sentence that explains why the light from galaxies has a Doppler shift toward the red end of the electromagnetic spectrum.

_____

_____

2. The information below describes galaxies. Match each letter and its information to the type of galaxy it describes. Write the letter in the box.

a. many young stars and areas of intense star formation

b. spherical halo of globular clusters and older, redder stars

c. probably form when the gravity of different spiral galaxies merges

d. little or no gas and dust and no internal structure

e. arms of stars, gas, and dust spiral outward from a thick central bulge

f. form from the gravitational pull of neighboring galaxies

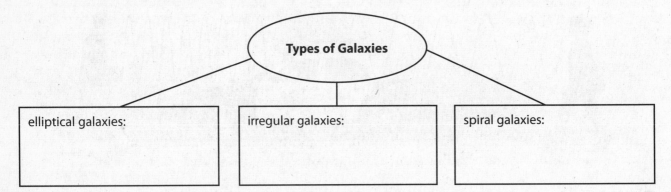

Types of Galaxies

elliptical galaxies: | irregular galaxies: | spiral galaxies:

3. What is one question that research on dark energy might help scientists answer?

_____

What do you think  **NOW?**

Reread the statements at the beginning of the lesson. Fill in the After column with an A if you agree with the statement or a D if you disagree. Did you change your mind?

Connect ED

Log on to ConnectED.mcgraw-hill.com and access your textbook to find this lesson's resources.

END OF LESSON

Minerals and Rocks

Minerals

·············· **Before You Read** ··············

What do you think? Read the two statements below and decide whether you agree or disagree with them. Place an A in the Before column if you agree with the statement or a D if you disagree. After you've read this lesson, reread the statements to see if you have changed your mind.

Before	Statement	After
	1. Minerals generally are identified by observing their color.	
	2. Minerals are made of crystals.	

·············· **Read to Learn** ··············

What is a mineral?

Do you ever drink mineral water? Maybe you take vitamins and minerals to stay healthy. The word *mineral* has many common meanings. For geologists, scientists who study Earth and the materials of which it is made, the word *mineral* has a very specific definition.

A **mineral** *is a naturally occurring, inorganic solid that has a crystal structure and a definite chemical composition.* In order for a substance to be classified as a mineral, it must have all five of the characteristics listed in this definition. Both coal and pyrite are shiny, hard substances that form deep inside Earth. But only one is a mineral. Coal formed from ancient plant material. Pyrite crystals are made of the elements iron and sulfur. One substance is a mineral, and one is not. ✔

Characteristics of Minerals

How can you determine which substance is the mineral? Consider each of the five characteristics of minerals.

Naturally Occurring To be classified as a mineral, a substance must form naturally. Materials made by people are not minerals. Diamonds that form deep beneath Earth's surface are minerals. Diamonds that are made in a laboratory are not minerals. However, manufactured diamonds may look similar to naturally occurring diamonds.

Key Concepts 🔑

- How do minerals form?
- What properties can be used to identify minerals?
- What are some uses of minerals in everyday life?

▸ **Mark the Text**

Identify the Main Ideas
Write notes next to each paragraph to summarize the main ideas. On a separate piece of paper, organize these notes into two columns. Place each main idea in the left column. List the details to support it in the right column. Use your table to review the lesson.

✔ **Reading Check**

1. Name the five characteristics that define a mineral.

_____

_____

_____

_____

Math Skills ×÷

A ratio compares numbers. For example, in the chemical formula for water, H_2O, the number *2* is called a subscript. The subscript tells you how many atoms of that element are in the formula. A symbol with no subscript means that element has one atom. So, the ratio of hydrogen (H) atoms to oxygen (O) atoms in H_2O is 2:1. This is read *two to one.*

2. Use Ratios Quartz has the formula SiO_2. What is the ratio of silicon (Si) atoms to oxygen (O) atoms in quartz?

_____

🗝 **Key Concept Check**
3. Describe How do minerals form?

_____

_____

_____

_____

_____

Inorganic A material that contains carbon and was once alive is organic. A mineral cannot be organic. This means that a mineral cannot have once been alive. Also, a mineral cannot contain anything that was once alive, such as plants.

Solid A mineral must be solid. Liquids and gases are not considered minerals. Solid ice is a mineral, but water is not.

Crystal Structure A mineral must have a crystal structure. *The atoms in a crystal are arranged in an orderly, repeating pattern called a* **crystal structure.** This organized structure produces smooth faces and sharp edges on a crystal.

Definite Chemical Composition A mineral is made of specific amounts of elements. A chemical formula shows the amount of each element in a mineral. For example, pyrite is made of the elements iron (Fe) and sulfur (S). There always must be one iron atom for every two sulfur atoms. Therefore, the chemical formula for pyrite is FeS_2.

Think again about coal and pyrite. The plants that turned into coal were once alive. Coal cannot be a mineral. Pyrite has all five characteristics of a mineral, so it is a mineral.

Mineral Formation

How do atoms form minerals? Atoms within a liquid join together and form a solid. **Crystallization** *is the process by which atoms form a solid with an orderly, repeating pattern.* Crystallization can happen in two main ways.

Crystallization from Magma Melted rock material is called magma. As magma cools, some of the atoms join together and form solid crystals. As the liquid magma continues to cool, more atoms are added to the surface of the crystals. The longer it takes the magma to cool, the larger the crystals become because atoms continue to be added to the crystals. Crystals grow large when the magma cools slowly. When magma cools quickly, the crystals that form remain small.

Crystallization from Water Many substances, such as salt, dissolve in water—especially if the water is warm. When water cools or evaporates, the particles of the dissolved substances come together again and crystallize. Gold crystals form this way. The orderly arrangement of atoms in the mineral gold is visible using a very powerful microscope. 🗝

Mineral Identification

Every mineral has a unique set of physical properties, or characteristics. These properties are used to identify minerals. By testing several properties, scientists can distinguish between similar minerals.

Density

If you pick up two mineral samples that are about the same size, one might feel heavier than the other. The heavier mineral has a higher density. It has more mass in the same volume. The densities of many minerals are similar, but a very high or a very low density can help identify a mineral.

Hardness

Scientists measure the hardness of a mineral by observing how easily it is scratched or how easily it scratches something else. The Mohs hardness scale, shown in the table below, ranks hardness from 1 to 10. On this scale, diamond is the hardest mineral, with a hardness value of 10. The softest mineral is talc, with a hardness of 1.

Mohs Hardness Scale for Minerals		
Mineral	**Hardness**	**Hardness of Common Objects**
talc	1 (softest)	
gypsum	2	fingernail (2.5)
calcite	3	copper wire or penny (3.5)
fluorite	4	wire nail (4.5)
apatite	5	glass, steel knife blade (5.5)
feldspar	6	streak plate (unglazed porcelain) (6.5)
quartz	7	
topaz	8	
corundum	9	
diamond	10 (hardest)	

Color and Streak

Some minerals can be identified by their unique color. The mineral malachite always has a certain green color. But the colors of most minerals vary. Quartz is a common mineral that has many different colors.

The colors of most minerals vary from sample to sample, but the color of a mineral's powder does not vary. *The color of a mineral's powder is called its* **streak.** You can observe streak by scratching the mineral across a tile of unglazed porcelain. Sometimes, the color of a mineral and the color of its streak are different. For example, the mineral hematite can have a red, brown, or black color, but its streak is always a dark, rusty red.

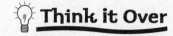

Think it Over

4. Apply A sample of mineral *A* and a sample of mineral *B* are about the same size. The mineral *B* sample is heavier than the mineral *A* sample. Which mineral has the higher density? How do you know?

_____

_____

_____

✓ **Visual Check**

5. Interpreting Tables Circle the minerals that are harder than glass.

Think it Over

6. Analyze Can you identify a mineral by its color alone? Explain your answer.

_____

_____

_____

Luster

Minerals reflect light in different ways. **Luster** *describes the way that a mineral's surface reflects light.* Some terms used to describe mineral luster are *metallic, glassy, earthy,* or *pearly.* Hematite can have either a metallic luster or a dull luster. Muscovite mica has a pearly luster. Quartz has a glassy luster.

Cleavage and Fracture

Sometimes the way a mineral breaks helps identify it. Minerals break in two ways. *If a mineral breaks along smooth, flat surfaces, it displays* **cleavage.** The mineral on the left in the figure below illustrates the property cleavage. It forms a flat surface where it breaks. A mineral can break along a single cleavage direction or along several directions. Muscovite mica has one cleavage direction and peels off in sheets. Halite has three cleavage directions and breaks into cubes.

A mineral that breaks along rough or irregular surfaces displays **fracture.** The mineral on the right in the figure below illustrates the property fracture.

Cleavage and Fracture

 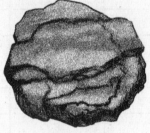

| Cleavage | Fracture |

Crystal Shape

Minerals <u>exhibit</u> many different crystal shapes. A mineral's atomic structure determines its crystal shape.

Crystal shapes can vary greatly. Crystals of hematite have no definite shape, or are shapeless. They are described as massive. Muscovite mica has diamond-shaped or six-sided crystals, but muscovite commonly occurs in flat, sheetlike layers. Amethyst, a type of quartz, has crystals shaped like pyramids.

Sometimes crystals grow so close to each other that the crystal shape is too small to see. If there is room for large crystals to grow, the crystal shape can be used to help identify the mineral. 🗝

✓ Visual Check

7. Contrast Highlight the surfaces formed by the breaks in the minerals. How does cleavage differ from fracture?

_____

_____

_____

ACADEMIC VOCABULARY

exhibit
(verb) to display, to present for the public to see

🗝 Key Concept Check

8. Identify What are the common properties used to identify minerals?

_____

_____

_____

_____

Unusual Properties

Some minerals have unusual properties that make them easy to identify. For example, halite tastes salty. Magnetite is magnetic and attracts steel objects. Calcite fizzes when acid touches it. A type of calcite called Iceland spar has a property called double refraction. Images viewed through a crystal of Iceland spar appear doubled.

Quartz crystals can produce an electric current when compressed. This property makes quartz crystals useful in radios, microphones, and watches. Several minerals display the property of fluorescence. Calcite and quartz glow under ultraviolet light. ✓

Minerals in Everyday Life

From the moment you wake in the morning until you fall asleep at night, you use materials made from minerals. For example, table salt contains the mineral halite. Toothpaste contains calcite or silica. Some cosmetics contain mica. Some minerals are valuable because we use them every day. We appreciate others simply for their beauty.

Did you know that beverage cans and car batteries are made from minerals? These items are made of metals. Most metals combine with other elements in the formation of a mineral. For example, aluminum can be removed from the mineral bauxite. The minerals must be processed to remove the metals from them. *Deposits of metallic or non-metallic minerals that can be produced at a profit are called* **ores.**

Some minerals, such as gemstones, are valuable because of their appearance. Gemstones have physical properties that make them valuable. They are usually harder than quartz. Gemstones often have intense colors and brilliant luster. The natural crystals are cut and polished. Emeralds are green gemstones often used in jewelry. ✓

> **✓ Reading Check**
>
> **9. Name** one unusual property of a mineral.
>
> _____
>
> _____
>
> _____

> **✓ Key Concept Check**
>
> **10. Describe** How are minerals used in everyday life?
>
> _____
>
> _____
>
> _____
>
> _____
>
> _____

Mini Glossary

cleavage: a break in a mineral along a smooth, flat surface

crystal structure: an orderly, repeating pattern of atoms in a crystal

crystallization: the process by which atoms form a solid with an orderly, repeating pattern

fracture: a break in a mineral along a rough or irregular surface

luster: the way that a mineral's surface reflects light

mineral: a naturally occurring, inorganic solid that has a crystal structure and a definite chemical composition

ore: a deposit of metallic or non-metallic minerals that can be produced at a profit

streak: the color of a mineral's powder

1. Review the terms and their definitions in the Mini Glossary. Write a sentence that describes two characteristics of a specific mineral discussed in the lesson.

_____

_____

2. Write the five characteristics of a mineral in the diagram below.

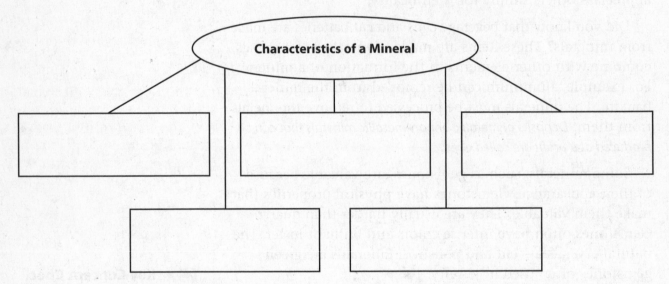

Characteristics of a Mineral

3. Write a sentence to explain why crystals formed from cooling magma vary in size.

_____

_____

What do you think NOW?

Reread the statements at the beginning of the lesson. Fill in the After column with an A if you agree with the statement or a D if you disagree. Did you change your mind?

 Connect ED

Log on to ConnectED.mcgraw-hill.com and access your textbook to find this lesson's resources.

END OF LESSON

Minerals and Rocks

Rocks

Copyright © Glencoe/McGraw-Hill, a division of The McGraw-Hill Companies, Inc.

·········· **Before You Read** ··············

What do you think? Read the two statements below and decide whether you agree or disagree with them. Place an A in the Before column if you agree with the statement or a D if you disagree. After you've read this lesson, reread the statements to see if you have changed your mind.

Before	Statement	After
	3. Once a rock forms, it lasts forever.	
	4. All rocks form when melted rock cools and changes into a solid.	

·············· **Read to Learn** ················

What is a rock?

Sometimes you can tell how an object was made simply by looking at the finished product. If someone serves you eggs for breakfast, you can tell whether they were fried or scrambled by looking at them. In much the same way, a geologist can tell how a rock formed just by looking at it. Two rocks can contain the same minerals. But because the rocks formed differently, they look different.

A **rock** is a naturally occurring solid mixture composed of minerals, smaller rock fragments, organic matter, or glass. The individual particles in rocks are called **grains.** The mineral grains in rocks give clues to understanding how the rocks formed. ✓

Classifying Rocks

Most of Earth's surface is made of rocks. Geologists classify the different kinds of rocks based on the way the rocks form. There are three major types of rocks: igneous, sedimentary, and metamorphic.

Igneous Rocks

There are more igneous rocks on Earth than there are metamorphic or sedimentary rocks. Most igneous rocks form deep below Earth's surface, but some form on Earth's surface. Igneous rocks might form in different places, but they all form in a similar way.

Key Concepts
- What characteristics can be used to classify rocks?
- How do the different types of rocks form?
- What are some uses of rocks in everyday life?

Study Coach

Building Vocabulary Write each vocabulary term in this lesson on an index card. Shuffle the cards. After you have studied the lesson, take turns picking cards with a partner. Each of you should define the terms using your own words.

✓ Reading Check

1. State What can geologists tell from studying mineral grains in rocks?

_____

_____

_____

_____

Formation of Igneous Rocks *Molten rock is called* **magma** *when it is inside Earth. Molten rock that erupts onto Earth's surface is called* **lava.** As magma or lava cools, mineral crystals begin to form. These minerals form the grains of a new igneous rock.

Texture and Composition Geologists classify igneous rocks according to texture and mineral composition. For rocks, **texture** *refers to grain size and how the grains are arranged.*

Lava at Earth's surface cools quickly. As a result, crystals do not have much time to form. The crystals are small, like the crystals in basalt. Geologists describe the texture of igneous rocks with small crystals as fine-grained. Locate the basalt in the figure below.

Deep below Earth's surface, magma cools slowly. The crystals have more time to grow. As a result, the crystals are larger, like the crystals in granite. Geologists describe the texture of igneous rocks with large crystals as coarse-grained. Locate the granite in the figure below.

Igneous rocks such as granite and basalt have different textures. They also have different compositions, or the minerals they contain. Granite contains mostly light-colored minerals such as quartz and potassium feldspar. Basalt is made of dark minerals such as pyroxene (pi RAHK seen) and olivine (AHL ih veen). Granite and basalt are both igneous rocks. However, because of the difference in composition, basalt looks darker than granite.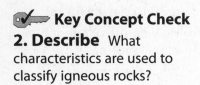

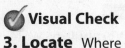

Key Concept Check

2. Describe What characteristics are used to classify igneous rocks?

_____

_____

_____

Visual Check

3. Locate Where does granite form?

_____

_____

_____

Environments of Igneous Rock Formation

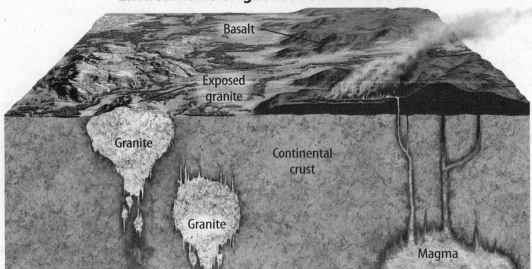

Basalt

Exposed granite

Granite

Continental crust

Granite

Magma

Sedimentary Rocks

Natural processes break down rocks exposed at Earth's surface. *Rock and mineral fragments that are loose or suspended in water are called* **sediment.**

You have read that magma is a source material for igneous rocks. The source material for sedimentary rocks is sediment.

Formation of Sedimentary Rocks Study the figure below. How can sediment become solid rock? **Lithification** *is the process through which sediment turns into rock.*

Usually, sediment is formed through weathering by water, ice, or wind. These agents also remove, or erode, the sediment. Eventually, the sediment is deposited in low areas called basins.

Layers build up as more sediment is deposited in the basins. The weight of this additional sediment compacts, or squeezes together, the lower layers. Dissolved minerals, usually quartz or calcite in water, cement the grains together and form sedimentary rocks such as sandstone and shale.

Dissolved solids also can crystallize directly from a water solution and form sedimentary rocks such as rock salt.

FOLDABLES

Make a folded table to compare types of rocks.

Rocks	Formation	Texture	Composition
Igneous			
Sedimentary			
Metamorphic			

Visual Check

4. Draw Circle the higher areas in the figure. Draw arrows to point out the basins.

Environments of Sedimentary Rock Formation

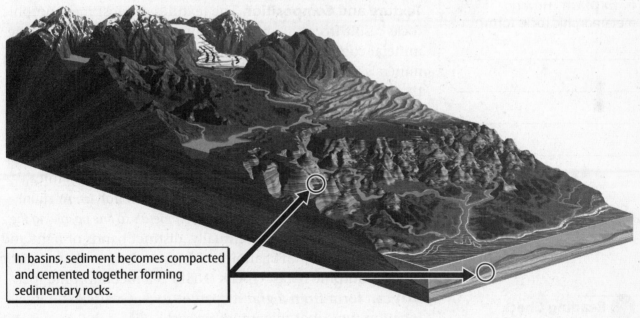

In basins, sediment becomes compacted and cemented together forming sedimentary rocks.

Similar to igneous rocks, sedimentary rocks can be described as fine-grained or course-grained. The shape of the grains also can be described as rounded or angular. Grains usually are angular when first broken but often become rounded during transport. This rounded texture of the grains can help distinguish sedimentary rocks from some igneous rocks. ✓

The composition of a sedimentary rock depends on the minerals in the sediment from which it formed. Sandstone is a sedimentary rock that usually is made of quartz grains. Shale is made from much smaller grains of quartz and clay minerals.

Metamorphic Rocks

Sometimes rocks change into different rocks without erosion or melting. Extreme high temperatures and pressure cause these changes. The original rocks are called parent rocks. The new rocks that form are called metamorphic rocks.

Formation of Metamorphic Rocks Metamorphic rocks form when parent rocks are squeezed, heated, or exposed to hot fluids. The rocks do not melt. They remain solid, but the texture, and sometimes the mineral composition of the parent rock, change. This process is metamorphism. ⌐✓

Texture and Composition The textures of most metamorphic rocks result from increases in temperature and pressure. The mineral composition of metamorphic rocks might result from minerals in the parent rock. Or, the minerals might grow in the new metamorphic rock. You read about gemstones in Lesson 1. Many gemstones are minerals that formed as a result of metamorphism.

Foliated Metamorphic Rocks Recall that crystals form in a variety of shapes. Minerals with flat shapes, such as mica, produce a texture called foliation. **Foliation** (foh lee AY shun) *results when uneven pressures cause flat minerals to line up, giving the rock a layered appearance.* Eventually, distinct bands of light and dark minerals form. Foliation is the most obvious characteristic of metamorphic rocks. Gneiss (NISE) is a metamorphic rock that can form from a granite parent rock. Gneiss exhibits foliation typical of metamorphic rocks. ✓

Nonfoliated Metamorphic Rocks Marble, a metamorphic rock that forms from limestone, does not exhibit foliation. The grains in marble are not flattened like the grains in gneiss. The calcite crystals that make up marble became blocklike and square when exposed to high temperatures and pressure. Marble has a nonfoliated texture.

✓ Reading Check

5. Contrast How are the grains of sedimentary rock different from the grains of igneous rock?

_____

_____

_____

_____

⌐ Key Concept Check

6. Explain How do metamorphic rocks form?

_____

_____

_____

_____

✓ Reading Check

7. Point Out What causes foliation?

_____

_____

_____

_____

Rocks in Everyday Life

Rocks are abundant natural resources. People use them in many ways, based on the physical characteristics of the rocks. Some igneous rocks are hard and durable, such as granite. You might see a fountain constructed from granite. The igneous rock pumice is soft but contains small pieces of hard glass, making it a useful polishing and cleaning product. ✓

Natural layering makes sedimentary rock a high-quality building stone. People use both sandstone and limestone in buildings. Builders use limestone to make cement, which they then use to construct highways and other structures.

Foliated metamorphic rocks split into flat pieces. Slate makes durable, fireproof roofing shingles.

Artists use other metamorphic rocks in their art. Because marble is soft enough to carve, artists often use it to make detailed sculptures. 🗝️

✓ **Reading Check**

8. Name three characteristics of rocks that make them useful to people.

_____

_____

_____

🗝️ **Key Concept Check**

9. Point Out What are some everyday uses for rocks?

_____

_____

_____

Mini Glossary

foliation (foh lee AY shun): the layered appearance of a rock that results when uneven pressures cause flat minerals to line up

grain: an individual particle in a rock

lava: molten rock that erupts onto Earth's surface

lithification: the process through which sediment turns into rock

magma: molten rock when it is inside Earth

rock: a naturally occurring solid mixture composed of minerals, smaller rock fragments, organic matter, or glass

sediment: rock and mineral fragments that are loose or suspended in water

texture: grain size and how the grains are arranged

1. Review the terms and their definitions in the Mini Glossary. Write a sentence to explain how sedimentary rock differs from igneous rock.

_____

_____

2. The boxes on the left describe different ways that rocks form. Write the type of rock that forms in each box on the right.

Parent rocks are squeezed, heated, or exposed to hot fluids.	type of rock:
Rock and mineral fragments are deposited in basins, compacted, and cemented by dissolved minerals.	type of rock:
Magma or lava cools and crystallizes.	type of rock:

3. Identify a term that you had difficulty defining as you worked with your index cards. Write the term and its definition in your own words.

_____

_____

What do you think NOW?

Reread the statements at the beginning of the lesson. Fill in the After column with an A if you agree with the statement or a D if you disagree. Did you change your mind?

Connect ED

Log on to ConnectED.mcgraw-hill.com and access your textbook to find this lesson's resources.

END OF LESSON

Minerals and Rocks

The Rock Cycle

Copyright © Glencoe/McGraw-Hill, a division of The McGraw-Hill Companies, Inc.

················ **Before You Read** ···············

What do you think? Read the two statements below and decide whether you agree or disagree with them. Place an A in the Before column if you agree with the statement or a D if you disagree. After you've read this lesson, reread the statements to see if you have changed your mind.

Before	Statement	After
	5. All rock types are related through the rock cycle.	
	6. Rocks move at a slow and constant rate through the rock cycle.	

Key Concepts

- How do surface processes contribute to the rock cycle?
- How is the rock cycle related to plate tectonics?

················ **Read to Learn** ···············

What is the rock cycle?

Do you have a recycling program at school? Or do you recycle at home? When materials such as paper or metal are recycled they are used over again, but not always for the same things. The metal from the beverage can you recycled yesterday might end up in a baseball bat.

Recycling also occurs naturally on Earth. The rock material that formed Earth 4.6 billion years ago is still here. However, much of it has changed many times throughout Earth's history. *The series of processes that continually change one rock type into another is called the* **rock cycle.**

As materials move through the rock cycle, they can take the form of igneous rocks, sedimentary rocks, or metamorphic rocks. At times, the material might not be rock at all. It might be sediment, magma, or lava. As Earth materials move through the rock cycle, both their form and their location on Earth change. ✔

Processes of the Rock Cycle

Mineral and rock formation are important processes in the rock cycle. The rock cycle is continuous. It has no beginning or end. Some processes take place on Earth's surface. Others take place deep beneath Earth's surface.

◀ **Study Coach**

Create a Quiz Write a quiz question for each paragraph. Answer the question with information from the paragraph. Then work with a partner to quiz each other.

✔ **Reading Check**

1. Compare How is the rock cycle similar to recycling?

_____

_____

_____

_____

Visual Check

2. Describe a path through the rock cycle that would result in the formation of a metamorphic rock.

_____

_____

_____

_____

_____

Cooling and Crystallization

The figure above shows the processes in the rock cycle. Rocks continually transform from one type to another as they move slowly through the cycle. Melted rock is present both on and below Earth's surface. _When lava erupts and cools and crystallizes on Earth's surface, the igneous rock that forms is called_ **extrusive rock.** _When magma cools and crystallizes inside Earth, the igneous rock that forms is called_ **intrusive rock.**

Uplift

If intrusive rocks form deep within Earth, how are they ever exposed at the surface? **Uplift** _is the process that moves large amounts of rock up to Earth's surface and to higher elevations._ Earth's tectonic activity drives uplift. Uplift is often part of mountain building.

Weathering and Erosion

Uplift brings rocks to Earth's surface. There they are exposed to the environment. Glaciers, wind, and rain, along with the activities of some organisms, break down exposed rocks. The same glaciers, wind, and rain also carry sediment to low-lying areas, called basins, by the process of erosion.

Deposition

Eventually, glaciers, wind, and water slow down enough that they can no longer transport the sediment. *The process of laying down sediment in a new location is called* **deposition.** Deposition forms layers of sediment. As time passes, more and more layers are deposited. 🗝️

Compaction and Cementation

The weight of top layers of sediment pushes the grains of the bottom layers closer together. This process is called compaction. Sedimentary rocks have tiny spaces, called pores, between the grains. Pores sometimes contain water and dissolved minerals. When these minerals crystallize, they cement the grains together. The rock cycle figure on the previous page shows the path of sediment from weathering and erosion to compaction and cementation.

Temperature and Pressure

Rocks subjected to high temperatures and pressure far below Earth's surface undergo metamorphism. For example, as temperature and pressure increase, the sedimentary rock called shale changes to the metamorphic rock called slate. As temperature and pressure continue to increase, slate changes to phyllite, then to schist, and finally to gneiss. If the temperature is high enough, the rock melts and becomes magma. Igneous rocks form as the magma cools, and the material continues through the rock cycle.

Rocks and Plate Tectonics

The theory of plate tectonics states that Earth's surface is broken into rigid plates. The plates move as a result of Earth's internal thermal energy and convection in the mantle. The theory explains the movement of continents. It also explains earthquakes, volcanoes, and the formation of new crust. These events occur at plate boundaries where plates interact.

Igneous rock forms where volcanoes occur and where plates move apart. Where plates collide, rocks are subjected to intense pressure and can undergo metamorphism. Colliding plates can also uplift rock or push rock deep below Earth's surface, where it melts and forms magma. At Earth's surface, uplifted rocks are exposed and weathered. Weathered rock forms sediment, which eventually can form sedimentary rock.

Processes within Earth that move tectonic plates also drive part of the rock cycle. The rock cycle also includes surface processes. As long as these processes exist, the rock cycle will continue. 🗝️

🗝️ Key Concept Check

3. Explain How are surface processes involved in the rock cycle?

_____

_____

_____

_____

FOLDABLES

Make a two-column chart book to organize your notes on rock formation.

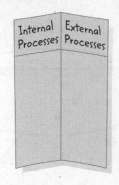

🗝️ Key Concept Check

4. Analyze How is the rock cycle related to plate tectonics?

_____

_____

_____

_____

Mini Glossary

deposition: the process of laying down sediment in a new location

extrusive rock: igneous rock that forms when lava erupts, cools, and crystallizes on Earth's surface

intrusive rock: igneous rock that forms when magma cools and crystallizes inside Earth

rock cycle: the series of processes that continually change one rock type into another

uplift: the process that moves large amounts of rock up to Earth's surface and to higher elevations

1. Review the terms and their definitions in the Mini Glossary. Write a sentence that explains the importance of uplift in the rock cycle.

_____

_____

2. Write the following processes of the rock cycle in the diagram in the order they might occur: compaction and cementation, deposition, uplift, sedimentary rock.

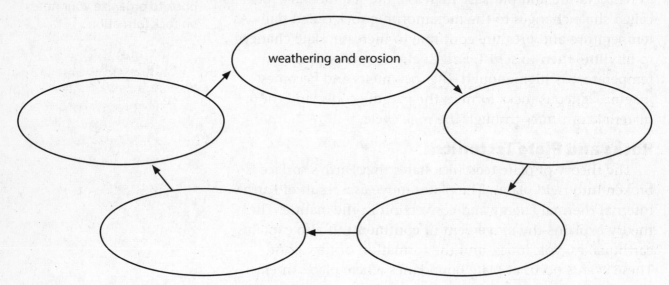

weathering and erosion

3. Explain how deposition in a basin leads to the formation of sedimentary rock.

_____

_____

What do you think NOW?

Reread the statements at the beginning of the lesson. Fill in the After column with an A if you agree with the statement or a D if you disagree. Did you change your mind?

 Connect ED

Log on to ConnectED.mcgraw-hill.com and access your textbook to find this lesson's resources.

END OF LESSON

Plate Tectonics

The Continental Drift Hypothesis

···········**Before You Read**···········

Before	Statement	After
	What do you think? Read the two statements below and decide whether you agree or disagree with them. Place an A in the Before column if you agree with the statement or a D if you disagree. After you've read this lesson, reread the statements to see if you have changed your mind.	

Before	Statement	After
	1. India has always been north of the equator.	
	2. All the continents once formed one supercontinent.	

···········**Read to Learn**···········

Pangaea

Nearly 100 years ago, a scientist named Alfred Wegener (VAY guh nuhr) began an investigation. He wanted to know if Earth's continents had always been in the same place, or if they had moved. Wegener proposed that *all the continents were once part of a supercontinent called* **Pangaea** (pan JEE uh). Over time, Pangaea broke apart, and the continents slowly drifted to their present locations. Wegener proposed the hypothesis of continental drift. *The* **continental drift** *hypothesis suggested that continents are in constant motion on the surface of Earth.*

Wegener looked at the coastlines of continents that are now separated by oceans. He saw similarities in their shapes. For instance, Africa and South America seemed to fit together like the pieces of a puzzle, as shown below.

Continental shelf

Key Concepts

- What evidence supports continental drift?
- Why did scientists question the continental drift hypothesis?

Mark the Text

Identify the Main Ideas
Highlight two or three phrases in each paragraph that summarize the information presented. After you have finished the lesson, review the highlighted text.

FOLDABLES

Make a half-book. Use it to organize your notes on the continental drift hypothesis.

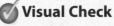
Evidence for the Continental Drift Hypothesis

✓ Visual Check

1. Identify With a pen or pencil, trace the area where Africa and South America match up.

Copyright © Glencoe/McGraw-Hill, a division of The McGraw-Hill Companies, Inc.

Evidence That Continents Move

Wegener knew that he needed evidence to support his hypothesis of continental drift. The most obvious evidence was how the continents fit together like pieces of a puzzle. But other scientists were doubtful of his hypothesis. Wegener needed more evidence.

Fossil Clues

There are many animals and plants that live only on one continent. For example, lions live in Africa but not in South America. Because oceans separate the continents, animals cannot travel from one continent to another by natural means. However, <u>fossils</u> of similar organisms have been discovered on several continents that are now separated by oceans.

Fossils of a plant called *Glossopteris* (glahs AHP tur us) have been discovered in rocks from South America, Africa, India, Antarctica, and Australia. Today these continents are far apart and separated by oceans. The plant's seeds could not have traveled across the oceans.

The figure below shows how some of the continents were joined as part of Pangaea 250 million years ago. The lighter area on the map shows where *Glossopteris* fossils have been found. Notice that the plant once grew in parts of five continents—South America, Africa, India, Antarctica, and Australia. Because these plants grew in a swampy environment, this region, including Antarctica, was different from how it is today. Most of Antarctica is covered in ice sheets. No swampy environments are found there now.

Copyright © Glencoe/McGraw-Hill, a division of The McGraw-Hill Companies, Inc.

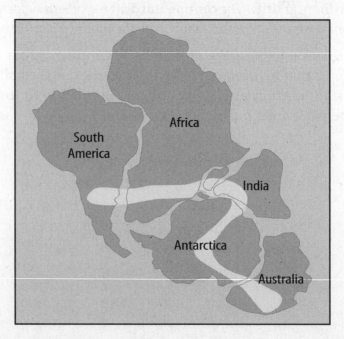

REVIEW VOCABULARY

fossil

the naturally preserved remains, imprints, or traces of organisms that lived long ago

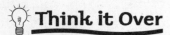

Think it Over

2. Specify Where did *Glossopteris* probably grow?

_____

_____

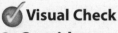

Visual Check

3. Consider Which of the continents would not support *Glossopteris* growth today?

_____

_____

Climate Clues

Other fossil evidence supported continental drift. Coal beds are in Antarctica, a polar climate today. Yet coal formed from fossilized plants that lived long ago in warm, wet climates. This meant that Antarctica must have been warmer and wetter when these plants were alive. Is it possible that Antarctica was at one time closer to the equator? Did Antarctica move to a colder climate near the South Pole?

Another climate clue used by Wegener to support continental drift came from glaciers. When Wegener pieced Pangaea together, he proposed that South America, Africa, India, and Australia were located closer to the South Pole 250 million years ago. He suggested that a large ice sheet covered much of the continents, as shown below. When the ice sheet melted as Pangaea spread apart, it left rock and sediment behind. Wegener studied the similarities of these sediments. ✓

Wegener also studied glacial grooves. Glacial grooves are deep scratches in rocks made as ice sheets move across the land. Wegener found glacial grooves on many different continents. By studying these grooves, he was able to determine the direction that the ice sheet moved across the joined continents.

✓ **Reading Check**

4. State Why did Wegener suggest that continents in the southern hemisphere had a colder climate long ago?

_____

_____

_____

Present

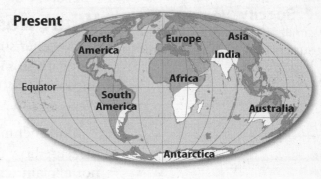

Past

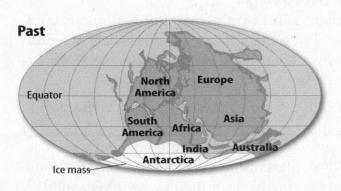

Rock Clues

Some of the evidence used by Wegener to support his idea of continental drift came from rock formations on different continents. The rock formations and mountain ranges seemed to have formed in the same way at the same time. Today geologists know that there were large-scale volcanic eruptions on the western coast of Africa and on the eastern coast of South America hundreds of millions of years ago. Geologists have studied rocks from these eruptions. They found that the volcanic rocks from both continents were identical in chemistry and age.

✓ **Visual Check**

5. Name the areas on Earth where there is evidence of ancient glaciers.

_____

_____

_____

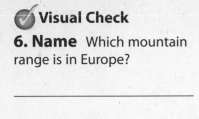
6. Name Which mountain range is in Europe?

_____

_____

Mountain Chains More evidence came from the rocks that make up two mountain chains in Europe and North America. Locate the caledonian mountain range and the Appalachian Mountains in the figure below on the left. The caledonian mountain range is in northern Europe, and the Appalachian Mountains are in eastern North America.

Mountain Ranges

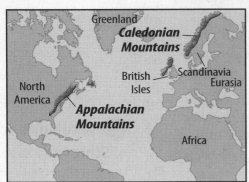
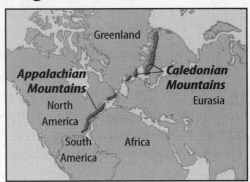

Rock Types Rocks in these two mountain chains are similar in age and structure. Both are also composed of the same rock types. If you could place North America and Europe next to each other, these mountain chains would meet. They would form one long, continuous mountain belt, shown in the figure above on the right.

What was missing?

Wegener supported his continental drift hypothesis until his death in 1930. Wegener's ideas were not widely accepted until nearly 40 years later. Why were scientists skeptical of Wegener's hypothesis?

Continental drift is a slow process. Wegener could not measure how fast the continents moved. Wegener also could not explain what forces caused the continents to move. The mantle under the continents and seafloor was made of solid rock. How could continents push their way through solid rock? Wegener needed more scientific evidence to prove his hypothesis.

The evidence for drifting continents was hidden on the seafloor. During Wegener's lifetime, scientists did not have the tools to determine what happened beneath the oceans. Wegener also could not have known what the seafloor looked like. The evidence needed to prove continental drift was not discovered until long after Wegener's death.

Key Concept Check

7. Specify How were similar rock types used to support the continental drift hypothesis?

_____

_____

_____

Key Concept Check

8. Explain Why did scientists argue against Wegener's continental drift hypothesis?

_____

_____

_____

Mini Glossary

continental drift: a hypothesis suggesting that continents are in constant motion on the surface of Earth

Pangaea (pan JEE uh): one supercontinent that all the continents were once part of

1. Review the terms and their definitions in the Mini Glossary. Write two sentences that explain how Pangaea and continental drift are related.

_____

_____

2. Draw three types of evidence on the map that support Wegener's hypothesis of continental drift. Highlight each drawing with a different-colored marker. Make a key next to the map that shows what each color means.

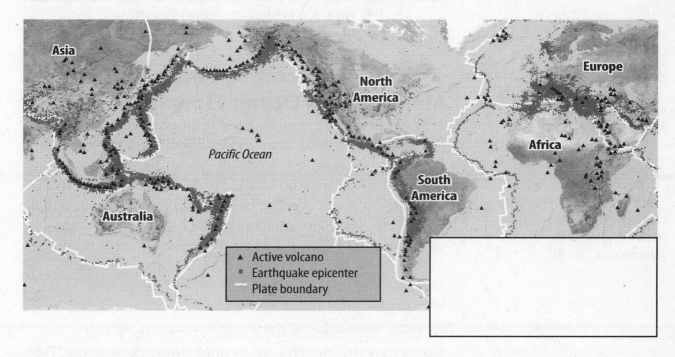

▲ Active volcano
● Earthquake epicenter
— Plate boundary

3. How did highlighting one or two phrases in each paragraph help you understand the hypothesis of continental drift?

_____

What do you think NOW?

Reread the statements at the beginning of the lesson. Fill in the After column with an A if you agree with the statement or a D if you disagree. Did you change your mind?

Connect ED

Log on to ConnectED.mcgraw-hill.com and access your textbook to find this lesson's resources.

END OF LESSON

Plate Tectonics

Development of a Theory

Copyright © Glencoe/McGraw-Hill, a division of The McGraw-Hill Companies, Inc.

•••••••••••••Before You Read•••••••••••••

Before	Statement	After
	3. The seafloor is flat.	
	4. Volcanic activity occurs only on the seafloor.	

What do you think? Read the two statements below and decide whether you agree or disagree with them. Place an A in the Before column if you agree with the statement or a D if you disagree. After you've read this lesson, reread the statements to see if you have changed your mind.

Key Concepts
- What is seafloor spreading?
- What evidence is used to support seafloor spreading?

Study Coach

Two-Column Notes As you read, organize your notes in two columns. In the left column, write the main idea of each paragraph. In the right column, write details that support each main idea. Review your notes to help you remember the details of the lesson.

••••••••••••••Read to Learn••••••••••••••

Mapping the Ocean Floor

Scientists began exploring the seafloor in greater detail during the late 1940s. They used a device called an echo sounder to measure the depths of the ocean floor. An echo sounder produces sound waves that travel from a ship to the seafloor. The waves echo, or bounce, off the seafloor and back to the ship. The echo sounder records the time it takes the echo to return. When the ocean is deeper, the time it takes for the sound waves to bounce back is longer. Scientists calculated ocean depths and used these data to create topographic maps of the seafloor.

These new topographic maps showed large mountain ranges that stretched for many miles along the seafloor. *The mountain ranges in the middle of the oceans are called* **mid-ocean ridges.** Mid-ocean ridges, shown in the figure below, are much longer than any mountain range on land.

✓ Visual Check

1. Identify Circle the area on the map that shows the mid-ocean ridge.

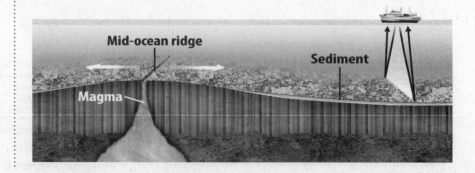

Seafloor Spreading

By the 1960s, scientists had discovered a new process to help explain continental drift. This process is called seafloor spreading. **Seafloor spreading** *is the process by which new oceanic crust forms along a mid-ocean ridge and older oceanic crust moves away from the ridge.*

When the seafloor spreads, Earth's mantle melts and forms magma. The liquid magma is less dense than the solid mantle. The magma rises through cracks in the crust along the mid-ocean ridge. When magma reaches Earth's surface, it is called lava.

As the lava cools and crystallizes on the seafloor, it forms a type of rock called basalt. Oceanic crust is mostly basalt. Because the lava erupts into water, it cools rapidly. The rapidly cooling lava forms rounded structures called pillow lava.

As the seafloor spreads apart, new crust that is forming pushes the older crust away from the mid-ocean ridge. The mid-ocean ridge, at the center of this formation, is shown below. The closer the crust is to a mid-ocean ridge, the younger the oceanic crust is. Scientists concluded that as the seafloor spreads, the continents must be moving. Seafloor spreading is the mechanism that explains Wegener's hypothesis of continental drift.

FOLDABLES

Make a layered book to record your notes and illustrate seafloor spreading.

Seafloor Spreading

Key Concept Check

2. Identify What is seafloor spreading?

_____

_____

Visual Check

3. Interpret Propose a pattern that exists in rocks on either side of the mid-ocean ridge.

_____

_____

_____

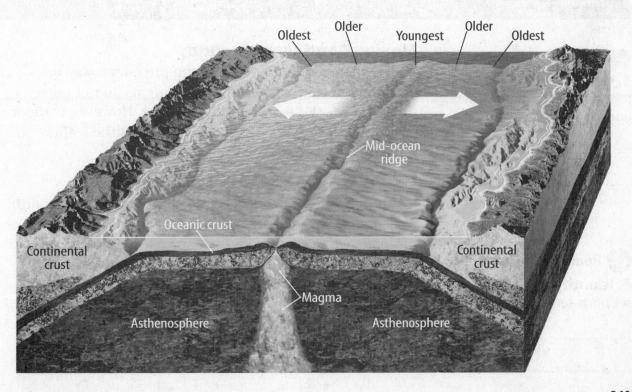

Topography of the Seafloor

What determines the topography of the ocean floor? One factor is seafloor spreading. The rugged mountains that make up the mid-ocean ridge system can form in two different ways. Some form as large amounts of lava erupt from the center of the ridge. That lava cools and builds up around the ridge. Others form as the lava cools and forms new crust that cracks. The rocks move up or down along these cracks and form jagged mountains. ✓

Sediment also determines the topography of the ocean floor. Close to a mid-ocean ridge, the crust is young, and there is not much sediment. However, farther from the ridge, sediment becomes thick enough to make the seafloor smooth. This deep, smooth part of the ocean floor, shown below, is called the abyssal (uh BIH sul) plain.

Mid-ocean ridge

Abyssal plain

Sediment

Oceanic crust

Continental crust

Magma

Moving Continents Around

The theory of seafloor spreading provides a way to explain how continents move. Continents do not move through the solid mantle or the seafloor. However, seafloor spreading suggests that continents move as the seafloor spreads along a mid-ocean ridge.

Development of a Theory

Just as evidence was needed to support continental drift, evidence was needed to support seafloor spreading. Some of the evidence to support seafloor spreading came from rocks on the ocean floor that were not covered with sediment. Scientists studied the magnetic signatures of minerals in these rocks. They discovered two important things. First, Earth's magnetic field changes. Second, these changes appear in rocks that make up the ocean floor. ✓

Reading Essentials

Magnetic Reversals

Earth's iron-rich, liquid outer core is like a giant magnet that creates Earth's magnetic field. The direction of this magnetic field is not always the same. Today's magnetic field is described as having normal polarity. **Normal polarity** *is a state in which magnetized objects, such as compass needles, will orient themselves to point north.*

Sometimes a **magnetic reversal** *occurs and the magnetic field reverses direction.* The opposite of normal polarity is reversed polarity. **Reversed polarity** *is a state in which magnetized objects reverse direction and orient themselves to point south.*

Magnetic reversals have occurred hundreds of times in Earth's past. They occur every few hundred thousand to every few million years.

Rocks Reveal Magnetic Signature

Ocean crust contains large amounts of basalt. Basalt contains iron-rich minerals that are magnetic. Each mineral acts like a small magnet. The figure below shows how magnetic minerals align themselves with Earth's magnetic field. When lava erupts along a mid-ocean ridge, it cools, crystallizes, and permanently records the direction of Earth's magnetic field at the time of the eruption. Scientists have discovered parallel patterns in the magnetic signature of rocks on either side of mid-ocean ridges. For example, in the figure below, notice the normal pattern exists closest to either side of the mid-ocean ridge. Likewise, the reversed polarity pattern exists at about the same distance on either side of the mid-ocean ridge.

Reading Check

7. Identify Does Earth's magnetic field currently have normal or reversed polarity?

_____

_____

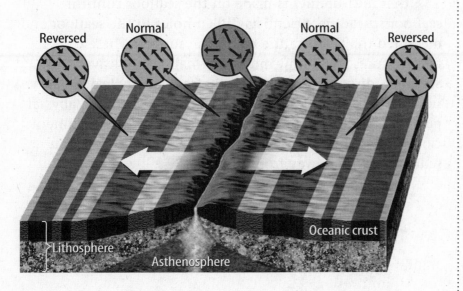

Visual Check

8. Describe the pattern in the magnetic stripes shown in the image to the left.

_____

_____

Evidence to Support the Theory

To support the theory of seafloor spreading, scientists collected data about the magnetic minerals in rocks from the seafloor. They used a magnetometer (mag nuh TAH muh tur) to measure and record the magnetic signature of these rocks. The data collected showed parallel magnetic stripes on either side of the mid-ocean ridge, as shown below. What do these stripes mean?

Each pair of magnetic stripes is similar in composition, age, and magnetic character. Each stripe also records whether Earth's magnetic field was in a period of normal or reversed polarity when the crust formed. Notice that the stripes on either side of the ridge are the same. This pattern supports the idea that ocean crust forms along mid-ocean ridges and is carried away from the center of the ridges. ✓

Reading Check

9. Discuss How do magnetic minerals help support the theory of seafloor spreading?

_____

_____

_____

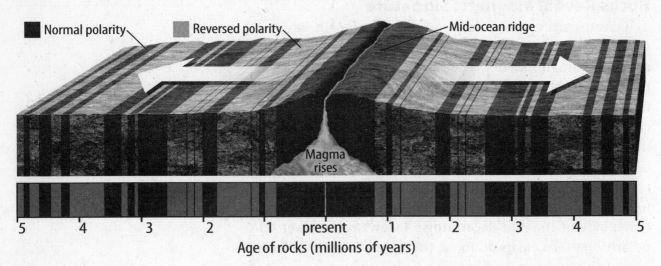

■ Normal polarity　　■ Reversed polarity　　Mid-ocean ridge

Magma rises

Age of rocks (millions of years)
5　4　3　2　1　present　1　2　3　4　5

Visual Check

10. Determine What was the polarity of Earth's magnetic field 4 million years ago?

_____

_____

Reading Check

11. Locate Where does more thermal energy leave Earth—near mid-ocean ridges or beneath abyssal plains?

_____

_____

Other measurements made on the seafloor confirm seafloor spreading. Scientists drilled holes in the seafloor and measured the temperature below the surface. These temperatures show how much thermal energy leaves Earth. Scientists discovered that more thermal energy leaves Earth near mid-ocean ridges than is released from beneath abyssal plains. In addition, studies of sediment show that sediment closest to a mid-ocean ridge is younger and thinner than sediment farther away from the ridge. ✓

Mini Glossary

magnetic reversal: when a magnetic field reverses direction

mid-ocean ridge: a mountain range in the middle of the ocean

normal polarity: a state in which magnetized objects, such as compass needles, will orient themselves to point north

reversed polarity: a state in which magnetized objects reverse direction and orient themselves to point south

seafloor spreading: the process by which new oceanic crust forms along a mid-ocean ridge and older oceanic crust moves away from the ridge

1. Review the terms and their definitions in the Mini Glossary. Write one or two original sentences to explain seafloor spreading.

_____

_____

2. Use words from the word bank to complete the events chain. Use each word only once.

crust   lava   mid-ocean ridge   new   oceanic   old   vents

Magma rises through cracks in the _____ along a _____.

⬇

_____ erupts from volcanic _____ in the ridge.

⬇

Lava cools and hardens to form new _____ crust.

⬇

_____ ocean crust pushes _____ crust away from the ridge.

3. What is the difference between normal polarity and reversed polarity?

_____

_____

What do you think NOW?

Reread the statements at the beginning of the lesson. Fill in the After column with an A if you agree with the statement or a D if you disagree. Did you change your mind?

Log on to ConnectED.mcgraw-hill.com and access your textbook to find this lesson's resources.

END OF LESSON

Plate Tectonics

The Theory of Plate Tectonics

Key Concepts 🔑

- What is the theory of plate tectonics?
- What are the three types of plate boundaries?
- Why do tectonic plates move?

Study Coach

Make an Outline Use the main heads in this lesson as the main points of your outline. Complete the outline with details found in the lesson. Study the lesson by reviewing your outline.

🔑 **Key Concept Check**

1. State What is plate tectonics?

_____

_____

_____

·············· **Before You Read** ··············

What do you think? Read the two statements below and decide whether you agree or disagree with them. Place an A in the Before column if you agree with the statement or a D if you disagree. After you've read this lesson, reread the statements to see if you have changed your mind.

Before	Statement	After
	5. Continents drift across a molten mantle.	
	6. Mountain ranges can form when continents collide.	

·············· **Read to Learn** ··············

The Plate Tectonics Theory

When you blow into a balloon, the balloon expands. Its surface area also increases. As more air is added to the balloon, the balloon gets larger. Similarly, if ocean crust continually forms at mid-ocean ridges and is never destroyed, Earth's surface should be expanding. But measurements of the planet show that Earth is not getting larger. How can this be explained?

Geologists proposed a more complete theory in the late 1960s. It was called plate tectonics theory. The theory of **plate tectonics** states that *Earth's surface is made of rigid slabs of rock, or plates, that move with respect to each other*, or in relation to each other. This new theory suggested that Earth's surface, the lithosphere, is divided into large pieces of rock. These pieces are called plates. Each plate moves slowly over Earth's hot and semiplastic mantle. 🔑

The word *tectonic* describes the forces that shape Earth's surface and the rock structures that form as a result. Plate tectonics explains why earthquakes occur and volcanoes erupt. When plates separate on the seafloor, earthquakes result and a mid-ocean ridge forms. When plates come together, one plate can move under the other. This causes earthquakes and creates a chain of volcanoes. When plates slide past each other, earthquakes can result.

Earth's Tectonic Plates

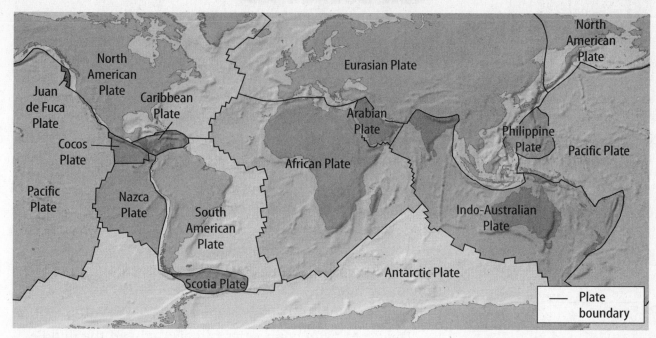

Tectonic Plates

Earth's surface is divided into rigid plates that move relative to one another. Look at the map above. It shows Earth's major plates and their boundaries. Notice how some boundaries are in the middle of the oceans. Many of these boundaries are located at mid-ocean ridges. The Pacific Plate is the largest plate. The Juan de Fuca is one of the smallest plates.

Earth's outermost layers are cold and rigid compared to the layers within Earth's interior. *The cold and rigid outermost rock layer is called the* **lithosphere.** The crust and the solid, uppermost mantle form the lithosphere.

The lithosphere varies in thickness. It is thin below mid-ocean ridges. It is thick below continents. Earth's tectonic plates are large pieces of lithosphere. These plates fit together like the pieces of a giant jigsaw puzzle.

Directly below the lithosphere is a very hot part of the mantle. This layer of Earth is called the asthenosphere (as THEE nuh sfihr). Even though it is solid, the asthenosphere behaves like a <u>plastic</u> material because it is so hot.

The asthenosphere flows below Earth's plates and enables the plates to move. The ways in which the lithosphere and asthenosphere interact help explain plate tectonics. ✓

SCIENCE USE V. COMMON USE ···
plastic
Science Use capable of being molded or changing shape without breaking

Common Use any of numerous organic, synthetic, or processed materials made into objects

✓ **Reading Check**
3. Identify What are Earth's outermost layers called?

_____

_____

Plate Boundaries

Imagine placing two books side by side. Imagine that each book is a tectonic plate. The place where the edges of the books meet represents a plate boundary. How many ways can you move the books along a set of boundaries? You can pull the books away from each other. You can push the books together. You can slide the books past each other. Earth's tectonic plates move in much the same way as you can move these books. ✓

Divergent Plate Boundaries

A **divergent plate boundary** *forms where two tectonic plates separate. Divergent* means "moving apart." Mid-ocean ridges are located along divergent plate boundaries. When the seafloor spreads at a mid-ocean ridge, lava erupts. As the lava cools and hardens, it forms new oceanic crust. As this process continues, the plates move away from each other.

Divergent plate boundaries can also exist in the middle of a continent. At these boundaries, continents pull apart and a rift valley forms. The East African Rift is one example of a continental rift.

Transform Plate Boundaries

The San Andreas Fault in California is a transform plate boundary. *A* **transform plate boundary** *forms where two tectonic plates slide past each other.* As they move past each other, the plates might get stuck and stop moving. Stress builds up where the plates are stuck. When this stress is too great, the rocks break and suddenly move apart. The result is a rapid release of energy in the form of an earthquake.

Convergent Plate Boundaries

A **convergent plate boundary** *forms where two plates collide. The denser plate sinks below the more buoyant plate in a process* called **subduction.** A subduction zone is the area where a denser plate descends into Earth along a convergent plate boundary. The two types of convergent plate boundaries are ocean-to-continent and continent-to-continent. ✓

Ocean-to-Continent Boundary When a dense oceanic plate and a less-dense continental plate collide, the oceanic plate subducts, or sinks, under the edge of the continental plate. This creates a deep ocean trench and a line of volcanoes forms on the edge of the continent. This process can also occur when two oceanic plates collide. An older and denser oceanic plate will subduct beneath a younger oceanic plate. A deep ocean trench forms, along with a line of volcanoes.

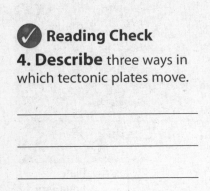

Reading Check

4. Describe three ways in which tectonic plates move.

_____

_____

_____

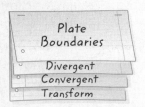

Make a layered book to organize your notes on the three types of plate boundaries.

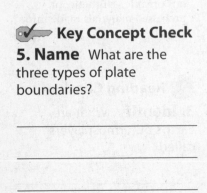

Key Concept Check

5. Name What are the three types of plate boundaries?

_____

_____

_____

Copyright © Glencoe/McGraw-Hill, a division of The McGraw-Hill Companies, Inc.

Continent-to-Continent Boundary Convergent plate boundaries also form when two continental plates collide. When this happens, neither plate is subducted. The less-dense plate folds and deforms, forming mountains such as the Himalayas in India.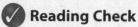

Evidence for Plate Tectonics

When Wegener proposed the continental drift hypothesis, technology was not available to measure how fast the continents moved. Remember that continents move apart or come together at speeds of only a few centimeters per year. This is about the length of a small paper clip.

Today, scientists can measure how fast continents move. A network of satellites orbiting Earth is used to monitor plate motion. By keeping track of the distance between satellites and Earth, it is possible to determine how fast a tectonic plate moves. This network of satellites is called the Global Positioning System (GPS).

The theory of plate tectonics explains why earthquakes and volcanoes are more common in some places than in others. Recall that when plates separate, collide, or slide past each other, stress builds. When this stress suddenly releases, earthquakes can result.

Volcanoes can also form along a mid-ocean ridge or continental rifts. They also form where plates collide along a subduction zone. Mountains can form where two continents collide. The map below shows that most earthquakes and volcanoes occur along tectonic plate boundaries.

 **Reading Check**

6. Identify Along what type of convergent plate boundary did the Himalayas form? (Circle the correct answer.)

a. ocean-to-ocean

b. continent-to-continent

c. ocean-to-continent

Key Concept Check

7. Explain How are earthquakes and volcanoes related to the theory of plate tectonics?

_____

_____

_____

Visual Check

8. Interpret Do earthquakes and volcanoes occur anywhere away from plate boundaries? If so, where?

_____

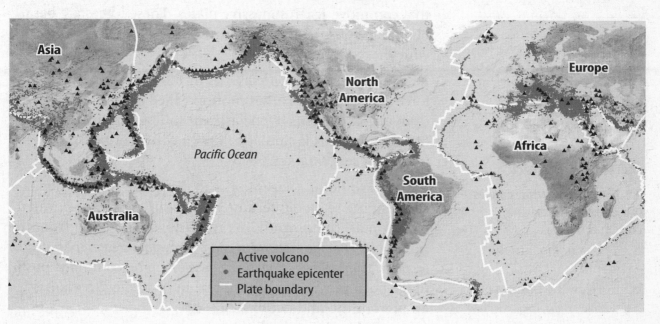

▲ Active volcano
● Earthquake epicenter
— Plate boundary

Plate Motion

You learned that the main objection to Wegener's continental drift hypothesis was that he could not explain how or why Earth's continents move. Scientists now understand that continents move because the asthenosphere moves underneath the rigid lithosphere.

Convection Currents

The circulation of material caused by differences in temperature and density is called **convection.** For example, the upstairs floors of homes are often warmer because hot air rises. Hot air is less dense than cold air. As the cold air sinks, the hot air rises. ✅

Convection in the mantle is related to plate tectonic activity. The warmth for convection comes from radioactive elements inside Earth, such as uranium, thorium, and potassium. When materials such as solid rock are heated, they expand and become less dense. Heated mantle material rises and comes in contact with Earth's crust. Thermal energy is transferred from hot mantle material to the colder surface above. As the mantle cools, it becomes denser and sinks, forming a convection current. These currents in the asthenosphere act like a conveyor belt moving the lithosphere above it. Therefore, tectonic plates move in response to the heating and cooling of mantle material. ⬤✓➡

Forces Causing Plate Motion

How can something as large as the Pacific Plate move? Convection currents in the mantle produce enormous forces that can move Earth's massive plates. These forces are basal drag, ridge push, and slab pull. Scientists' opinions differ on which force is strongest.

Basal Drag Convection currents in the mantle produce a force on plates that causes motion called basal drag. Convection currents in the asthenosphere can drag the lithosphere. This is similar to how a conveyor belt moves items at a supermarket.

Ridge Push Recall that mid-ocean ridges are higher than the surrounding seafloor. Because mid-ocean ridges are elevated, gravity pulls the surrounding rocks down and away from the ridge. *Rising mantle material at mid-ocean ridges creates the potential for plates to move away from the ridge with a force called* **ridge push.** Ridge push moves the lithosphere in opposite directions away from the mid-ocean ridge. ✅

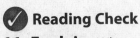

Reading Check

9. Explain What causes convection?

_____

_____

_____

🔑 Key Concept Check

10. Cause and Effect Why do tectonic plates move?

_____

_____

_____

_____

✅ Reading Check

11. Explain What causes ridge push? (Circle the correct answer.)

a. a plate going into the mantle

b. force on the bottom of a plate

c. movement along a mid-ocean ridge

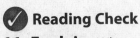

Copyright © Glencoe/McGraw-Hill, a division of The McGraw-Hill Companies, Inc.

Slab Pull You learned that when tectonic plates collide, the denser plate will sink into the mantle along a subduction zone. This sinking plate is called a slab. Because the slab is old and cold, it is denser than the surrounding mantle. Therefore, it sinks. *As a slab sinks, it pulls on the rest of the plate with a force called* **slab pull.** This is similar to pushing a tablecloth over the edge of a table. When enough of the cloth slides over the edge, it will pull the rest of the cloth off the table.

A Theory in Progress

Plate tectonics is often said to be the unifying theory in geology. It explains the connection between continental drift and the formation and destruction of crust along plate boundaries. It also helps explain why earthquakes and volcanoes occur and why mountains form.

The investigation that Wegener began nearly a century ago is still being updated. Several questions remain.

- Why is Earth the only planet in the solar system that has plate tectonic activity? No other planet in our solar system is known to have active tectonic plates.

- Why do some earthquakes and volcanoes occur far from plate boundaries? Perhaps it is because plates are not perfectly rigid. Different thicknesses and weaknesses exist within plates. Also, the mantle is much more active than scientists originally understood.

- What forces actually dominate plate motions? Currently accepted models suggest that convection currents occur in the mantle. However, there is no way to measure or observe them.

- How will scientists answer these questions? One topic of interest is creating 3-D images of seismic wave velocities in a subduction zone. This technology is called anisotropy. It might help scientists better understand the processes that occur within the mantle and along plate boundaries.

Math Skills

The plates along the Mid-Atlantic Ridge spread at an average rate of 2.5 cm/y. How long will it take the plates to spread 1 m? Use proportions to find the answer.

a. Convert the distance to the same unit.

$$1 \text{ m} = 100 \text{ cm}$$

b. Set up a proportion:

$$\frac{2.5 \text{ cm}}{1 \text{ y}} = \frac{100 \text{ cm}}{x \text{ y}}$$

c. Cross-multiply and solve for x as follows:

$$2.5 \text{ cm} \times x\text{y} = 100 \text{ cm} \times 1\text{y}$$

d. Divide both sides by 2.5 cm.

$$x\text{ y} = \frac{100 \text{ cm y}}{2.5 \text{ cm}}$$

$$x = 40 \text{ y}$$

12. Use Proportions The Eurasian Plate travels the slowest, at about 0.7 cm/y. How long would it take the plate to travel 3 m?

_____

Reading Check

13. Explain Why does the theory of plate tectonics continue to change?

_____

_____

Mini Glossary

convection: the circulation of material caused by differences in temperature and density

convergent plate boundary: forms where two plates collide

divergent plate boundary: forms where two tectonic plates separate

lithosphere: the cold and rigid outermost rock layer of Earth

plate tectonics: the theory that states that Earth's surface is made of rigid slabs of rocks, or plates, that move with respect to each other

ridge push: when the rising mantle material at mid-ocean ridges creates the potential for plates to move away from the ridge with a force

slab pull: when a slab sinks and pulls on the rest of the plate with a force

subduction: the process by which the denser plate sinks below the more buoyant plate when two plates collide

transform plate boundary: forms where two tectonic plates slide past each other

1. Review the terms and their definitions in the Mini Glossary. Choose one term and explain what it means in your own words.

_____

_____

2. Use what you have learned about plate tectonics to complete the concept map.

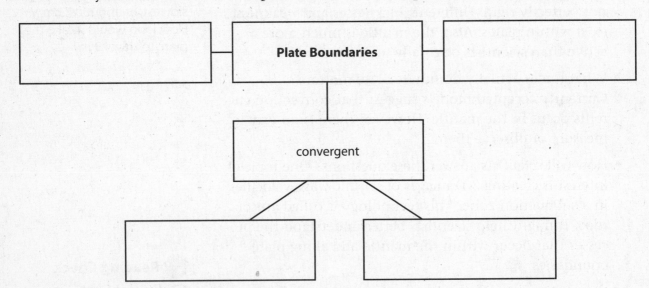

What do you think NOW?

Reread the statements at the beginning of the lesson. Fill in the After column with an A if you agree with the statement or a D if you disagree. Did you change your mind?

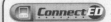

Log on to ConnectED.mcgraw-hill.com and access your textbook to find this lesson's resources.

END OF LESSON

Earthquakes and Volcanoes

Earthquakes

·············· **Before You Read** ··············

What do you think? Read the three statements below and decide whether you agree or disagree with them. Place an A in the Before column if you agree with the statement or a D if you disagree. After you've read this lesson, reread the statements to see if you have changed your mind.

Before	Statement	After
	1. Earth's crust is broken into rigid slabs of rock that move, causing earthquakes and volcanic eruptions.	
	2. Earthquakes create energy waves that travel through Earth.	
	3. All earthquakes occur on plate boundaries.	

·············· **Read to Learn** ··············

What are earthquakes?

Imagine bending a stick until it breaks. When the stick snaps, it vibrates, releasing energy. Earthquakes release energy in a similar way. **Earthquakes** *are the vibrations in the ground that result from movement along breaks in Earth's lithosphere.* These breaks are called faults.

Why do rocks move along a fault? The forces that move tectonic plates also push and pull rocks along a fault. If these forces become large enough, the blocks of rock on either side of the fault can move past each other. The rocks might move vertically—up or down—or horizontally—sideways.

The size of an earthquake depends on the amount of force applied to the fault. The greater the force applied to a fault, the greater the chance of a large and destructive earthquake occurring.

Earthquakes can cause billions of dollars in damage. Injuries and fatalities often occur during earthquakes. Earthquakes are common in the state of California. In 1994, the Northridge earthquake along the San Andreas Fault in California caused $20 billion in damage.

Key Concepts 🔑
- What is an earthquake?
- Where do earthquakes occur?
- How do scientists monitor earthquake activity?

> **Study Coach**

Sticky-Note Questions As you read the lesson, write questions on sticky-notes. Stick each question next to the paragraph that relates to the question. When you finish reading the lesson, discuss your questions with your teacher or with a partner.

🔑 **Key Concept Check**
1. Define What is an earthquake?

_____

_____

_____

Where do earthquakes occur?

Few earthquakes occur in the middle of a continent. Most earthquakes occur in the oceans and along the edges of continents where tectonic plates meet.

Earthquakes and Plate Boundaries

Stress builds up along plate boundaries. Earthquakes result from the buildup and release of this stress along the active plate boundaries. The deepest and strongest earthquakes occur along convergent plate boundaries. At a convergent plate boundary, plates collide. The denser oceanic plate subducts, or drops down, into the mantle. These earthquakes release great amounts of energy.

Shallow earthquakes commonly occur where plates separate along a divergent plate boundary or along a transform plate boundary. Earthquakes occur at varying depths where continents collide. Continental collisions form large, deformed mountain ranges.

Rock Deformation

Force, or pressure, applied along plate boundaries can cause a body of rock to bend and change shape. This is called rock deformation. Over time, the rocks can break and move.

Faults

A **fault** *is a break in Earth's lithosphere where one block of rock moves toward, away from, or past another.* When rocks move in any direction along a fault, an earthquake occurs. The table below describes three types of faults. The forces applied to a fault determine the direction the rocks move.

Copyright © Glencoe/McGraw-Hill, a division of The McGraw-Hill Companies, Inc.

Types of Faults		
Fault Name	**Location**	**Movement**
Strike-slip	transform plate boundaries	Two blocks of rock slide horizontally past each other in opposite directions.
Normal	divergent plate boundaries	Forces pull two blocks of rock apart. One block drops down relative to the other.
Reverse	convergent plate boundaries	Forces push two blocks of rock together. One block moves up relative to the other.

Key Concept Check

2. Specify Where do most earthquakes occur?

_____

_____

_____

Reading Check

3. Define What is a fault?

_____

_____

_____

Interpreting Tables

4. Identify What are the three types of faults?

_____

_____

_____

Earthquake Focus and Epicenter

When rocks move along a fault, they release energy. *Energy that travels as vibrations on and in Earth is called **seismic waves**. Seismic waves originate where rocks first move along the fault, at a location inside Earth called the **focus**.* Earthquakes can occur anywhere between Earth's surface and depths greater than 600 km. In a news report, you might hear a reporter identify the earthquake's epicenter. *The **epicenter** is the location on Earth's surface directly above an earthquake's focus.*

Seismic Waves

During an earthquake, there is a rapid release of energy along a fault. This release of energy produces seismic waves. The waves travel outward in all directions through rock, much like ripples in water. As the waves travel, they transfer energy through the ground and produce the motion associated with an earthquake. The energy released is strongest near the epicenter. As seismic waves move away from the epicenter, their energy and intensity decrease. The greater the distance from an earthquake's epicenter, the less the ground moves.

Types of Seismic Waves

During an earthquake, particles in the ground can move back and forth or up and down. Particles can also move in an elliptical motion parallel to the direction the seismic wave travels. Scientists use wave motion, wave speed, and the type of material the wave travels through to classify seismic waves. The three types of seismic waves are primary waves, secondary waves, and surface waves.

Primary waves, *also called P-waves, cause particles in the ground to move in a push-pull motion, similar to a coiled spring.* P-waves move faster than any other seismic waves. They are the first waves detected and recorded after an earthquake.

Secondary waves, *also called S-waves, cause particles to move up and down at right angles relative to the direction that the wave travels.* These waves move like a coiled spring when it is shaken side-to-side and up and down at the same time.

Surface waves *cause particles in the ground to move up and down in a rolling motion,* similar to ocean waves. Surface waves travel only on Earth's surface. P-waves and S-waves can travel through Earth's interior. However, scientists have discovered that S-waves cannot travel through liquid. ✔

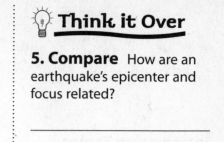

Think it Over

5. Compare How are an earthquake's epicenter and focus related?

_____

_____

_____

FOLDABLES

Make a trifold book to organize your notes about the types of plate movement and resulting activities along these plate boundaries.

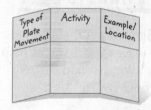

✔ **Reading Check**

6. Describe the three types of seismic waves.

_____

_____

_____

_____

_____

Properties of Seismic Waves	
Seismic Waves	**Description**
Primary waves (P-waves)	• Cause rock particles to vibrate in same direction as waves travel • Fastest seismic waves • First waves that seismometers detect • Travel through solids and liquids
Secondary waves (S-waves)	• Cause rock particles to vibrate perpendicular to direction that waves travel • Slower than P-waves; faster than surface waves • Second waves that seismometers detect • Travel only through solids
Surface waves	• Cause rock particles to move in a rolling or elliptical motion in the same direction that waves travel • Slowest seismic waves • Cause the most damage at Earth's surface

Interpreting Tables

7. Predict what will happen to an S-wave when it reaches Earth's liquid outer core.

_____

_____

💡 **Think it Over**

8. Specify What properties of P-waves and S-waves help seismologists map Earth's interior?

_____

_____

_____

✅ **Reading Check**

9. Explain How did scientists discover that Earth's outer core is liquid?

_____

_____

_____

Mapping Earth's Interior

Scientists who study earthquakes are called **seismologists** (size MAH luh justs). They use properties of seismic waves, as described in the table above, to map Earth's interior. P-waves and S-waves change speed and direction as they travel through different materials. Seismologists measure the speed and direction of waves as they move through Earth at different depths. Using these measurements, seismologists can determine the materials that make up Earth's layers.

Inner and Outer Core Seismologists discovered that S-waves cannot travel through the outer core. This discovery proved that Earth's outer core is liquid, unlike its solid inner core. By analyzing the speed of P-waves, seismologists also discovered that the inner and outer cores are mostly iron and nickel. ✅

The Mantle Seismologists have used seismic waves to model convection currents in the mantle. The speeds of seismic waves depend on the temperature, pressure, and chemistry of the rocks that the waves travel through. Seismic waves tend to travel slower as they move through hot material, such as in areas of the mantle below mid-ocean ridges. Seismic waves travel faster in cooler areas of the mantle near subduction zones.

Locating an Earthquake's Epicenter

An instrument called a **seismometer** *(size MAH muh ter) measures and records ground motion and can be used to determine the distance seismic waves travel. A seismometer records ground motion as a* **seismogram,** *a graphical illustration of seismic waves.*

Seismologists use a method called triangulation to locate an earthquake's epicenter. This method uses the speeds and travel times of seismic waves to determine the distance to the earthquake's epicenter from at least three seismometers at different locations.

1. **Find the arrival time difference.** First, scientists determine the number of seconds between the arrival of the first P-wave and the first S-wave on the seismogram. This time difference is called lag time.

2. **Find the distance to the epicenter.** Next, seismologists plot the lag time against distance on a graph. This reveals the distance of the epicenter from the seismograph's location.

3. **Plot the distance on a map.** Seismologists determine the distance of the epicenter to seismographs in at least three different locations. The map below shows circles around the locations of three seismometer stations. The distance from each station to its circle measures the distance from that station to the earthquake's epicenter. The epicenter must lie somewhere on the circle around each station. Only one point lies on all three circles. The point where the three circles intersect is the epicenter.

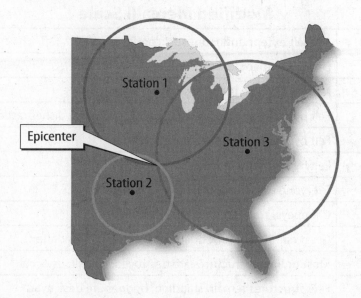

Station 1

Station 3

Epicenter

Station 2

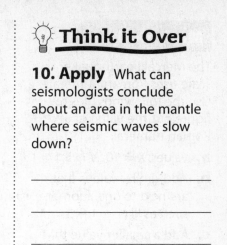
10. Apply What can seismologists conclude about an area in the mantle where seismic waves slow down?

_____

_____

_____

_____

 Visual Check

11. Compare Which station on the map is farthest from the earthquake's epicenter?

_____

The Mercalli earthquake scale uses Roman numerals. Use the following rules to figure out the value of a Roman numeral.

a. Values: X = 10; V = 5; I = 1

b. Add similar values that are next to one another, such as III (1 + 1 + 1 = 3).

c. Add a smaller value that comes after a larger value, such as XV (10 + 5 = 15).

d. Subtract a smaller value that precedes a larger value, such as IX (10 − 1 = 9).

e. Use the fewest possible numerals to express the value (X rather than VV).

12. Use Roman Numerals What is the value of the Roman numeral XVI? XIV?

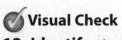

Visual Check

13. Identify At what point on the Modified Mercalli scale of earthquake measurement does it become dangerous to be indoors?

Determining Earthquake Magnitude

The Richter magnitude scale uses the amount of ground motion at a given distance from an earthquake to determine magnitude. Each increase of one unit on the Richter scale represents 10 times the amount of ground motion. For example, a magnitude 8 earthquake produces 10 times greater shaking than a magnitude 7 earthquake does and 100 times greater shaking than a magnitude 6 earthquake does (10 × 10).

The moment magnitude scale measures the total amount of energy released by an earthquake. For each increase of one unit on the scale, an earthquake releases 31.5 times more energy. For example, a magnitude 8 earthquake releases more than 992 times the amount of energy than that of a magnitude 6 earthquake (31.5 × 31.5).

Describing Earthquake Intensity

Another way to measure and describe an earthquake is to examine the amount of damage that results from the shaking. The Modified Mercalli scale measures the intensity of an earthquake based on descriptions of its effects on people and structures. The scale, shown below, ranges from I, an earthquake that people do not feel, to XII, an earthquake that destroys everything. The higher the number is, the greater the effects.

An area's geology also influences earthquake damage. The shaking produces more damage in areas covered by loose sediment than it does in places built on solid bedrock.

Modified Mercalli Scale	
I	Not felt except under unusual conditions.
II	Felt by few people; suspended objects swing.
III	Most noticeable indoors; strong vibrations.
IV	Felt by many people indoors but few outdoors; dishes rattle.
V	Felt by nearly everyone; dishes break.
VI	Felt by all; furniture shifts.
VII	Everyone runs outdoors; some chimneys break.
VIII	Chimneys, smokestacks, and walls fall.
IX	Great damage occurs; buildings shift off of foundations.
X	Most ordinary structures are destroyed; landslides occur.
XI	Few structures remain standing; bridges are destroyed.
XII	Damage is total.

Earthquake Risk

In the United States, the highest risk of earthquakes occurs near tectonic plate boundaries of the western states. The transform plate boundary in California and the convergent plate boundaries in Oregon, Washington, and Alaska have the highest earthquake risks. However, not all earthquakes occur near plate boundaries. Some of the largest earthquakes in the United States have occurred far from plate boundaries.

High-energy, destructive earthquakes are not very common. Only about ten earthquakes with a magnitude greater than 7.0 occur worldwide each year. Earthquakes with magnitudes greater than 9.0, such as the Indian Ocean earthquake in 2004, are rare.

Seismologists evaluate risk in several ways because earthquakes threaten people's lives and property. They study the probability that an earthquake will occur in an area. Seismologists study past earthquake activity, the geology around a fault, the population density, and the building design in an area.

Engineers use these risk assessments to design buildings that can withstand the shaking during an earthquake. City and state government officials use risk assessments to help plan and prepare for future earthquakes. ✔️🗝️

REVIEW VOCABULARY

convergent
tending to move toward one point or approaching each other

🗝️ **Key Concept Check**
14. Explain How do seismologists evaluate risk?

_____

_____

_____

Mini Glossary

earthquake: the vibrations in the ground that result from movement along breaks in Earth's lithosphere

epicenter: the location on Earth's surface directly above an earthquake's focus

fault: a break in Earth's lithosphere where one block of rock moves toward, away from, or past another

focus: a location inside Earth where rocks first move along a fault, starting seismic waves

primary wave: a seismic wave, also called a P-wave, that causes particles in the ground to move in a push-pull motion similar to a coiled spring

secondary wave: a seismic wave, also called an S-wave, that causes particles to move up and down at right angles relative to the direction the wave travels

seismic wave: energy that travels as vibrations on and in Earth

seismogram: a graphical illustration of seismic waves

seismologist (size MAH luh just): a scientist who studies earthquakes

seismometer (size MAH muh ter): an instrument that measures and records ground motion and can be used to determine the distance seismic waves travel

surface wave: a seismic wave that causes particles in the ground to move up and down in a rolling motion

1. Review the terms and their definitions in the Mini Glossary. Write a sentence that describes how primary waves differ from secondary waves.

_____

_____

2. Circle the set of arrows that shows how rock moves at a strike-slip fault.

Strike-Slip Fault

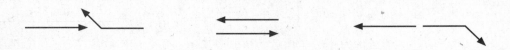

3. What three measurements are used by scientists to locate an earthquake's epicenter?

_____

_____

What do you think NOW?

Reread the statements at the beginning of the lesson. Fill in the After column with an A if you agree with the statement or a D if you disagree. Did you change your mind?

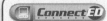 **Connect ED**

Log on to ConnectED.mcgraw-hill.com and access your textbook to find this lesson's resources.

END OF LESSON

Earthquakes and Volcanoes

Volcanoes

·············· **Before You Read** ··············

What do you think? Read the three statements below and decide whether you agree or disagree with them. Place an A in the Before column if you agree with the statement or a D if you disagree. After you've read this lesson, reread the statements to see if you have changed your mind.

Before	Statement	After
	4. Volcanoes can erupt anywhere on Earth.	
	5. Volcanic eruptions are rare.	
	6. Volcanic eruptions only affect people and places close to the volcano.	

·············· **Read to Learn** ··············

What is a volcano?

A **volcano** *is a vent in Earth's crust through which melted—or molten—rock flows. Molten rock below Earth's surface is called* **magma.** Have you heard of some famous volcanoes such as Mount St. Helens, Kilauea, or Mount Pinatubo? All of these volcanoes have erupted within the last 30 years. Volcanoes exist in many places around the world. Some places have more volcanoes than others. ✅

How do volcanoes form?

Volcanic eruptions constantly shape Earth's surface. They can form large mountains, create new crust, and destroy anything in their path. Scientists have learned that the movement of Earth's tectonic plates causes volcanoes to form and to erupt.

Convergent Boundaries

Volcanoes can form along convergent plate boundaries. When two plates collide, the denser plate sinks, or subducts, into the hot mantle. The thermal energy below the surface and fluids driven off the subducting plate melt the mantle and form magma. Magma is less dense than the mantle and rises through cracks in the crust. This forms a volcano. *Molten rock that erupts onto Earth's surface is called* **lava.**

Key Concepts

- How do volcanoes form?
- What factors contribute to the eruption style of a volcano?
- How are volcanoes classified?

Study Coach

Make Flash Cards Think of a quiz question for each paragraph. Write the question on one side of a flash card. Write the answer on the other side. Work with a partner to quiz each other using the flash cards.

✅ **Reading Check**

1. Define What is magma?

_____

_____

Divergent Boundaries

Lava also erupts along divergent plate boundaries. As the plates spread apart, magma rises through the vent or opening between them. More than 60 percent of all eruptions occur at divergent plate boundaries along mid-ocean ridges. There, the lava forms new oceanic crust.

Hot Spots

Not all volcanoes form on or near plate boundaries. *Volcanoes that are not associated with plate boundaries are called* **hot spots.** Geologists hypothesize that hot spots form above a rising current of hot mantle materials, called a plume.

Plumes do not move. As shown in the figure below, a volcano forms as a tectonic plate moves over the plume. As the moving plate carries a volcano away from the hot spot, the volcano becomes dormant, or inactive. As the plate continues to move, a chain of volcanoes forms. The oldest volcano will be the farthest away from the hot spot. The youngest volcano will be directly above the hot spot.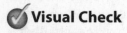

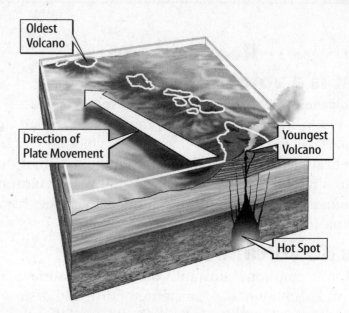

Oldest Volcano

Direction of Plate Movement

Youngest Volcano

Hot Spot

Where do volcanoes form?

The figure on the next page shows the world's active volcanoes. Notice that most volcanoes are close to plate boundaries.

Ring of Fire

In the figure, notice that volcanoes form a ring around most of the Pacific Ocean. Because of its earthquake and volcanic activity, the area surrounding the Pacific Ocean has earned the name Ring of Fire. ✔

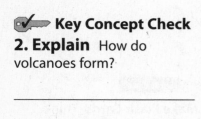

Key Concept Check

2. Explain How do volcanoes form?

_____

_____

_____

✔ **Visual Check**

3. Interpret As a hot spot forms a chain of volcanoes, what moves and what does not move?

_____

_____

_____

✔ **Reading Check**

4. Locate Where is the Ring of Fire?

_____

Volcanoes, Hot Spots, and Plate Boundaries

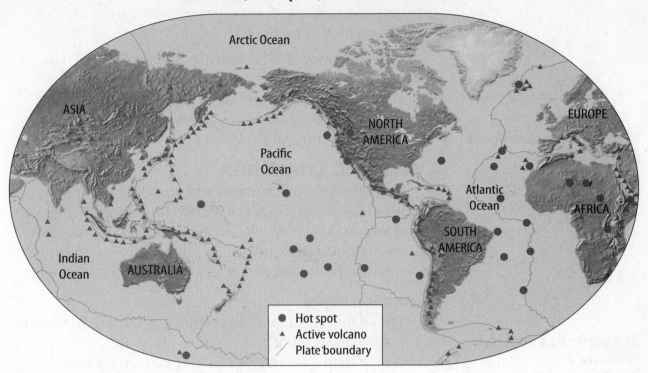

- ● Hot spot
- ▲ Active volcano
- / Plate boundary

Volcanoes in the United States

The United States has 60 active volcanoes. Most are part of the Ring of Fire. Alaska, Hawaii, Washington, Oregon, and northern California all have active volcanoes. Mount Redoubt in Alaska is an active volcano. Mount St. Helens in Washington is also an active volcano. It exploded with a violent eruption in 1980.

The United States Geological Survey (USGS) operates volcano observatories. Because many people live near volcanoes, scientists monitor earthquake activity, changes in the shape of volcanoes, and gas emissions. Scientists also study the history of past eruptions to determine the possibility of future eruptions.

Types of Volcanoes

Scientists classify volcanoes based on the shape and size of the volcano. The magma composition and eruption style of a volcano contribute to its shape.

Shield Volcanoes *The **shield volcanoes** are common along divergent plate boundaries and oceanic hot spots. They are large, with gentle slopes of basaltic lavas.*

Composite Volcanoes *The **composite volcanoes** are large, steep-sided volcanoes. They result from explosive eruptions of andesitic and rhyolitic lava and ash along convergent plate boundaries.*

 Visual Check

5. Locate the Ring of Fire on the map. Highlight it.

FOLDABLES

Make a pyramid book. Inside the pyramid, organize your notes about the three main types of volcanoes.

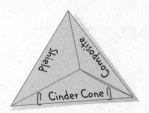

Cinder Cones *The cinder cones are small, steep-sided volcanoes that erupt gas-rich, basaltic lavas.* Cinder cones are made from mildly explosive eruptions.

Some volcanoes are classified as supervolcanoes. The Yellowstone Caldera in Wyoming is the result of a supervolcano. A large volcanic depression formed approximately 630,000 years ago when the summit was blown away during an explosive eruption. 🔑

Volcanic Eruptions

Sometimes magma surfaces and erupts as a lava flow. Other times, magma erupts explosively. Explosive eruptions send **volcanic ash**—*tiny particles of pulverized volcanic rock and glass*—high into the atmosphere. An example of an explosive eruption is the violent Mount St. Helens eruption in 1980.

Eruption Style

The chemical composition of the magma and the amount of <u>dissolved</u> gases in it contribute to the eruption style. The most abundant gas dissolved in magma is water vapor.

Magma Chemistry All magmas are made mainly of silica. The amount of silica in magma affects magma thickness and viscosity. **Viscosity** *is a liquid's ability to flow.*

Magma with a low silica content has low viscosity. It flows easily, like warm syrup. When this type of magma erupts, it flows as fluid lava. The lava cools, crystallizes, and forms the volcanic rock basalt. This type of lava commonly erupts along mid-ocean ridges and at oceanic hot spots, such as Hawaii.

Magma with a high silica content has high viscosity. It flows like sticky toothpaste. This type of magma forms when rocks high in silica content melt. The volcanic rocks andesite and rhyolite form when high silica magma erupts from subduction zones and continental hot spots. 🔑

Dissolved Gases All magmas contain dissolved gases—mainly water vapor and small amounts of carbon dioxide and sulfur dioxide. As magma moves toward the surface, pressure from the weight of the rock above decreases. As the pressure decreases, the gases can no longer stay dissolved in the magma. Bubbles begin to form. As the magma continues to rise, the bubbles get larger and the gas begins to escape. But gases cannot easily escape from lava with high viscosity. The combination of high-viscosity lava and large gas bubbles often results in explosive eruptions.

🔑 **Key Concept Check**

6. Summarize What determines the shape of a volcano?

_____

_____

ACADEMIC VOCABULARY
dissolve
(verb) to cause to disperse or disappear

🔑 **Key Concept Check**

7. Specify What factors affect the eruption style of a volcano?

_____

_____

_____

Effects of Volcanic Eruptions

Lava flows, ash fall, mudflows, and pyroclastic flows from volcanic eruptions can affect all life on Earth. Erupted materials enrich rock and soil with nutrients and help regulate climate. However, eruptions also destroy and kill.

Lava Flows Lava flows move so slowly that they are rarely deadly. But lava flows do cause damage. People living near Mount Etna in Sicily, Italy, flee its frequent eruptions.

Ash Fall Explosive eruptions can spew volcanic ash high into the air. Recall that ash is a mixture of particles of pulverized rock and glass. Ash can cause airplane engines to stop in mid-flight. Ash in the air can cause breathing problems. Large quantities of ash can affect climate by blocking sunlight and cooling Earth's atmosphere.

Mudflows The thermal energy produced during an eruption can melt snow and ice on a volcano's summit. This meltwater can then mix with mud and ash on the mountain to form mudflows. Mudflows, also called lahars, can sweep down the mountainside and bury everything below.

Pyroclastic Flow Explosive volcanoes can produce pyroclastic (pi roh KLAS tihk) flows—avalanches of hot gas, ash, and rock. These flows can travel at speeds of more than 100 km/h and reach temperatures above 1,000°C. A pyroclastic flow from Mount St. Helens killed 58 people.

Predicting Volcanic Eruptions

Volcanic eruptions can be predicted. Geologists study changes that could signal a brewing eruption. Moving magma can deform ground features, change a volcano's shape, or set off a series of earthquakes called an earthquake swarm. A volcano might emit more gas before it erupts. Ground and surface water near the volcano can become more acidic. Geologists study these events and photographs from airplanes and satellites to assess the danger.

Volcanic Eruptions and Climate Change

Volcanic eruptions can affect climate. Ash in the atmosphere blocks sunlight. High-altitude winds can move ash around the world. Also, sulfur-dioxide gas released from a volcano forms sulfuric acid droplets in the atmosphere. These droplets reflect sunlight back into space. Global temperatures decrease as less sunlight reaches Earth's surface. 🔑

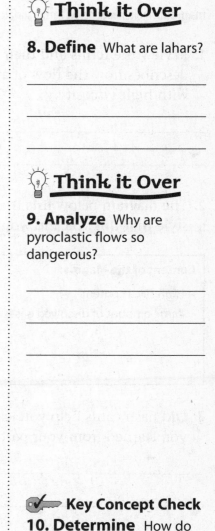

💡 Think it Over

8. Define What are lahars?

_____

_____

_____

💡 Think it Over

9. Analyze Why are pyroclastic flows so dangerous?

_____

_____

_____

🔑 Key Concept Check
10. Determine How do volcanic eruptions affect climate?

_____

_____

_____

Copyright © Glencoe/McGraw-Hill, a division of The McGraw-Hill Companies, Inc.

Mini Glossary

cinder cone: a small, steep-sided volcano that erupts gas-rich, basaltic lava

composite volcano: a large, steep-sided volcano that results from explosive eruptions of andesitic and rhyolitic lava and ash along convergent plate boundaries

hot spot: a volcano that is not associated with plate boundaries

lava: molten rock that erupts onto Earth's surface

magma: molten rock below Earth's surface

shield volcano: a large volcano with gentle slopes of basaltic lava that is common along divergent plate boundaries and oceanic hot spots

viscosity: a liquid's ability to flow

volcanic ash: tiny particles of pulverized volcanic rock and glass

volcano: a vent in Earth's crust through which melted—or molten—rock flows

1. Review the terms and their definitions in the Mini Glossary. Write a sentence that describes how the flow of a liquid with low viscosity differs from the flow of a liquid with high viscosity.

_____

_____

2. The diagram below lists the content of the magma in a volcano. Describe the eruption style that this volcano would probably have.

Content of the Magma:		Eruption Style:
• High silica content • Large amount of dissolved gases	→	

3. Did flash cards help you learn about volcanoes? On the lines below, list two facts that you learned from your partner's flash cards.

_____

_____

What do you think NOW?

Reread the statements at the beginning of the lesson. Fill in the After column with an A if you agree with the statement or a D if you disagree. Did you change your mind?

 **ConnectED**

Log on to ConnectED.mcgraw-hill.com and access your textbook to find this lesson's resources.

 END OF LESSON

Clues to Earth's Past

Fossils

············ **Before You Read** ·············

What do you think? Read the two statements below and decide whether you agree or disagree with them. Place an A in the Before column if you agree with the statement or a D if you disagree. After you've read this lesson, reread the statements to see if you have changed your mind.

Before	Statement	After
	1. Fossils are pieces of dead organisms.	
	2. Only bones can become fossils.	

············ **Read to Learn** ···············

Evidence of the Distant Past

Old family photos provide clues to a family's history. Each photo reveals something of the family's past. You can often guess the age of the photos based on the clothes people are wearing or the vehicles they are driving. The material the photographs are printed on also provides a clue about when the photograph was taken.

Just as old photos can provide clues about a family's past, rocks can provide clues about Earth's past. Some of the most obvious clues found in rocks are the remains or traces of ancient living things. **Fossils** *are the preserved remains or evidence of ancient living things*.

Catastrophism

Many fossils represent plants and animals that no longer live on Earth. Ideas about how these fossils formed have changed over time. Some early scientists hypothesized that great, sudden disasters killed the organisms that became fossils. These scientists explained Earth's history as a series of disastrous events that occurred over short periods of time. **Catastrophism** (kuh TAS truh fih zum) *is the idea that conditions and creatures on Earth change in quick, violent events*. Supernatural forces were thought to have caused many of these events. For example, ancient Romans thought that a supernatural being called Vulcan caused volcanic eruptions. ✓

Key Concepts 🔑

- What are fossils and how do they form?
- What can fossils reveal about Earth's past?

[Mark the Text]

Identify Main Ideas
Highlight each head and the details that support it. Use the highlighted information to review the lesson.

✓ **Reading Check**
1. Define What is catastrophism?

_____

_____

_____

Copyright © Glencoe/McGraw-Hill, a division of The McGraw-Hill Companies, Inc.

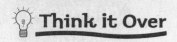
2. Explain Why did James Hutton think that Earth was more than a few thousand years old?

_____

_____

_____

ACADEMIC VOCABULARY

uniform

(adjective) having always the same form, manner, or degree; not varying or variable

 Reading Check

3. Define What is uniformitarianism?

_____

_____

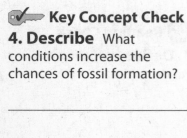 **Key Concept Check**

4. Describe What conditions increase the chances of fossil formation?

_____

_____

_____

Uniformitarianism

Most people who supported catastrophism thought that Earth was only a few thousand years old. In the 1700s, James Hutton rejected this idea. Hutton was a naturalist and farmer in Scotland. He noticed that the landscape on his farm changed over the years. Hutton thought that the processes responsible for changing the landscape on his farm could also change Earth's surface. For example, Hutton thought that erosion caused by streams could also wear down mountains. He realized that these processes would take a long time to change Earth's surface. Hutton proposed that Earth was much older than a few thousand years.

Hutton's ideas became part of the principle of uniformitarianism (yew nuh for muh TER ee uh nih zum). *The principle of* **uniformitarianism** *states that geologic processes that occur today are similar to those that have occurred in the past.* According to this view, Earth's surface is constantly being reshaped in a steady, <u>uniform</u> manner. ✓

Today, uniformitarianism is the basis for understanding Earth's past. But scientists also know that catastrophic events do occur. Volcanic eruptions and meteorite impacts can change Earth's surface quickly. These catastrophic events are not caused by supernatural beings or forces. They can be explained by natural processes.

Formation of Fossils

You learned that fossils are the remains or traces of ancient living organisms. Not all dead organisms become fossils. How are these remains preserved? Fossils form only under certain conditions.

Conditions for Fossil Formation

Most plants and animals are eaten or decay when they die. They leave no evidence that they ever lived. Think about the chances of an apple becoming a fossil. If the apple is on the ground for many months, it will decay into a soft, rotting lump. Eventually, insects and bacteria will eat it.

Some conditions increase the chances of fossil formation. An organism is more likely to become a fossil if it has hard parts, such as shells, teeth, or bones. The hard parts of organisms do not decay easily. An organism is also more likely to form a fossil if it is buried quickly after it dies. If layers of sand or mud bury an organism quickly, decay is slowed or stopped. The figure on the next page shows how a fish might become a fossil. ✓

Copyright © Glencoe/McGraw-Hill, a division of The McGraw-Hill Companies, Inc.

① ② ③

✅ **Visual Check**
5. Identify What parts of an organism become a fossil?

_____

_____

Fossils Come in All Sizes

When many people think of fossils, they think of dinosaur fossils. Many dinosaurs were large animals. Their large bones remained after they died. Not all fossils are large enough for you to see. Some fossils can be seen only by using a microscope. These tiny fossils are called microfossils. Some microfossils are about the size of a speck of dust. ✅

Types of Preservation

Fossils are preserved in different ways. Some are trapped in ice or embedded in rock. Others actually become rock. Some fossils are not body parts, but only evidence of once-living things.

Preserved Remains

Sometimes the actual remains of organisms are preserved as fossils. For this to happen, an organism must be completely enclosed in some material for a long period of time. Materials such as amber, tar, or ice trap the remains and prevent air or bacteria from causing decay. Usually, preserved remains are 10,000 or fewer years in age. However, insects preserved in amber can be millions of years old.

Carbon Films

Sometimes pressure on a buried organism is so great that all gases and liquids are released from the organism's tissues. The only thing left behind is the carbon. _A_ **carbon film** _is the fossilized carbon outline of an organism or part of an organism._ ✅

Carbon films are usually shiny black or dark brown. Fish, insects, and plant leaves are often preserved as carbon films.

Mineral Replacement

Copies of organisms can form from minerals in groundwater. Rock-forming minerals dissolved in groundwater enter the pore spaces or replace the tissues of dead organisms. Petrified wood is a mineral-replacement fossil. The mineral silica (SiO_2) filled in the spaces between the cell walls in a dead tree. The wood petrified when the SiO_2 hardened.

✅ **Reading Check**
6. Identify Which is a microfossil? (Circle the correct answer.)
a. a dinosaur leg bone
b. an ancient fish bone
c. a fossil about the size of a speck of dust

✅ **Reading Check**
7. Explain How does a carbon film form?

_____

_____

_____

_____

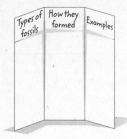

Make a tri-fold book to organize your notes about different types of fossils.

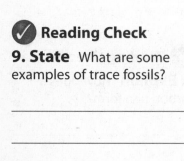

Visual Check

8. Sequence How is a cast made? (Circle the correct answer.)

a. A carbon print is filled with sediment.

b. A fossil impression is filled with sediment.

c. A fossil copy is filled with sediment.

Reading Check

9. State What are some examples of trace fossils?

_____

_____

_____

Molds

Sometimes a fossilized imprint, or impression, is all that is left of an organism. This type of fossil is called a mold. *A* **mold** *is the impression in a rock left by an ancient organism.* A mold can form when sediment hardens around a buried organism. As the organism decays over time, an impression of its shape remains in the sediment. The sediment eventually turns to rock.

Casts

Sometimes, after a mold forms, sediment fills the mold. This type of fossil is called a cast. *A* **cast** *is a fossil copy of an organism made when a mold of the organism is filled with sediment or mineral deposits.* The figure below shows how a fossil mold and cast form.

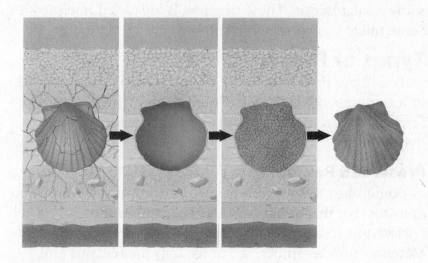

Trace Fossils

Some animals leave fossilized traces of their activities. *A* **trace fossil** *is the preserved evidence of the activity of an organism.* Trace fossils include tracks, footprints, and nests. Scientists use trace fossils to learn about animal characteristics. For example, dinosaur tracks give clues about the dinosaur's size, speed, and whether it was traveling alone or in a group. ✓

Ancient Environments

Paleontologists (pay lee ahn TAH luh justs) *are scientists who study fossils.* Paleontologists use the principle of uniformitarianism to learn about ancient organisms and the environments in which they lived. They compare fossils of ancient organisms with organisms living today. A trilobite fossil and a horseshoe crab look alike. Horseshoe crabs today live in shallow water on the ocean floor. Partly because trilobite fossils look like horseshoe crabs, paleontologists infer that trilobites also lived in shallow ocean water.

Shallow Seas

Today, Earth's continents are mostly above sea level. Many times in the past, the sea level rose and flooded Earth's continents. For example, the figure below shows a shallow ocean that covered much of North America about 450 million years ago. Fossils of organisms that lived in that shallow ocean help scientists reconstruct what the seafloor looked like in the past.

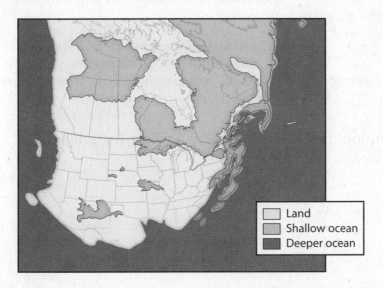

Land
Shallow ocean
Deeper ocean

Key Concept Check

10. Describe What can fossils tell us about ancient environments?

_____

_____

_____

_____

Visual Check

11. Illustrate Mark an *X* over each area on the map where there might have been ancient, shallow oceans.

Past Climates

You might have heard people talking about global climate change. There is evidence that Earth's present-day climate is warming. Fossils show that Earth's climate has warmed and cooled many times in the past.

Plant fossils are good indicators of climate change. Fossils of ferns and other tropical plants indicate that much of Earth was very warm 100 million years ago. Tropical swamps and forests covered much of the land. Dinosaurs lived on Earth during this period.

Millions of years later, the swamps and forests were gone. In some of these areas, coarse grasses grew. Huge sheets of ice called glaciers formed as the climate cooled. This ice spread over parts of North America, Europe, and Asia. Organisms that adapted to the cold climate survived. Fossils of organisms, such as the woolly mammoth, help scientists learn about this cold time in Earth's history.

Fossils of organisms, such as ferns and mammoths, help scientists learn about ancient organisms and past environments. In the following lessons, you will read how scientists use fossils and other clues, such as rock layers and radioactivity, to learn about the ages of Earth's rocks.

Key Concept Check

12. Describe What was Earth's climate like when dinosaurs lived?

_____

_____

_____

_____

Mini Glossary

carbon film: the fossilized carbon outline of an organism or part of an organism

cast: a fossil copy of an organism made when a mold of the organism is filled with sediment or mineral deposits

catastrophism (kuh TAS truh fih zum): the idea that conditions and creatures on Earth change in quick, violent events

fossil: the preserved remains or evidence of an ancient living thing

mold: the impression in a rock left by an ancient organism

paleontologist (pay lee ahn TAH luh jihst): a scientist who studies fossils

trace fossil: the preserved evidence of the activity of an organism

uniformitarianism (yew nuh for muh TER ee uh nih zum): the principle that states that geologic processes that occur today are similar to those that have occurred in the past

1. Review the terms and their definitions in the Mini Glossary. Explain the principle of uniformitarianism in your own words.

_____

_____

_____

2. Use what you have learned about fossils to complete the table.

Fossil Type	What They Are/How They Form
Preserved remains	Actual remains that become preserved in ice, tar, or amber
Carbon films	
Mineral-replacement fossils	Form when minerals replace organic remains
Molds and casts	
Trace fossils	

What do you think NOW?

Reread the statements at the beginning of the lesson. Fill in the After column with an A if you agree with the statement or a D if you disagree. Did you change your mind?

Log on to ConnectED.mcgraw-hill.com and access your textbook to find this lesson's resources.

END OF LESSON

Clues to Earth's Past

Relative-Age Dating

·············· **Before You Read** ··············

What do you think? Read the two statements below and decide whether you agree or disagree with them. Place an A in the Before column if you agree with the statement or a D if you disagree. After you've read this lesson, reread the statements to see if you have changed your mind.

Before	Statement	After
	3. Older rocks are always located below younger rocks.	
	4. Relative age means that scientists are relatively sure of the age.	

·············· **Read to Learn** ··············

Relative Ages of Rocks

You just remembered where you left the money that you have been looking for. It is in the pocket of the pants you wore to the movies last Saturday. Now imagine that the pants are in your pile of dirty laundry. How can you tell where your money is? There really is some order to your pile of clothes. Every time you add clothes to the pile, you place them on top. The clothes from last Saturday are on the bottom. That is where your money is!

There is order in a rock formation just as there is order in a pile of clothes. In many rock formations, the oldest rocks are in the bottom layer and the youngest rocks are in the top layer.

If you have brothers and sisters, you might describe your age by saying, "I'm older than my sister and younger than my brother." This tells how your age relates to others in your family. It is your relative age. Geologists are scientists who study Earth and rocks. They have developed a set of principles to compare the ages of rock layers. These principles help them organize rocks according to their relative ages. **Relative age** *is the age of rocks and geologic features compared with other rocks and features nearby.*

Key Concepts

- What does relative age mean?
- How can the positions of rock layers be used to determine the relative ages of rocks?

Study Coach

Ask Questions As you read, write a question about any topic you don't understand. When you finish reading the lesson, discuss your question with your teacher or another student.

Key Concept Check
1. Define How might you define your relative age?

_____

_____

_____

FOLDABLES

Make a five-tab book and use it to organize information about the principles of relative-age dating.

Superposition
Original Horizontality
Lateral Continuity
Cross-cutting Relationships
Inclusions

Reading Check

2. Explain How might rock layers be disturbed?

_____

_____

_____

Visual Check

3. Sequence Which rock layer is the oldest?

_____

_____

_____

Superposition

Your pile of dirty clothes demonstrates the first principle of relative-age dating—superposition. **Superposition** *is the principle that in undisturbed rock layers, the oldest rocks are on the bottom.*

Forces do sometimes disturb rock layers after they are deposited. But if no disturbance takes place, each layer of rocks is younger than the layer below it. The principle of superposition is shown in the top part of the figure below. Layer 1 in the figure is the oldest rock layer, while layer 4 is the youngest.

Original Horizontality

The second principle of relative-age dating is called original horizontality. It is shown in the middle part of the figure below. Again, layer 1 is the oldest rock layer and layer 4 is the youngest.

According to the principle of original horizontality, most rock-forming materials are deposited in horizontal layers. Sometimes rock layers are deformed or disturbed after they form. For example, the layers might be tilted or folded. When you see rocks that are tilted, remember that all layers were originally deposited horizontally. ✓

Lateral Continuity

Another principle of relative-age dating is that sediments are deposited in large, flat sheets. The sheets, or layers, continue in all lateral directions until they thin out or until they meet a barrier. This principle, shown in the bottom part of the figure below, is called the principle of lateral continuity. For example, a river might erode the layers, but the order of the layers does not change.

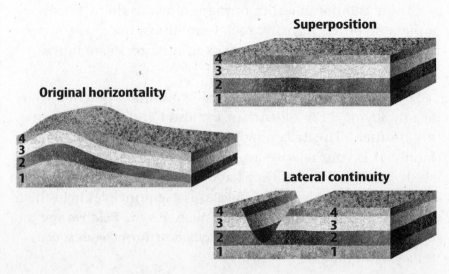

Inclusions

Sometimes, when rocks form, they contain pieces of other rocks. This can happen when part of an existing rock breaks off and falls into soft sediment or flowing magma. When the sediment or the magma becomes rock, the broken piece of rock becomes a part of it. *A piece of an older rock that becomes part of a new rock is called an* **inclusion.**

According to the principle of inclusions, if one rock contains pieces of another rock, the rock containing the pieces is younger than the pieces. The first part of the figure below shows sediments deposited in layers that have become rock. The vertical intrusion shown in the middle part of the figure below is called a dike. The dike formed when magma flowed into the rock layers. The dike is younger than the pieces of rock, or inclusions, inside it. ✔

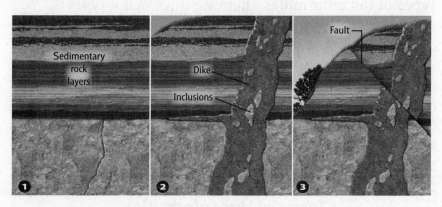

Cross-Cutting Relationships

Sometimes forces within Earth cause rock formations to break, or fracture. When rocks move along a fracture line, the fracture is called a fault.

According to the principle of cross-cutting relationships, if one geologic feature cuts across another feature, the feature that it cuts across is older. Notice in the figure above that both faults and dikes cut across existing rock. In the figure on the right, the fault cuts across rock layers and the dike. Scientists conclude that the dike is older than the fault because the fault is cutting across the dike. Both the fault and the dike are younger than the rock layers. ⚷

Copyright © Glencoe/McGraw-Hill, a division of The McGraw-Hill Companies, Inc.

✔ Reading Check

4. Define What are inclusions?

_____

_____

_____

✔ Visual Check

5. Sequence Is the dike older or younger than the fault? Explain your answer.

_____

_____

_____

⚷ Key Concept Check

6. Name What geologic principles are used in relative-age dating?

_____

_____

_____

_____

Unconformities

After rocks form, they are sometimes uplifted and exposed at Earth's surface. As soon as rocks are exposed, wind and rain start to weather and erode them. These eroded areas represent a gap in the rock record.

A Gap in Time Often, new rock layers are deposited on top of old, eroded rock layers. When this happens, an unconformity (un kun FOR muh tee) occurs. *An **unconformity** is a surface where rock has eroded away, producing a break, or gap, in the rock record.*

An unconformity is not a hole or a space in the rock. It is a surface on a layer of eroded rocks with younger rocks on top. An unconformity does represent a gap in time. It could represent a few hundred years, a million years, or even billions of years.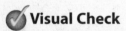

Types of Unconformities There are three major types of unconformities, as shown in the figure below. In a disconformity, younger sedimentary layers are deposited on top of older, horizontal sedimentary layers that have been eroded. In an angular unconformity, sedimentary layers are deposited on top of tilted or folded sedimentary layers that have been eroded. In a nonconformity, younger sedimentary layers are deposited on older igneous or metamorphic rock layers that have been eroded.

Key Concept Check

7. Explain How does an unconformity represent a gap in time?

_____

_____

_____

Visual Check

8. Explain What is the difference between a disconformity and a nonconformity?

_____

_____

_____

Types of Unconformities		
Disconformity	→	**Disconformity** Younger sedimentary rock / Older sedimentary rock
Angular Unconformity	→	**Unconformity** Younger sedimentary rock / Older sedimentary rock
Nonconformity	→	**Nonconformity** Younger sedimentary rock / Older sedimentary rock

Correlation

Rock layers contain clues about Earth. Geologists use these clues to build a record of Earth's geologic history. Many times the rock record is incomplete. For example, unconformities create gaps in the geologic record. Geologists fill in the gaps in the rock record by matching rock layers and fossils from separate locations. *Matching rocks and fossils from separate locations is called* **correlation** (kor uh LAY shun).

Matching Rock Layers

Another word for correlation is *connection*. Sometimes, geologists can connect rock layers simply by walking along rock formations and looking for similarities. Other times, soil might cover the rocks, or rocks might be eroded. When this happens, geologists correlate rocks by matching exposed rock layers in different locations. As shown in the figure below, geologists have used correlation to develop a historical record for part of the southwestern United States.

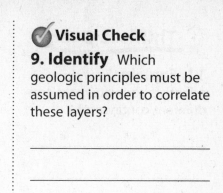

Visual Check

9. Identify Which geologic principles must be assumed in order to correlate these layers?

_____

_____

_____

_____

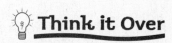

Think it Over

10. Apply Look at the figure. Which is older, the Moenkopi formation or the Navajo sandstone?

_____

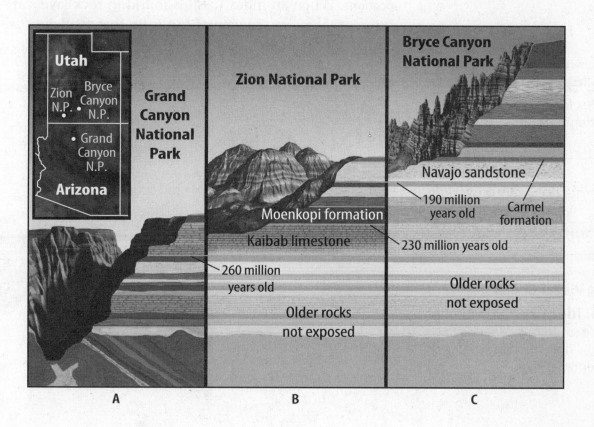

Reading Essentials

Clues to Earth's Past **285**

Copyright © Glencoe/McGraw-Hill, a division of The McGraw-Hill Companies, Inc.

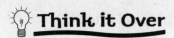
11. Explain How do scientists correlate rocks on different continents?

_____

_____

_____

_____

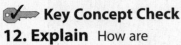 **Key Concept Check**

12. Explain How are index fossils useful in relative dating?

_____

_____

_____

_____

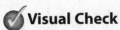 **Visual Check**

13. Identify Circle the tail on each trilobite species to show how trilobites changed over time.

Index Fossils

Some correlated rock formations are within a few hundred kilometers of one another, such as those in some national parks. They are correlated based on similarities in rock type, structure, and fossil evidence. If scientists want to learn the relative ages of rock formations that are very far apart or on different continents, they often use fossils. If two or more rock formations contain fossils of about the same age, scientists can infer that the formations are also about the same age.

Not all fossils are useful in determining the relative ages of rock formations. Fossils of species that lived on Earth for hundreds of millions of years are not helpful. They represent time spans that are too long.

The most useful fossils represent species such as certain trilobites. These trilobites existed for only a short time in many different areas on Earth. The fossils of these trilobites are index fossils. **Index fossils** *represent species that existed on Earth for a short length of time, were abundant, and inhabited many locations.* When an index fossil is found in rock layers at different locations, geologists can infer that the layers are of similar age. Trilobites, shown in the figure below, are examples of index fossils.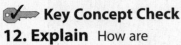

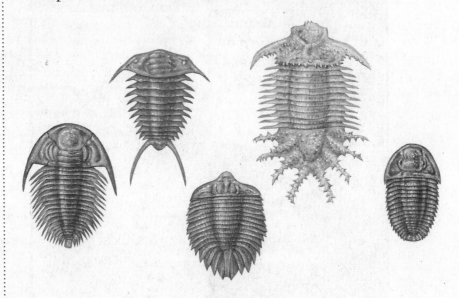

Mini Glossary

correlation (kor uh LAY shun): matching rocks and fossils from separate locations

inclusion: a piece of an older rock that becomes part of a new rock

index fossil: a fossil that represents a species that existed on Earth for a short length of time, was abundant, and inhabited many locations

relative age: the age of rocks and geologic features compared with other rocks and features nearby

superposition: the principle that in undisturbed rock layers, the oldest rocks are on the bottom

unconformity (un kun FOR muh tee): a surface where rock has eroded away, producing a break, or gap, in the rock record

1. Review the terms and their definitions in the Mini Glossary. Explain the principle of superposition in your own words.

_____

_____

2. Use what you have learned about relative dating methods to complete the table.

Relative Dating Principles			
Superposition	**Lateral Continuity**		**Cross-Cutting Relationships**
	Sediment is deposited in large, continuous sheets.	Most rock-forming material was originally deposited in horizontal layers.	

3. Use the principle of inclusion to explain whether a rock that contains pieces of another rock is older or younger than the pieces.

_____

_____

What do you think NOW?

Reread the statements at the beginning of the lesson. Fill in the After column with an A if you agree with the statement or a D if you disagree. Did you change your mind?

Log on to ConnectED.mcgraw-hill.com and access your textbook to find this lesson's resources.

END OF LESSON

CHAPTER 16
LESSON 3

Clues to Earth's Past

Absolute-Age Dating

Key Concepts
- What does absolute age mean?
- How can radioactive decay be used to date rocks?

Study Coach

Build Vocabulary Make flash cards to help you learn the important terms in this lesson. Write each bolded word on one side of a card and the definition on the other side. Use your flash cards to review this lesson.

Key Concept Check
1. Contrast How is absolute age different from relative age?

____

____

____

Before You Read

What do you think? Read the two statements below and decide whether you agree or disagree with them. Place an A in the Before column if you agree with the statement or a D if you disagree. After you've read this lesson, reread the statements to see if you have changed your mind.

Before	Statement	After
	5. Absolute age means that scientists are sure of the age.	
	6. Scientists use radioactive decay to determine the ages of some rocks.	

Read to Learn

Absolute Ages of Rocks

Remember that you read in Lesson 2 that you have a relative age. You might be older than your sister and younger than your brother. Or, you might be the youngest in your family.

You also can describe your age in years. For example, you might be 13 years old. This is not your relative age. It is your age in numbers, or your numerical age.

Relative age helps scientists compare the ages of rock layers. But scientists also look for more specific information. Scientists can describe the ages of some kinds of rocks using numbers. Scientists use the term **absolute age** *to mean the numerical age, in years, of a rock or an object.*

Scientists have only been able to determine absolute ages of rocks and other objects for about 100 years. Early in the twentieth century, radioactivity was discovered. Radioactivity is the release of energy from unstable atoms. The release of radioactive energy can be used to make images called X-rays. How can radioactivity be used to determine the absolute age of rocks? To answer this question, you need to know about the internal structure of the atoms that make up elements.

Atoms

You are probably familiar with the periodic table of the elements. Each element is made up of atoms. An atom is the smallest part of an element that has all the properties of the element. Each atom contains smaller particles. They are protons, neutrons, and electrons. Protons and neutrons are in an atom's nucleus. Electrons surround the nucleus.

Isotopes

All atoms of a given element have the same number of protons. For example, all hydrogen atoms have one proton. But an element's atoms can have different numbers of neutrons. The three atoms in the figure below are all hydrogen atoms. Notice that each hydrogen atom has only one proton. But one has no neutrons, one has one neutron, and the other has two neutrons. The three different forms of hydrogen atoms are called isotopes (I suh tohps). **Isotopes** *are atoms of the same element that have different numbers of neutrons.* ✓

Hydrogen   Hydrogen-2   Hydrogen-3
nucleus   nucleus   nucleus

Radioactive Decay

Most isotopes are stable. Stable isotopes do not change under normal conditions. Some isotopes are unstable. Unstable isotopes are called radioactive isotopes. Radioactive isotopes decay, or change, over time. As they decay, they release energy and form new, stable atoms. **Radioactive decay** *is the process by which an unstable element naturally changes into another element that is stable.* The unstable isotope that decays is called the parent isotope. The new element that forms is called the daughter isotope. As shown below, the atoms of an unstable isotope of hydrogen (parent) decay into atoms of a stable isotope of helium (daughter).

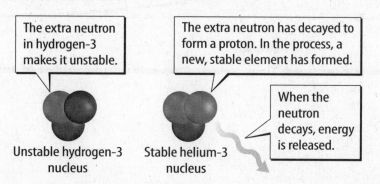

The extra neutron in hydrogen-3 makes it unstable.

The extra neutron has decayed to form a proton. In the process, a new, stable element has formed.

When the neutron decays, energy is released.

Unstable hydrogen-3 nucleus

Stable helium-3 nucleus

✔ **Reading Check**

2. Explain How do an element's isotopes differ?

_____

_____

_____

✔ **Visual Check**

3. Describe How do the hydrogen atoms in the figure differ?

_____

_____

_____

✔ **Visual Check**

4. Identify Which isotope in the figure is the parent? Which is the daughter?

_____

_____

_____

Half-Life

The radioactive isotopes of different elements decay to form daughter isotopes at different rates. But the rate of decay is constant for a given isotope. This rate of decay is measured in time units called half-lives. *An isotope's* **half-life** *is the time required for half of the parent isotopes to decay into daughter isotopes.* Half-lives of radioactive isotopes can be as short as a few microseconds or as long as billions of years. ✔

The graph below shows how half-life is measured. As time passes, parent isotopes decay and form daughter isotopes. That means the ratio between the number of parent and daughter isotopes is always changing.

When half of the parent isotopes have decayed into daughter isotopes, the isotope has reached one half-life. At this point, there are equal numbers of parent and daughter isotopes.

After two half-lives, one-half of the remaining parent isotopes have decayed. At this point, only one-quarter of the original parent isotopes remain. So, 25 percent of the isotopes are parent and 75 percent are daughter.

After three half-lives, another half of the remaining parent isotopes have decayed into daughter isotopes. Radioactive decay continues until nearly all the parent isotopes have decayed into stable daughter isotopes.

✔ Reading Check

5. Define What is half-life?

_____

_____

_____

✔ Visual Check

6. State What percentages of parent isotopes and daughter isotopes will there be after four half-lives?

_____

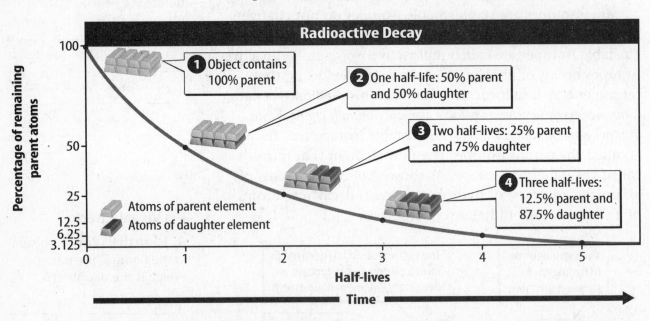

Radioactive Decay

Percentage of remaining parent atoms

100

50

25
12.5
6.25
3.125

0   1   2   3   4   5

Half-lives

Time

1 Object contains 100% parent

2 One half-life: 50% parent and 50% daughter

3 Two half-lives: 25% parent and 75% daughter

4 Three half-lives: 12.5% parent and 87.5% daughter

Atoms of parent element
Atoms of daughter element

Radiometric Ages

Radioactive isotopes decay at a constant rate. Scientists use them like clocks to measure the age of the material that contains them. The process scientists use is called radiometric dating. In radiometric dating, scientists select a sample of the material they want to date. They then measure the ratio of the amount of parent isotope to the amount of daughter product in the sample. From this ratio, scientists can determine the material's age. They make these precise measurements in laboratories. ✓

Radiocarbon Dating

One important radioactive isotope used for dating is an isotope of carbon called radiocarbon. Radiocarbon is also known as carbon-14, or C-14, because there are 14 particles in its nucleus—six protons and eight neutrons.

Radiocarbon forms in Earth's upper atmosphere. There, it mixes with a stable isotope of carbon called carbon-12, or C-12. The ratio of C-14 to C-12 in the atmosphere is constant.

All living organisms use carbon as they build and repair tissues. As long as an organism is alive, the ratio of C-14 to C-12 in its tissues is identical to the ratio in the atmosphere. However, when an organism dies, it stops taking in C-14.

The C-14 already present in the organism starts to decay into nitrogen-14. As the dead organism's C-14 decays, the ratio of C-14 to C-12 changes. Scientists measure the ratio of C-14 to C-12 in the remains of the dead organism to determine how much time has passed since the organism died.

Carbon-14 has a half-life of 5,730 years. That means radiocarbon dating is useful for measuring the age of the remains of organisms that died up to about 50,000 years ago. In older remains, too much of the carbon-14 has decayed to be measured accurately.

Dating Rocks

Radiocarbon dating is useful only for dating organic material. Organic material is material from once-living organisms. Organic material includes bones, wood, parchment, and charcoal.

Most rocks do not contain organic material. Even most fossils are no longer organic. In fossils, living tissue has been replaced by rock-forming minerals. For dating rocks, geologists use different kinds of radioactive isotopes.

✓ **Reading Check**

7. Explain What is measured in radiometric dating?

_____

_____

_____

💡 **Think it Over**

8. Contrast What happens to the C-14 in a living organism? What happens to the C-14 in a dead organism?

_____

_____

_____

_____

FOLDABLES

Make a two-tab book to compare how absolute ages of organic materials and rocks are determined.

Dating Organic Material

Dating Rocks

Dating Igneous Rock Another common isotope used in radiometric dating is uranium-235 (U-235). Radioactive isotopes can be trapped in the minerals of igneous rocks that crystallize from hot, molten magma. U-235 trapped in the minerals immediately begins to decay to lead-207 (Pb-207).

Scientists measure the ratio of U-235 to Pb-207 in a mineral to determine how much time has passed since the mineral formed. This provides the age of the rock that contains the mineral.

Dating Sedimentary Rock To be dated by radiometric means, a rock must have U-235 or other radioactive isotopes trapped inside it. Radiometric dating is not as useful for dating sedimentary rocks. The grains in sedimentary rocks come from a variety of weathered rocks in different locations. The radioactive isotopes in the grains record the age of the grains that make up the sedimentary rock. This does not help scientists learn when the sediment was deposited. For this reason, sedimentary rock cannot be as easily dated using radiometric dating as igneous rock.

Different Types of Isotopes Isotopes with short half-lives are useful for dating relatively young rocks. Isotopes with longer half-lives are useful for dating much older rocks. The half-life of uranium-235 is 704 million years old. This makes it useful for dating rocks that are very old.

Often, geologists use combinations of several isotopes to measure the age of a rock. Using combinations helps make the measurements more accurate. The table below shows some of the most useful radioactive isotopes used for dating rocks. Notice that all are isotopes with long half-lives.

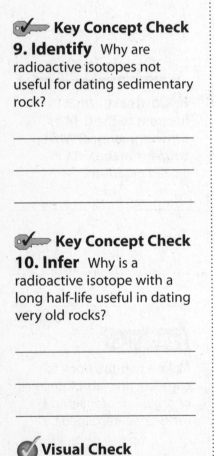

Key Concept Check

9. Identify Why are radioactive isotopes not useful for dating sedimentary rock?

_____

_____

_____

Key Concept Check

10. Infer Why is a radioactive isotope with a long half-life useful in dating very old rocks?

_____

_____

_____

Visual Check

11. Identify Which isotope in the table has the longest half-life?

_____

Radioactive Isotopes Used for Dating Rocks		
Parent Isotope	**Half-Life**	**Daughter Product**
Uranium-235	704 million years	lead-207
Potassium-40	1.25 billion years	argon-40
Uranium-238	4.5 billion years	lead-206
Thorium-232	14.0 billion years	lead-208
Rubidium-87	48.8 billion years	strontium-87

The Age of Earth

Since the discovery of radiometric dating, geologists have tried to find Earth's oldest rocks. The oldest rock formation dated by geologists using radiometric means is in Canada. It is estimated to be between 4.03 billion and 4.28 billion years old. However, individual crystals of the mineral zircon have been found in rocks in Australia. These zircon crystals have been dated at 4.4 billion years old. ✔

For a long time, people have searched for ways to determine the age of Earth. With rocks and minerals more than 4 billion years old, scientists know that Earth must be at least that old. Radiometric dating of rocks from the Moon and meteorites indicates that Earth is 4.54 billion years old. Scientists accept this age because evidence suggests that Earth, the Moon, and meteorites all formed at about the same time.

Radiometric dating, the relative order of rock layers, and fossils all help scientists understand Earth's long history. Understanding Earth's history can help scientists understand changes taking place on Earth today. Scientists can also use what they have learned to predict changes that are likely to take place in the future.

✔ Reading Check

12. Identify How old is the oldest rock formation geologists have dated?

_____

_____

_____

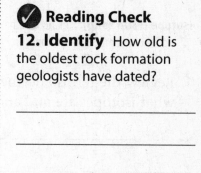

Math Skills

The answer to a problem involving measurement cannot be more precise than the measurement with the fewest number of significant digits. For example if you begin with 36 grams (2 significant digits) of U-235, how much U-235 will remain after 2 half-lives?

a. After the first half-life,
$$\frac{36 \text{ g}}{2} = 18 \text{ g of U-235}$$
remain.

b. After the second half-life,
$$\frac{18 \text{ g}}{2} = 9.0 \text{ g of U-235}$$
remain. Add the zero to retain two significant digits.

13. Use Significant Digits The half-life of rubidium-87 (Rb-87) is 48.8 billion years. What is the length of three half-lives of Rb-87?

Mini Glossary

absolute age: the numerical age, in years, of a rock or object

half-life: the time required for half of the parent isotopes to decay into daughter isotopes

isotope (I suh tohp): an atom of the same element that has different numbers of neutrons

radioactive decay: the process by which an unstable element naturally changes into another element that is stable

1. Review the terms and their definitions in the Mini Glossary. Write a sentence explaining what isotopes are and another sentence that names three radioactive isotopes.

   _____

   _____

   _____

2. Use what you have learned about carbon-14 to complete the events chain map.

   An animal is alive. → The ratio of C-14 to C-12 is _____.

   The animal dies. → Carbon-14 _____

   while carbon-12 _____. → Time of death is determined

   by measuring the ratio of _____ to _____.

3. In this lesson, you made vocabulary cards as you read. How did this strategy help you remember the meanings of the important words?

   _____

   _____

   _____

What do you think NOW?

Reread the statements at the beginning of the lesson. Fill in the After column with an A if you agree with the statement or a D if you disagree. Did you change your mind?

Connect ED

Log on to ConnectED.mcgraw-hill.com and access your textbook to find this lesson's resources.

END OF LESSON

Geologic Time

Geologic History and the Evolution of Life

···············**Before You Read**··············

What do you think? Read the two statements below and decide whether you agree or disagree with them. Place an A in the Before column if you agree with the statement or a D if you disagree. After you've read this lesson, reread the statements to see if you have changed your mind.

Before	Statement	After
	1. All geologic eras are the same length of time.	
	2. Meteorite impacts cause all extinction events.	

·················**Read to Learn**················

Developing a Geologic Time Line

Geologists have developed a time line of Earth's past called the geologic time scale. The scale divides the 4.6 billion years of Earth's history into time units.

Units in the Geologic Time Scale

Eons *are the longest units of geologic time. Eons are subdivided into smaller units of time called* **eras.** *Eras are subdivided into* **periods.** *Periods are subdivided into* **epochs** (EH pocks). Notice in the geologic time line below that the units of time are not equal. Epochs are not shown on the time line.

Key Concepts

- How was the geologic time scale developed?
- What are some causes of mass extinctions?
- How is evolution affected by environmental change?

 Mark the Text

Identify Main Ideas
Highlight each head and the information that explains it to help you review this lesson.

✓ **Visual Check**

1. Interpret The Jurassic period was part of what era?

_____

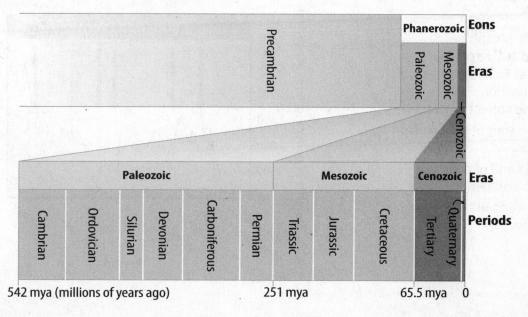

542 mya (millions of years ago)   251 mya   65.5 mya   0

Copyright © Glencoe/McGraw-Hill, a division of The McGraw-Hill Companies, Inc.

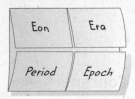
The Time Scale and Fossils

As geologists developed the geologic time scale, they chose boundaries between time units based on what they observed in Earth's rock layers. Different layers contained different fossils. For example, older rocks contained only fossils of small, simple life-forms. Younger rocks contained these fossils, too. But younger rocks also contained fossils of more-complex organisms, such as dinosaurs.

Major Divisions in the Geologic Time Scale

While studying the fossils in rock layers, geologists often saw sudden changes in the types of fossils within the layers. Sometimes, fossils in one rock layer did not appear in the rock layers right above it. It seemed as though the organisms that lived during that period had disappeared suddenly. Geologists used these sudden changes in the fossil record to mark divisions in geologic time. Because the lengths of time between changes were different, the geologic time scale is divided into unequal units of time.

The geologic time scale is a work in progress. The boundaries, or lengths in time, can change as scientists make new discoveries.

Responses to Change

Sudden changes in the fossil record represent times when large populations of species of organisms died or became extinct. *A* **mass extinction** *is the extinction of many species on Earth within a short period of time.* There have been several mass extinction events in Earth's history. Five events are shown in the graph below. In each one, the number of genera—groups of species—decreased sharply.

Key Concept Check

2. Explain Why are fossils important in the development of the geologic time scale?

_____

_____

_____

_____

✓ **Visual Check**

3. Read a Graph
When was Earth's greatest mass extinction event? (Circle the correct answer.)

a. at the start of the Silurian period

b. at the end of the Devonian period

c. at the end of the Permian period

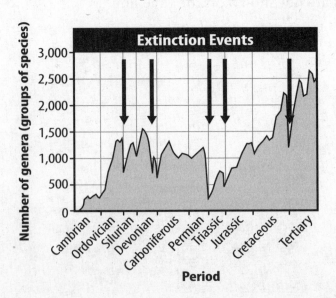

Changes in Climate

All organisms depend on the environment for the food and the other resources that they need to live. If the environment changes quickly and organisms cannot survive in the new conditions, they die.

A rapid change in climate can cause a mass extinction. Climate can change when gas and dust from volcanoes block sunlight and reduce temperatures. The results of a meteorite crashing into Earth would also block sunlight and change climate.

The impact of a meteorite 65.5 million years ago might have caused the extinction of the dinosaurs. Evidence of this impact is in a clay layer around the world. Rocks in this layer contain the element iridium. Iridium is rare in Earth rocks. However, iridium is common in meteorites. No dinosaur fossils have been found in rocks above the iridium layer. 🔑

Geography and Evolution

When environments change, some species of organisms do not adapt. They become extinct. However, other species do adapt to changes in the environment. Evolution is the change in species over time as they adapt to their environments. Sudden changes in the environment can affect evolution. The slow movement of Earth's tectonic plates can also affect evolution.

Land Bridges When continents collide or when sea level drops, landmasses can join together. *A* **land bridge** *connects two continents that were previously separated*. Over time, organisms move across land bridges and evolve as they adapt to new environments.

Geographic Isolation The movement of tectonic plates or other slow geologic events can cause geographic areas to move apart. When areas separate, populations of organisms can become isolated. **Geographic isolation** *is the separation of a population of organisms from the rest of its species due to some physical barrier, such as a mountain range or an ocean.*

Separated populations of species evolve in different ways as they adapt to different environments. For example, a population of squirrels was gradually separated as the Grand Canyon formed. The squirrels on one side of the canyon became adapted to a slightly different environment from the squirrels on the other side of the canyon. Each group evolved in a different way. 🔑

🔑 **Key Concept Check**

4. Describe a possible event that could cause a mass extinction.

_____

_____

_____

💡 **Think it Over**

5. Predict If temperatures on Earth decrease, what changes might occur in a species of squirrel over time as a result of this change in climate?

_____

_____

_____

_____

🔑 **Key Concept Check**

6. Summarize How can geographic isolation affect evolution?

_____

_____

_____

_____

Precambrian Time

Life has been evolving on Earth for billions of years. The oldest fossil evidence of life on Earth is in rocks that are about 3.5 billion years old. These ancient life-forms were simple, unicellular organisms, much like bacteria on Earth today.

The oldest fossils of multicellular organisms are about 600 million years old. Early geologists did not know about these rare fossils. They hypothesized that multicellular life first appeared in the Cambrian (KAM bree un) period, at the beginning of the Phanerozoic (fan er oh ZOH ihk) eon 542 million years ago (mya). Time before the Cambrian period was called Precambrian time, as shown below. Scientists today know that Precambrian time is nearly 90 percent of Earth's history.

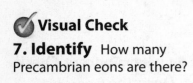
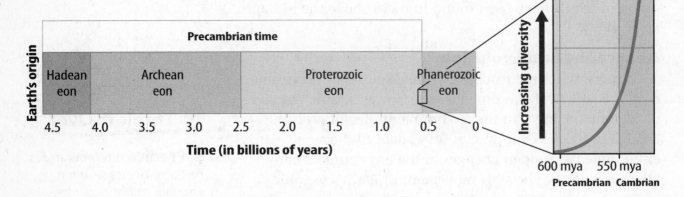

Precambrian Life

The rare fossils of multicellular life-forms in Precambrian rocks are from organisms that had soft bodies. These organisms were not like any organisms that live on Earth today. Precambrian life-forms lived 600 mya at the bottom of the sea. Many of these species became extinct at the end of Precambrian time.

Cambrian Explosion

Precambrian life led to a sudden appearance of new types of multicellular life-forms in the Cambrian period. This sudden appearance of new, complex life-forms is often called the Cambrian explosion.

Some Cambrian life-forms, such as trilobites, were the first organisms to have hard body parts. Because of their hard parts, trilobites were more easily preserved than were organisms with only soft body parts. More evidence of trilobites is in the fossil record. Scientists hypothesize that some of the trilobites are distant ancestors of organisms that are alive today.

Reading Check

8. Explain What was the Cambrian explosion?

Mini Glossary

eon: the longest unit of geologic time

epoch (EH pock): a subdivision of a geologic period

era: a subdivision of a geologic eon

geographic isolation: the separation of a population of organisms from the rest of its species due to some physical barrier, such as a mountain range or an ocean

land bridge: land that connects two continents that were previously separated

mass extinction: the extinction of many species on Earth within a short period of time

period: a subdivision of a geologic era

1. Review the terms and their definitions in the Mini Glossary. Use the term *mass extinction* in an original sentence.

_____

_____

2. Use what you have learned about geologic time to complete the table.

Unit of Time	Definition
Eon	
Era	
Period	
Epoch	

What do you think NOW?

Reread the statements at the beginning of the lesson. Fill in the After column with an A if you agree with the statement or a D if you disagree. Did you change your mind?

Log on to ConnectED.mcgraw-hill.com and access your textbook to find this lesson's resources.

END OF LESSON

Geologic Time

The Paleozoic Era

Copyright © Glencoe/McGraw-Hill, a division of The McGraw-Hill Companies, Inc.

Key Concepts 🗝

- What major geologic events occurred during the Paleozoic era?
- What does fossil evidence reveal about the Paleozoic era?

Study Coach ▶

Answer Questions Write each Key Concept question on half a sheet of paper. As you read, write phrases or sentences that answer each question. Use your answers to review the lesson.

············· **Before You Read** ··············

What do you think? Read the two statements below and decide whether you agree or disagree with them. Place an A in the Before column if you agree with the statement or a D if you disagree. After you've read this lesson, reread the statements to see if you have changed your mind.

Before	Statement	After
	3. North America was once on the equator.	
	4. All of Earth's continents were part of a huge supercontinent 250 million years ago.	

············· **Read to Learn** ···············

Early Paleozoic

The Phanerozoic eon has three eras. *The* **Paleozoic** (pay lee uh ZOH ihk) **era** *is the oldest era of the Phanerozoic eon. The* **Mesozoic** (mez uh ZOH ihk) **era** *is the middle era of the Phanerozoic eon. The* **Cenozoic** (sen uh ZOH ihk) **era** *is the youngest era of the Phanerozoic eon.*

As shown below, the Paleozoic era lasted for more than half of the Phanerozoic eon. The Paleozoic era is divided into three parts: early, middle, and late. The Cambrian and Ordovician periods make up the Early Paleozoic.

✓ **Visual Check**

1. Sequence Which is the youngest period of the Paleozoic era?

_____

_____

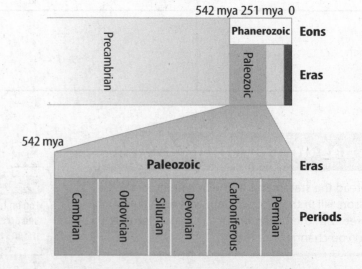

The Age of Invertebrates

The organisms from the Cambrian explosion were invertebrates (ihn VUR tuh brayts). Invertebrates are animals without backbones.

The invertebrates of the Cambrian lived only in the oceans. So many kinds of invertebrates lived in the Early Paleozoic oceans that this time is often called the age of invertebrates.

Geology of the Early Paleozoic

If you could have visited Earth during the Early Paleozoic, it would have seemed unfamiliar to you. There was no life on land. All life was in the oceans.

The shapes and locations of Earth's continents also would have been unfamiliar to you if you could have visited Earth during the Early Paleozoic. The landmass that would become North America was on the equator.

Earth's climate was warm during the Early Paleozoic. Rising seas flooded the continents. The water formed many shallow, inland seas. *An* **inland sea** *is a body of water formed when ocean water floods continents.* Most of North America was covered by an inland sea. ✓

Middle Paleozoic

The Early Paleozoic ended with a mass extinction event, but many invertebrates survived. Huge coral reefs along the edges of the continents were home to new forms of life during the Middle Paleozoic. The Middle Paleozoic consists of the Silurian (suh LOOR ee un) period, 441–416 mya, and the Devonian (dih VOH nee un) period, 416–359 mya.

Soon, vertebrates evolved. Vertebrates are animals with backbones.

The Age of Fishes

Some of the earliest vertebrates were fishes. So many types of fishes lived during the Middle Paleozoic that this time is often called the age of fishes. Bony armor covered the bodies of some fishes, such as the *Dunkleosteus*. This fish was a top Devonian predator.

On land, cockroaches, dragonflies, and other insects evolved. Earth's first plants grew. These early plants were small and lived in water.

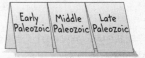

FOLDABLES®

Make a horizontal three-tab book to record information about changes during the Paleozoic era.

Early Paleozoic | Middle Paleozoic | Late Paleozoic

✓ **Reading Check**

2. Explain How do inland seas form?

_____

_____

_____

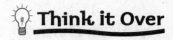 **Think it Over**

3. Identify When was the age of fishes? (Circle the correct answer.)

a. the Early Paleozoic

b. the Middle Paleozoic

c. the Late Paleozoic

Geology of the Middle Paleozoic

Middle Paleozoic rocks contain evidence of major collisions between moving continents. These collisions created mountain ranges. When several landmasses collided with the eastern coast of North America, the Appalachian (ap uh LAY chun) Mountains began to form. By the end of the Paleozoic era, these mountains were probably as high as the Himalayas are today. 🔑

Late Paleozoic

Like the Early Paleozoic, the Middle Paleozoic ended with a mass extinction event. Many of the invertebrates that lived in the seas and some land animals disappeared.

The Age of Amphibians

In the Late Paleozoic, some fishlike organisms spent part of their lives on land. Amphibians had lungs and could breathe air. *Tiktaalik* (tihk TAH lihk) was one of the earliest of amphibians. Amphibians were so common in the Late Paleozoic that this time is known as the age of amphibians.

Amphibians were adapted to land in several ways. They had lungs and could breathe air. They had thick skin, which kept them from drying out too quickly. They also had strong limbs, which helped them move around on land. These early amphibians, like amphibians today, had to return to the water to mate and lay eggs. 🔑

Reptile species evolved toward the end of the Paleozoic era. Reptiles were the first animals that did not need water for reproduction. Like the eggs of reptiles today, their eggs had tough, leathery shells that kept them from drying out. Because of this, the entire life of a reptile could be spent on land.

Coal Swamps

During the Late Paleozoic, dense, tropical forests grew in swamps along shallow inland seas. When trees and other plants died, they sank into the swamps. *A* **coal swamp** *is an oxygen-poor environment where, over time, plant material changes into coal.* The coal swamps of the Carboniferous (car buhn IF er us) period, 359–299 mya, and the Permian period, 299–251 mya, are the major sources of coal that we use today.

🔑 Key Concept Check

4. Explain How did the Appalachian Mountains form?

_____

_____

_____

_____

🔑 Key Concept Check

5. Explain What adaptations enabled amphibians to live on land?

_____

_____

_____

_____

💡 **Think it Over**

6. Contrast Describe one major difference between reptiles and amphibians.

_____

_____

_____

Formation of Pangaea

Geologic evidence shows that many continental collisions occurred during the Late Paleozoic. As continents moved closer together, new mountain ranges formed. By the end of the Paleozoic era, Earth's continents had formed a giant supercontinent called Pangaea. *A* **supercontinent** *is an ancient landmass that separated into present-day continents.* Pangaea formed from land masses that came together close to Earth's equator, as shown below. As Pangaea formed, coal swamps dried up. Earth's climate became cooler and drier.

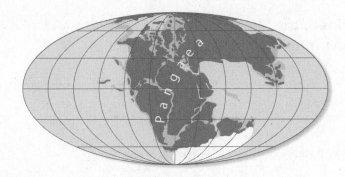

The Permian Mass Extinction

The largest mass extinction in Earth's history occurred at the end of the Paleozoic era. Fossil evidence indicates that 95 percent of life-forms in the oceans and 70 percent of all life on land became extinct. This extinction event is called the Permian mass extinction.

Scientists debate what caused this mass extinction. The formation of Pangaea likely decreased the amount of space where ocean organisms could live. The formation of Pangaea also would have changed ocean currents, making the center part of Pangaea drier. But Pangaea formed over many millions of years. The extinction event occurred more suddenly.

Some scientists hypothesize that the impact of a large meteorite caused an extreme climate change. Other scientists propose that huge volcanic eruptions changed Earth's climate. Both a meteorite impact and erupting volcanoes would have caused large amounts of ash, dust, and rock to enter the atmosphere. This debris would have blocked sunlight, reducing temperatures and destroying food webs.

Whatever the cause, Earth had fewer species after the Permian mass extinction. Only a few species adapted to the changes and survived this mass extinction.

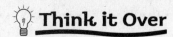

Think it Over

7. Define What was Pangaea? (Circle the correct answer.)

a. a supercontinent

b. a type of coal swamp

c. a Paleozoic amphibian

Visual Check

8. Sketch Use a highlighter to outline the area that is the continent of Africa today.

Key Concept Check

9. Specify What does fossil evidence reveal about the end of the Paleozoic era?

_____

_____

_____

_____

Mini Glossary

Cenozoic (sen uh ZOH ihk) era: the youngest era of the Phanerozoic eon

coal swamp: an oxygen-poor environment where, over time, plant material changes into coal

inland sea: a body of water that forms when oceans flood continents

Mesozoic (mez uh ZOH ihk) era: the middle era of the Phanerozoic eon

Paleozoic (pay lee uh ZOH ihk) era: the oldest era of the Phanerozoic eon

supercontinent: an ancient landmass that separated into present-day continents

1. Review the terms and their definitions in the Mini Glossary. Put the terms *Mesozoic era, Cenozoic era,* and *Paleozoic era* in order from oldest to youngest.

_____

_____

2. Use what you have learned about the Paleozoic era to complete the table.

Paleozoic Event	When It Happened—Early, Middle, or Late
Pangaea formed.	
Insects evolved on land.	Middle
Vertebrates began to evolve.	
Inland seas covered much of Earth.	Early
Reptiles appeared on Earth.	
The Appalachian Mountains formed.	Middle
This time is called the age of fishes.	
Coal swamps formed in many places.	Late
The largest mass extinction occurred.	
Amphibians appeared on Earth.	Late

3. How did answering the Key Concept questions help you learn about the Paleozoic era?

_____

What do you think NOW?

Reread the statements at the beginning of the lesson. Fill in the After column with an A if you agree with the statement or a D if you disagree. Did you change your mind?

Log on to ConnectED.mcgraw-hill.com and access your textbook to find this lesson's resources.

END OF LESSON

Geologic Time

The Mesozoic Era

·············· **Before You Read** ··············

What do you think? Read the two statements below and decide whether you agree or disagree with them. Place an A in the Before column if you agree with the statement or a D if you disagree. After you've read this lesson, reread the statements to see if you have changed your mind.

Before	Statement	After
	5. All large Mesozoic vertebrates were dinosaurs.	
	6. Dinosaurs disappeared in a large mass extinction event.	

··············· **Read to Learn** ···············

Geology of the Mesozoic Era

When people imagine what Earth looked like millions of years ago, they often picture dinosaurs. Dinosaurs lived during the Mesozoic era. This era of geologic time lasted from 251 mya to 65.5 mya. As shown in the figure below, the Mesozoic era is divided into three periods: the Triassic (tri A sihk), the Jurassic (joo RA sihk), and the Cretaceous (krih TAY shus).

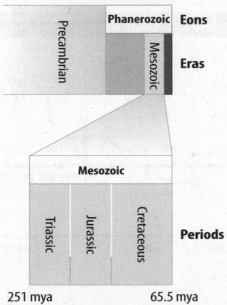

Mesozoic Era

Eons — Phanerozoic

Eras — Precambrian / Mesozoic

Periods — Mesozoic: Triassic, Jurassic, Cretaceous

251 mya — 65.5 mya

Key Concepts 🔑

- What major geologic events occurred during the Mesozoic era?
- What does fossil evidence reveal about the Mesozoic era?

▸ Study Coach

Make Flash Cards Change each main head into a question. Write each question on one side of a small index card. Write the answer on the other side of the card. Quiz yourself until you know all of the answers.

✓ **Visual Check**

1. Name List the periods of the Mesozoic era from oldest to youngest.

_____

_____

_____

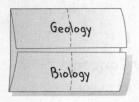
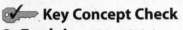
Breakup of Pangaea

Recall that Pangaea was a supercontinent that formed at the end of the Paleozoic era. Pangaea began to break apart in the Late Triassic period. Eventually, Pangaea split into two landmasses—Gondwanaland (gahn DWAH nuh land) and Laurasia (la RAY SHZah).

Gondwanaland was the southern continent. It included the future continents of Africa, Antarctica, Australia, and South America. Laurasia was the northern continent. It included the future continents of North America, Europe, and Asia.

Return of Shallow Seas

Much of Earth during the Mesozoic era was covered by lush, tropical forests and warm ocean waters. This is because the climate of the Mesozoic era was much warmer than the climate of the Paleozoic era. Many species were adapted to this type of environment. It was so warm that, for most of the Mesozoic era, there were no ice caps, even at the poles. With no glaciers, more water filled the oceans.

Some of this water flowed onto the continents as Pangaea began to split apart. This created narrow channels that grew larger as the continents moved apart. Eventually, the channels became oceans. The Atlantic Ocean began to form at this time.

The graph below shows the rise in sea level during the Mesozoic era. Toward the end of the era, sea level was so high that inland seas covered much of Earth's continents. These warm, shallow waters provided environments in which new organisms evolved.

Key Concept Check

2. Explain When did the Atlantic Ocean begin to form?

_____

_____

_____

Interpret a Graph

3. Identify In which period was sea level at its highest?

_____

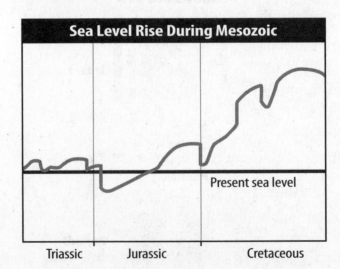

Sea Level Rise During Mesozoic

Present sea level

Triassic   Jurassic   Cretaceous

Mesozoic North America

Along North America's eastern coast and the Gulf of Mexico, sea level rose and fell over millions of years. As this happened, much seawater <u>evaporated</u>, leaving large salt deposits behind. Some of these salt deposits are sources of salt today. Other salt deposits later became traps for oil. Today, salt traps in the Gulf of Mexico are an important source of oil.

Throughout the Mesozoic era, the North American continent moved slowly and steadily westward. Its western edge collided with several small landmasses carried on an ancient oceanic plate. As this plate subducted beneath the North American continent, the crust buckled inland, slowly pushing up the Rocky Mountains. In the dry Southwest, windblown sand formed huge dunes. In the middle of the continent, a warm inland sea formed. ✔

Mesozoic Life

The species that survived the Permian mass extinction event lived in a world with few species. Vast amounts of space were open for animals and plants to inhabit. New types of cone-bearing trees, such as pines and cycads, began to appear. Toward the end of the Mesozoic era, the first flowering plants evolved. The dinosaurs were the dominant vertebrates living on land. Hundreds of species existed.

Dinosaurs

Scientists today disagree about classifying dinosaurs as reptiles. Dinosaurs share a common ancestor with present-day reptiles, such as crocodiles. However, dinosaurs differ from present-day reptiles in their hip structure. *Dinosaurs were dominant Mesozoic land vertebrates that walked with legs positioned directly below their hips.* This means that dinosaurs walked upright. In contrast, the legs of a crocodile stick out sideways from its body. As a result, a crocodile appears to drag itself along the ground.

Scientists hypothesize that some dinosaurs are more closely related to present-day birds than they are to present-day reptiles. Some dinosaur fossils show evidence of a feathery covering. *Archaeopteryx* (ar kee AHP tuh rihks), for example, was a small bird the size of a pigeon. It had wings and feathers. It also had claws and teeth. Many scientists suggest that it was an ancestor to birds. Future discoveries might help answer the question about why some dinosaurs are more closely related to present-day birds than they are to present-day reptiles.

✔ **Key Concept Check**
4. Explain How did the Rocky Mountains form?

_____

_____

_____

_____

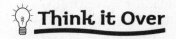 **Think it Over**

5. Make Connections Which animals on Earth today are most closely related to dinosaurs of the Mesozoic era? (Circle the correct answer.)

a. birds

b. mammals

c. reptiles

Other Mesozoic Vertebrates

Dinosaurs dominated the land during the Mesozoic era. But fossils show that other large vertebrates swam in the seas and flew in the air. **Plesiosaurs** (PLY zee oh sorz) *were Mesozoic marine reptiles with small heads, long necks, and flippers.* Through much of the Mesozoic, these reptiles dominated the oceans. Some were as long as 14 m.

Other Mesozoic reptiles could fly. **Pterosaurs** (TER oh sorz) *were Mesozoic flying reptiles with large, batlike wings.* The *Quetzalcoatlus* (kwetz oh koh AHT lus) was one of the largest pterosaurs. It had a wingspread of nearly 12 m. Though pterosaurs could fly, they were not birds. Recall that birds are more closely related to dinosaurs.

Appearance of Mammals

Dinosaurs and reptiles dominated the Mesozoic era, but another type of vertebrate also lived during this time— mammals. Mammals evolved early in the Mesozoic era. They were small in size. Few were larger than present-day cats.

Cretaceous Extinction Event

The Mesozoic era ended 65.5 mya with a mass extinction called the Cretaceous extinction event. Scientists propose that a large meteorite impact was the main cause of this extinction. This crash would have produced enough dust to block sunlight for a long time. There is evidence that volcanoes also erupted at the same time. These eruptions would have added more dust to the atmosphere. Without light, plants died. Without plants, animals died. Dinosaur species and other large Mesozoic vertebrate species could not survive in the changed environment. They became extinct.

<hr>

✏️ Key Concept Check

6. Contrast How could you distinguish fossils of plesiosaurs and pterosaurs from fossils of dinosaurs?

_____

_____

_____

💡 Think it Over

7. Generalize How could the impact of a meteorite help cause a mass extinction event?

_____

_____

_____

_____

Mini Glossary

dinosaur: a dominant Mesozoic land vertebrate that walked with legs positioned directly below its hips

plesiosaur (PLY zee oh sor): a Mesozoic marine reptile with a small head, long neck, and flippers

pterosaur (TER oh sor): a Mesozoic flying reptile with large, batlike wings

1. Review the terms and their definitions in the Mini Glossary. Write a sentence describing one group of Mesozoic animals.

_____

_____

2. In the center oval, list the similarities among the three groups shown.

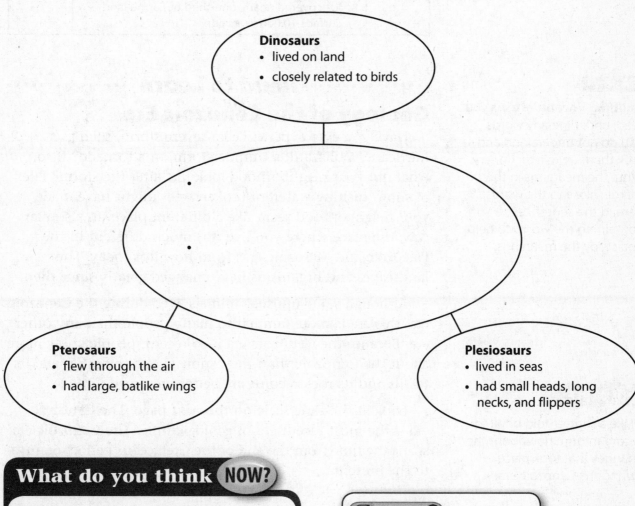

Dinosaurs
- lived on land
- closely related to birds

Pterosaurs
- flew through the air
- had large, batlike wings

Plesiosaurs
- lived in seas
- had small heads, long necks, and flippers

What do you think NOW?

Reread the statements at the beginning of the lesson. Fill in the After column with an A if you agree with the statement or a D if you disagree. Did you change your mind?

 Connect ED

Log on to ConnectED.mcgraw-hill.com and access your textbook to find this lesson's resources.

 END OF LESSON

Geologic Time

The Cenozoic Era

Copyright © Glencoe/McGraw-Hill, a division of The McGraw-Hill Companies, Inc.

Key Concepts 🔑

- What major geologic events occurred during the Cenozoic era?
- What does fossil evidence reveal about the Cenozoic era?

Mark the Text

Building Vocabulary As you read, underline any words you do not understand and look them up in a dictionary. Write the meanings in the margin close to the words. Reread the sentences containing the words to help you study the meanings.

FOLDABLES®

Make a shutter-fold book to record information about the changes that took place during the Cenozoic era.

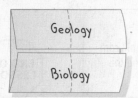

·············· **Before You Read** ··············

What do you think? Read the two statements below and decide whether you agree or disagree with them. Place an A in the Before column if you agree with the statement or a D if you disagree. After you've read this lesson, reread the statements to see if you have changed your mind.

Before	Statement	After
	7. Mammals evolved after dinosaurs became extinct.	
	8. Ice covered nearly one-third of Earth's land surface 10,000 years ago.	

·············· **Read to Learn** ··············

Geology of the Cenozoic Era

Have you ever experienced a severe storm, such as a hurricane, a blizzard, a thunderstorm, or a tornado? If so, what did your neighborhood look like after the storm? Piles of snow, rushing water, or broken trees might have made your neighborhood seem like a different place. In a similar way, the place where you live was much different in the Paleozoic and Mesozoic eras from how it is today. The landscapes and organisms have changed greatly since then.

Although some unusual animals lived during the Cenozoic era, this era is more familiar to many people than any other era. Because the Cenozoic era is so recent, people know more about the Cenozoic than they know about any other era. Its fossils and its rock records are better preserved.

Look at the time scale on the next page. The Cenozoic era is the most recent era of geologic time. The Cenozoic era spans the time from the end of the Cretaceous period, 65 mya, to the present.

Cenozoic Era

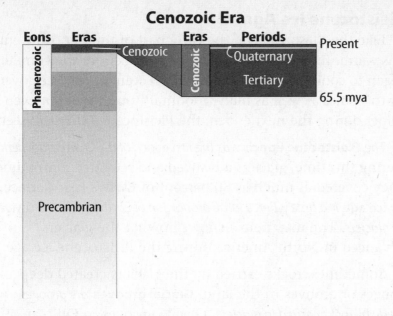

Eons | Eras | Eras | Periods

Phanerozoic

Cenozoic

Cenozoic

Quaternary

Tertiary

Present

65.5 mya

Precambrian

Geologists divide the Cenozoic era into two periods—the Tertiary (TUR shee ayr ee) period and the Quaternary (KWAH tur nayr ee) period. These two periods are further divided into seven epochs. *The most recent epoch, the* **Holocene** (HOH luh seen) **epoch,** *began 10,000 years ago.* You live in the Holocene epoch.

Cenozoic Mountain Building

Earth's continents continued to move apart during the Cenozoic era. The Atlantic Ocean continued to widen. As the continents moved, some landmasses collided. Early in the Tertiary period, India crashed into Asia. This collision began to push up the Himalayas—the highest mountains on Earth today. At about the same time, Africa began to push into Europe. This collision formed the Alps—another mountain range. Both the Himalayas and the Alps continue to get higher today.

Recall that during the Mesozoic era, the western coast of North America pushed against the seafloor next to it and formed the Rocky Mountains. These plates are still colliding today, causing the Rockies to continue to rise. Plate collisions during the Cenozoic era also caused other mountain ranges—the Cascades and the Sierra Nevadas—to begin forming along the western coast of North America. On the eastern coast, there was little tectonic activity. The Appalachian Mountains, which formed during the Paleozoic era, continue to erode, or wear away. ✓

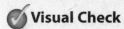

Visual Check

1. Identify the two periods of the Cenozoic era.

_____

_____

Math Skills ÷

The Cenozoic era began 65.5 mya. What percentage of the Cenozoic era is taken up by the Quaternary period, which began 2.6 mya? To calculate the percentage of a part to the whole, perform the following steps:

a. Express the problem as a fraction.

$$\frac{2.6 \text{ mya}}{65.5 \text{ mya}}$$

b. Convert the fraction to a decimal.

2.6 mya ÷ 65.5 mya = 0.040

c. Multiply by 100 and add %.

0.040 × 100 = 4.0%

2. Use Percentages

What percent of the Cenozoic era is represented by the Tertiary period, which lasted from 65.5 mya to 2.6 mya? [Hint: Subtract to find the length of the Tertiary period.]

_____

Reading Check

3. Explain Why are the Appalachian Mountains relatively small today?

_____

_____

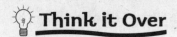

Think it Over

4. Contrast How was the climate of the Pleistocene epoch different from the climate of the early Cenozoic era?

_____

_____

Think it Over

5. Summarize Why did sea level drop during the Pleistocene ice age?

_____

_____

_____

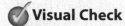 **Visual Check**

6. Estimate Approximately what percentage of the United States was covered with ice?

_____

_____

Pleistocene Ice Age

Like the Mesozoic era, the early part of the Cenozoic era was warm. In the middle of the Tertiary period, the climate began to cool. By the Pliocene (PLY oh seen) epoch, ice covered Earth's poles as well as many mountaintops. Earth got even colder during the next epoch, the Pleistocene (PLY stoh seen).

The **Pleistocene epoch** _was the first epoch of the Quaternary period._ During this time, glaciers advanced and retreated many times. They covered as much as 30 percent of Earth's land surface. _An_ **ice age** _is a time when a large proportion of Earth's surface is covered by glaciers._ The map below shows how far the glaciers advanced in North America during the Pleistocene ice age.

Sometimes, rocks carried by the glaciers created deep gouges or grooves in the land. **Glacial grooves** _are grooves made by rocks carried in glaciers._ Glacial grooves in Ohio are evidence that glaciers advanced far into North America during the Pleistocene ice age.

The glaciers contained huge amounts of water. This water came from the oceans. Glaciers trapped so much water that sea level dropped. As sea level dropped, the inland seas that had covered the continents drained away, exposing dry land. When sea level was at its lowest, the Florida peninsula was about twice as wide as it is today.

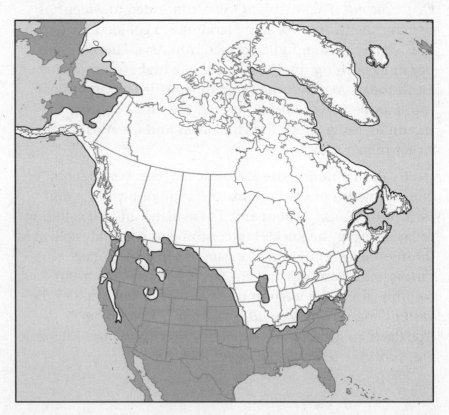

Cenozoic Life—The Age of Mammals

The mass extinction event at the end of the Mesozoic era meant that there was more space for the surviving species. Flowering plants, including grasses, evolved and began to dominate the land. Flowering trees and grasses provided new food sources. The new food sources helped make it possible for many animals, including mammals, to evolve. Mammals were so successful during the Cenozoic era that this era is sometimes called the age of mammals.

Mega-Mammals

Recall that mammals were small during the Mesozoic era. Many new types of mammals appeared during the Cenozoic era. Some were very large, such as those shown below. *The large mammals of the Cenozoic era are called* **mega-mammals.**

Some of the largest mega-mammals lived during the Oligocene and Miocene periods, from 34 mya to 5 mya. Others, such as woolly mammoths, giant sloths, and saber-toothed cats, lived during the cool climate of the Pliocene and Pleistocene periods. These periods lasted from 5 mya to 10,000 years ago. Many fossils of these animals have been discovered. A few preserved mammoth bodies also have been discovered in glacial ice. All of the Cenozoic mega-mammals shown below are extinct.

 Key Concept Check

7. Explain How do scientists know that mega-mammals lived during the Cenozoic era?

_____

_____

Visual Check

8. Compare the sizes of Cenozoic era animals in the figure to the human. Make a general statement about their sizes.

_____

_____

_____

Isolated Continents and Land Bridges

The mega-mammal lived in the Americas, Europe, and Asia. Different mammal species evolved in Australia during this time. This is mostly because of the movement of Earth's tectonic plates. Recall that land bridges can connect continents that were once separated. Also recall that when continents are separated, members of a species that once lived together can become geographically isolated.

Most mammals that live in Australia today are marsupials (mar SOO pee ulz). Marsupials are mammals, such as kangaroos, that carry their young in pouches. Some scientists suggest that marsupials did not evolve in Australia. Instead, scientists <u>hypothesize</u> that the ancestors of marsupials migrated to Australia from South America. This migration would have happened when South America and Australia were both connected to Antarctica by land bridges, as shown below. After the ancestors of marsupials arrived in Australia, Australia moved away from the other landmasses. Sea level rose and water covered the land bridges. Over time, the marsupials evolved into the types of marsupials that live in Australia today. ✓

Copyright © Glencoe/McGraw-Hill, a division of The McGraw-Hill Companies, Inc.

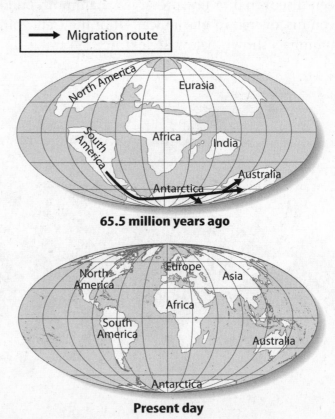

65.5 million years ago

Present day

ACADEMIC VOCABULARY

hypothesize
(verb) to make an assumption about something that is not positively known

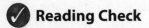 **Reading Check**

9. Contrast What major geologic events affected the evolution of marsupials in Australia?

_____

_____

_____

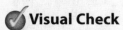 **Visual Check**

10. Analyze Would the migration that took place 65.5 mya be possible today? Explain.

_____

_____

_____

_____

Rise of Humans

The oldest fossil remains of human ancestors have been discovered in Africa. The fossil evidence suggests that Africa is where humans first evolved. These fossils are nearly 6 million years old. Scientists discovered a skeleton of a 3.2-million-year-old human ancestor. They named the skeleton *Lucy.*

Modern humans, called *Homo sapiens,* didn't evolve until the Pleistocene epoch. Early *Homo sapiens* migrated to Europe, Asia, and eventually North America. Early humans likely migrated to North America from Asia, using a land bridge that connected the continents during the Pleistocene ice age. Today, this land bridge is covered with water.

Pleistocene Extinctions

Earth's climate changed at the close of the Pleistocene epoch 10,000 years ago. The Holocene epoch was warmer and drier. Forests replaced grasses. The mega-mammals that lived during the Pleistocene became extinct. Some scientists suggest that mega-mammal species did not adapt to the changes in the environment.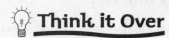

Future Changes

Evidence suggests that present-day Earth is undergoing a global-warming climate change. Many scientists hypothesize that humans have contributed to this change because of their use of coal, oil, and other fossil fuels over the past few centuries.

Think it Over

11. Specify In what epoch did modern humans first appear on Earth?

_____

Key Concept Check
12. Describe How did climate change at the end of the Pleistocene epoch?

_____

_____

Mini Glossary

glacial groove: a groove made by rocks carried in glaciers

Holocene (HOH luh seen) epoch: the most recent epoch, which began 10,000 years ago

ice age: a time when a large proportion of Earth's surface is covered by glaciers

mega-mammal: a large mammal of the Cenozoic era

Pleistocene (PLY stoh seen) epoch: the first epoch of the Quaternary period

1. Review the terms and their definitions in the Mini Glossary. Write a sentence that explains the order in which the Pleistocene and the Holocene epochs occurred.

_____

2. Complete the diagram by listing different things that happened during the Cenozoic era.

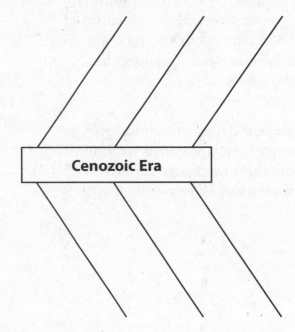

Cenozoic Era

3. The oldest well-preserved skeleton of a human ancestor was found in Africa. How old is the skeleton and what was it named?

_____

_____

What do you think NOW?

Reread the statements at the beginning of the lesson. Fill in the After column with an A if you agree with the statement or a D if you disagree. Did you change your mind?

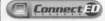 **Connect ED**

Log on to ConnectED.mcgraw-hill.com and access your textbook to find this lesson's resources.

END OF LESSON

Interactions Within Ecosystems

Ecosystems

············ **Before You Read** ··············

What do you think? Read the two statements below and decide whether you agree or disagree with them. Place an A in the Before column if you agree with the statement or a D if you disagree. After you've read this lesson, reread the statements to see if you have changed your mind.

Before	Statement	After
	1. In symbiosis, two species cooperate in a way that benefits both species.	
	2. Overpopulation can be damaging to an ecosystem.	

············· **Read to Learn** ··············

Abiotic and Biotic Factors

A productive garden requires sunlight, soil, water, and plants. Compost enriches the soil in a garden and insects and birds pollinate garden flowers. Gardens are home to birds, rabbits, beetles, and many other organisms that feed on plants. A garden is an example of an ecosystem.

Ecosystems contain all the nonliving and living parts of the environment in a given area. The nonliving parts of an ecosystem are called abiotic factors. They include sunlight, water, soil, and air. The abiotic factors, such as those in this figure, determine what kinds of organisms can live in the ecosystem. For example, dragonflies can survive only in wetland ecosystems. The living or once-living parts of an ecosystem are called biotic factors. These factors include living organisms, wastes from living organisms, and the decayed remains of dead organisms.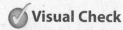

Wetland Ecosystem

Key Concepts

• How can you describe an ecosystem?

• In what ways do living organisms interact?

• How do population changes affect ecosystems?

Mark the Text

Building Vocabulary As you read, underline the words and phrases that you do not understand. When you finish reading, discuss these words and phrases with another student or your teacher.

✅ **Visual Check**

1. Examine What abiotic factors can you identify in this ecosystem?

_____

_____

🔑 **Key Concept Check**

2. Summarize How can you describe an ecosystem?

_____

_____

_____

Habitats

Every species lives in a particular habitat. *A **habitat** is the place within an ecosystem that provides the biotic and abiotic factors an organism needs to survive and reproduce.* For example, a habitat for dragonflies includes plants that grow in shallow water. The leaves of the plant provide a place for adult dragonflies to lay eggs. The underwater stems and leaves provide shelter for young dragonflies.

Populations and Communities

Each species of dragonfly in a wetland ecosystem forms a population. *A **population** is all the organisms of the same species that live in the same area at the same time.* Populations in an African savanna ecosystem include giraffes, kudus, wildebeests, zebras, and the other species shown in the figure below. *All the populations living in an ecosystem at the same time form a **community.***

☑ **Visual Check**
3. Identify Circle three biotic factors in this ecosystem.

Savanna Ecosystem

Interactions of Living Things

More than one population can live in the same habitat. For example, giraffes and two types of antelope—kudus and steenboks—feed on trees that grow in the African savanna. How can these three populations share the same habitat? Each species uses resources, such as water, food, and shelter, in its habitat. However, each species uses the resources in a different way.

A **niche** (NICH) *is the way a species interacts with abiotic and biotic factors to obtain food, find shelter, and fulfill other needs.* As shown in the figure on the previous page, giraffes feed at the tops of the trees. Kudus eat from mid-level branches. Steenboks feed on the lowest branches of trees and shrubs. Although all of these animals live in the same area, they use resources in their habitat in different ways. Giraffes, kudus, and steenboks have different niches.

Predation

A predator is an organism that hunts and kills other organisms for food. Prey is an organism that a predator catches and eats. **Predation** *is the act of one organism, the predator, feeding on another organism, its prey.* An African lion, like the one in the figure on the previous page, is a predator that eats zebras and other savanna species. These animals are the lion's prey. ✓

Symbiosis

Another type of interaction between organisms is symbiosis. **Symbiosis** *is a close, long-term relationship between two species that usually involves an exchange of food or energy.* There are three types of symbiosis: mutualism, commensalism, and parasitism.

Mutualism Both species benefit from the relationship. For example, honeybees pollinate acacia flowers as they collect nectar, helping the acacias produce fertile seeds.

Commensalism One species benefits from the relationship. The other species is neither harmed nor benefited. For example, a bird nesting in a tree has a place to raise its young. It neither harms nor benefits the tree.

Parasitism One species (the parasite) benefits and the other (the host) is harmed. For example, the roots of the *Striga* plant grow into the host plant, robbing the host plant of water and nutrients. ✓

Copyright © Glencoe/McGraw-Hill, a division of The McGraw-Hill Companies, Inc.

FOLDABLES

Create a horizontal four-door book to organize your notes on ecosystems.

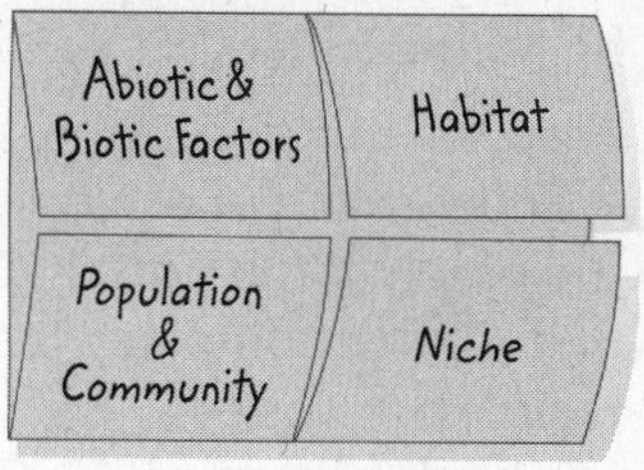

| Abiotic & Biotic Factors | Habitat |
| Population & Community | Niche |

✓ **Reading Check**

4. Apply Name one other organism that is a predator. What is its prey?

_____

_____

🔑 **Key Concept Check**

5. Describe What is one way that living things interact?

_____

_____

_____

_____

Competition

Organisms that share the same habitat often compete for resources. Competition describes interactions between two or more organisms that need the same resource at the same time. For example, trees compete for sunlight, and shade from tall trees can slow the growth of younger trees. Wolves compete with ravens for meat from the animals that wolves kill.

Population Changes

The number of individuals in a population is always changing. Populations get larger when offspring are produced or when new individuals move into a community. Populations get smaller when individuals die or leave.

Changes in the abiotic factors of an ecosystem affect population size. If a drought reduces plant growth, less food is available for plant eaters. Less food can lead to a decrease in the size of plant-eater populations. Interactions between organisms also affect population size. As shown in the graph below, predators help control the size of prey populations.

Population density is the size of a population compared to the amount of space available. A high population density means individuals live closer together. This can increase competition and make it easier for disease to be transmitted from one individual to another.

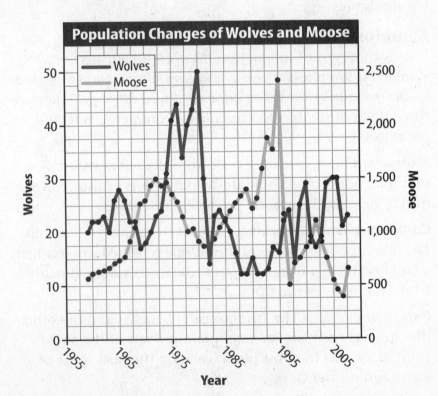

Population Changes of Wolves and Moose

— Wolves
— Moose

Overpopulation

In summer, an adult moose eats about 18 kg of leaves and twigs every day. As long as food is available, a moose population can continue to grow. But there is a limit to the resources an ecosystem can provide. **Carrying capacity** *is the largest number of individuals of one species that an ecosystem can support over time.* A habitat's carrying capacity depends on the abiotic and biotic factors that are present. ✓

What happens if a population exceeds its carrying capacity? The area becomes overpopulated. When a moose population gets too large, the moose eat so much that they can damage or kill plant life. The destruction of plants destroys habitat for the moose and other species. When food becomes scarce, individuals might leave the area, starve, or sicken and die.

Changes in an ecosystem can increase or decrease its carrying capacity for a particular species. Drought, flood, or the arrival of a competing species can reduce carrying capacity. On the other hand, good growing conditions or the disappearance of a competing species can increase carrying capacity. ✓

Extinction

If all the members of a population die or move away from an area, that population becomes extinct. If all populations of a species disappear from Earth, the entire species becomes extinct.

Extinction of one population can affect other populations. What if all the individuals in a moose population died or moved from an area? Carrying capacity for wolves in that population's area could decrease because no moose would be available as a source of food. Some plant-eating animals would no longer be competing with moose for food. The carrying capacity for those populations of plant eaters could increase. ✓━

Copyright © Glencoe/McGraw-Hill, a division of The McGraw-Hill Companies, Inc.

✓ **Reading Check**

8. Define What is carrying capacity?

_____

_____

_____

✓ **Reading Check**

9. Identify two factors that can decrease an ecosystem's carrying capacity.

_____

_____

✓━ **Key Concept Check**

10. Explain How do population changes affect ecosystems?

_____

_____

_____

_____

Mini Glossary

carrying capacity: the largest number of individuals of one species that an ecosystem can support over time

community: all the populations living in an ecosystem at the same time

habitat: the place within an ecosystem that provides the biotic and abiotic factors an organism needs to survive and reproduce

niche (NICH): the way a species interacts with abiotic and biotic factors to obtain food, find shelter, and fulfill other needs

population: all the organisms of the same species that live in the same area at the same time

predation: the act of one organism, the predator, feeding on another organism, its prey

symbiosis: a close, long-term relationship between two species that usually involves an exchange of food or energy

1. Review the terms and their definitions in the Mini Glossary. Write a sentence explaining the relationship between population and carrying capacity.

   _____

   _____

2. Suppose an ecosystem had a good growing season with above average rainfall. In the population boxes, write *increase* or *decrease* to identify the effects on populations living in the ecosystem.

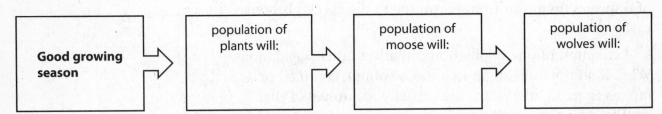

| Good growing season | → | population of plants will: | → | population of moose will: | → | population of wolves will: |

3. Explain why competition exists in a habitat.

   _____

   _____

What do you think NOW?

Reread the statements at the beginning of the lesson. Fill in the After column with an A if you agree with the statement or a D if you disagree. Did you change your mind?

Log on to ConnectED.mcgraw-hill.com and access your textbook to find this lesson's resources.

END OF LESSON

Interactions Within Ecosystems

Energy and Matter

Copyright © Glencoe/McGraw-Hill, a division of The McGraw-Hill Companies, Inc.

·············· Before You Read ··············

What do you think? Read the two statements below and decide whether you agree or disagree with them. Place an A in the Before column if you agree with the statement or a D if you disagree. After you've read this lesson, reread the statements to see if you have changed your mind.

Before	Statement	After
	3. Sunlight provides the energy at the base of all food chains on Earth.	
	4. A detritivore is a type of carnivore.	

·············· Read to Learn ··············

Food Energy

Organisms need a constant supply of energy, in the form of food, to stay alive. How a species obtains energy is an important part of its niche.

Producers *are organisms that use an outside energy source, such as the Sun, and produce their own food.* Producers include green plants, algae, and some kinds of bacteria. Most producers make energy-rich compounds through photosynthesis. Recall that photosynthesis is the chemical process that uses carbon dioxide, water, and light energy—usually from the Sun—to produce glucose (a type of sugar) and oxygen.

Some producers make energy-rich compounds through chemosynthesis, a chemical process similar to photosynthesis. Chemosynthesis uses a chemical such as hydrogen sulfide or methane, instead of light, to produce glucose. Producers that use chemosynthesis include bacteria that live in hot springs or near deep-sea thermal vents. ✔

Consumers

Organisms that cannot make their own food are **consumers.** Consumers obtain energy and nutrients by consuming other organisms or compounds produced by other organisms. There are four types of consumers: herbivores, omnivores, carnivores, and detritivores.

Key Concepts

- How does energy move through an ecosystem?
- How does matter move through an ecosystem?

Study Coach

Make Flash Cards For each head in this lesson, write a question on one side of a flash card and the answer on the other side. Quiz yourself until you know all of the answers.

✔ **Reading Check**

1. Differentiate How do chemosynthesis and photosynthesis differ?

_____

_____

_____

_____

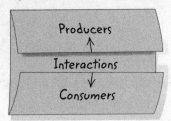

Make a horizontal shutterfold book to organize your notes on producers and consumers.

Producers
↑
Interactions
↓
Consumers

Herbivores Consumers that eat only producers are herbivores. For example, insects called aphids obtain energy and nutrients by sucking fluids from plants. Herbivores include butterflies, snails, mice, rabbits, fruit-eating bats, gorillas, and cows.

Omnivores Consumers that eat both producers and consumers are omnivores. For example, raccoons will eat almost anything, from fruit and insects to fish and garbage. Omnivores include corals, crickets, ants, bears, robins, and humans.

Carnivores Consumers are carnivores if they eat herbivores, omnivores, and other carnivores. For example, moles eat small animals such as earthworms. Carnivores include scorpions, octopuses, sharks, tuna, frogs, insect-eating bats, and owls.

Detritivores *A* **detritivore** (duh TRI tuh vor) *consumes dead organisms and wastes produced by living organisms*. For example, termites feed on decaying plant materials. Detritivores also include wood lice and earthworms.

Scavengers are detritivores that eat animals killed by carnivores or omnivores. Examples include hyenas, jackals, and vultures.

Decomposers are microscopic detritivores. They cause the decay of dead organisms or wastes produced by living organisms. Most decomposers are fungi and bacteria. ✓

The Flow of Energy

Recall that producers use energy from the environment and make their own food. This is the first step in the flow of energy through an ecosystem.

Most producers use photosynthesis and convert the energy from sunlight into chemical energy stored in food molecules. Others use chemosynthesis. Once energy from the environment is converted into food energy, it can be transferred to other organisms. ✓

In an ecosystem, food energy is transferred from one organism to another through feeding relationships. Food chains and food webs are models that describe how energy is transferred through an ecosystem.

✓ **Reading Check**

2. Explain How do organisms obtain energy?

_____

_____

_____

_____

✓ **Reading Check**

3. Describe How do most producers obtain energy?

_____

_____

_____

Food Chains

A food chain is a simple model that shows how energy moves from a producer to one or more consumers through feeding relationships. Every food chain begins with a producer because producers are the source of all food energy in an ecosystem.

Food Webs

Most ecosystems contain many food chains. *A food web is a model of energy transfer that can show how the food chains in a community are interconnected.* The figure below shows a food web for a rain-forest community. One food chain in this food web shows the transfer of energy from the Sun, to berries, to a sloth.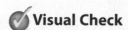

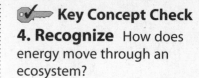 **Key Concept Check**

4. Recognize How does energy move through an ecosystem?

_____

_____

_____

✓ **Visual Check**

5. Identify Circle two food sources for the golden lion tamarin.

Energy Flow Through Ecosystems

Interactions Within Ecosystems **325**

Energy Pyramids

Most food chains have at least three links, but no more than five. Why? Think about a blackberry plant. It converts light energy from the Sun to chemical energy stored in the plant's tissues. The plant uses some of that energy to perform life processes, including growing blackberries. Some of the stored energy is lost as thermal energy.

When a robin eats blackberries, only part of the energy stored in the plant is transferred to the bird. The robin uses some of the energy for its own life processes, and again some of the energy is lost as thermal energy. A falcon that eats the robin receives even less energy for its life processes.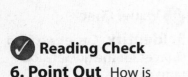

An **energy pyramid** *is a model that shows the amount of energy available in each link of a food chain.* As shown in the figure below, an energy pyramid shows that each step in a food chain contains less energy than the previous steps. The loss of energy at each level of an energy pyramid helps explain why there are always more producers than carnivores in a community. ✓

Copyright © Glencoe/McGraw-Hill, a division of The McGraw-Hill Companies, Inc.

An Energy Pyramid

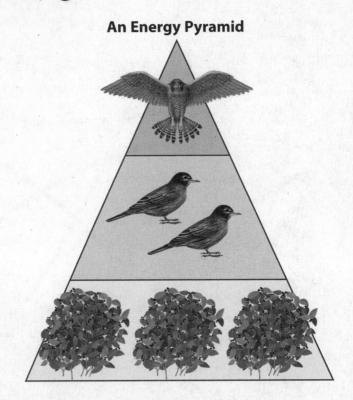

✅ **Reading Check**

6. Point Out How is energy lost in a food chain?

_____

_____

✅ **Reading Check**

7. Explain Why are there always more producers than carnivores in a community?

_____

_____

_____

_____

✅ **Visual Check**

8. Identify Which level in this pyramid has the greatest amount of available energy?

_____

_____

Cycling Materials

Living organisms need more than a constant supply of energy. They also need matter to build cells and tissues. Almost all matter on Earth today has been here since the planet formed. The law of conservation of matter states that matter cannot be created or destroyed, but it can change form. Matter is recycled through ecosystems, changing form along the way. Three of the most important pathways of matter moving through an ecosystem are described by the nitrogen cycle, the water cycle, and the oxygen-carbon dioxide cycle.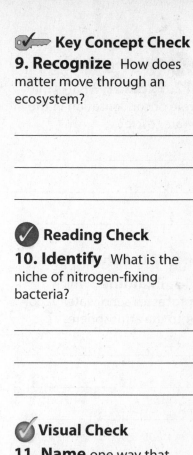

Nitrogen Cycle

Proteins are essential to all life. An important component of every protein molecule is the element nitrogen. Nitrogen gas makes up about 78 percent of Earth's atmosphere, but most organisms cannot obtain nitrogen from the air. Nitrogen-fixing bacteria live in the soil or on the roots of plants. These bacteria convert nitrogen gas into compounds that plants and other producers can absorb. ✓

Nitrogen changes form as it cycles through an ecosystem. The nitrogen cycle describes how nitrogen moves from the atmosphere, to the soil, into the bodies of living organisms, and back to the atmosphere. The figure below illustrates the nitrogen cycle.

🔑 Key Concept Check

9. Recognize How does matter move through an ecosystem?

_____

_____

_____

✓ Reading Check

10. Identify What is the niche of nitrogen-fixing bacteria?

_____

_____

_____

✓ Visual Check

11. Name one way that nitrogen compounds get into the soil.

_____

_____

The Nitrogen Cycle

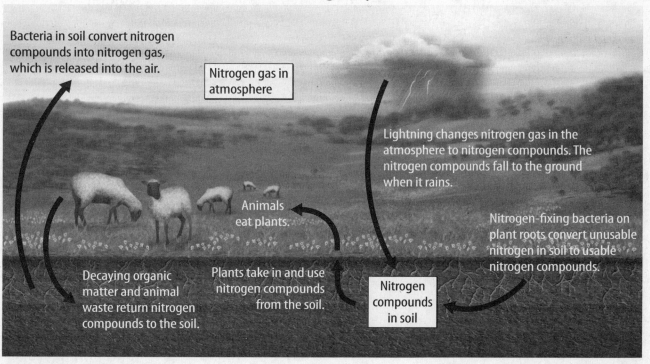

Bacteria in soil convert nitrogen compounds into nitrogen gas, which is released into the air.

Nitrogen gas in atmosphere

Lightning changes nitrogen gas in the atmosphere to nitrogen compounds. The nitrogen compounds fall to the ground when it rains.

Animals eat plants.

Nitrogen-fixing bacteria on plant roots convert unusable nitrogen in soil to usable nitrogen compounds.

Decaying organic matter and animal waste return nitrogen compounds to the soil.

Plants take in and use nitrogen compounds from the soil.

Nitrogen compounds in soil

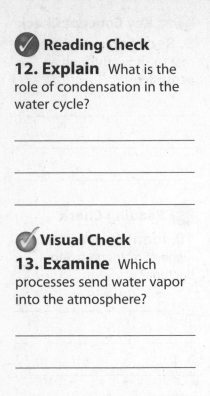

Reading Check

12. Explain What is the role of condensation in the water cycle?

_____

_____

_____

Visual Check

13. Examine Which processes send water vapor into the atmosphere?

_____

_____

_____

Water Cycle

Life as it is on Earth could not exist without water. Water is required for every process that takes place in cells and tissues, including cellular respiration, photosynthesis, and digestion. Recall that cellular respiration is a series of chemical reactions that convert the energy in food molecules into a usable source of energy called ATP.

The figure below illustrates the water cycle. All the freshwater on Earth's surface and in the bodies of living organisms is recycled through the water cycle.

Water evaporates from Earth's surface and rises into the atmosphere as water vapor. Water vapor also is released from the leaves of plants in the process called transpiration and from animals when they exhale. When water vapor comes into contact with cooler air, it condenses and forms clouds. Clouds condense further, and it rains or snows. Water returns to Earth's surface as precipitation. Plants and other organisms absorb water from the soil. Animals drink water or get water from the food they eat.

The Water Cycle

The Oxygen and Carbon Dioxide Cycles

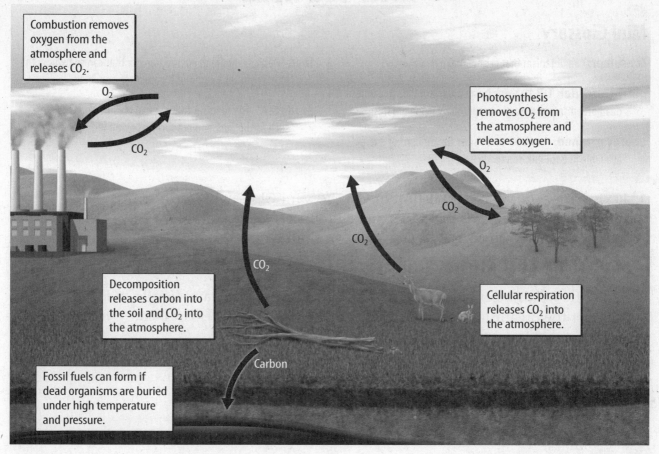

Combustion removes oxygen from the atmosphere and releases CO_2.

O_2

CO_2

Photosynthesis removes CO_2 from the atmosphere and releases oxygen.

O_2

CO_2

CO_2

CO_2

Decomposition releases carbon into the soil and CO_2 into the atmosphere.

CO_2

Cellular respiration releases CO_2 into the atmosphere.

Carbon

Fossil fuels can form if dead organisms are buried under high temperature and pressure.

The Oxygen and Carbon Dioxide Cycles

Oxygen and carbon dioxide also cycle through Earth's ecosystems. The cells of most organisms, including those in all plants and animals, require oxygen for cellular respiration. Cells use the energy released during cellular respiration for life processes.

Cellular respiration releases carbon dioxide into the atmosphere, where it can be taken in by plant leaves and used for photosynthesis. Photosynthesis releases oxygen into the atmosphere, where it can be taken in by animals, plants, and other organisms. As shown in the figure above, photosynthesis and cellular respiration are important processes in the cycling of carbon dioxide and oxygen through the living and nonliving parts of ecosystems. The oxygen and carbon dioxide cycle includes processes such as the formation and combustion of fossil fuels. ✓

✓ Visual Check

14. State What role does combustion play in the oxygen and carbon dioxide cycle?

_____

_____

_____

✓ Reading Check

15. Compare How do cellular respiration and photosynthesis compare?

_____

_____

_____

Mini Glossary

consumer: an organism that cannot make its own food

detritivore (duh TRI tuh vor): an organism that consumes dead organisms and wastes produced by living organisms

energy pyramid: a model that shows the amount of energy available in each link of a food chain

food web: a model of energy transfer that can show how the food chains in a community are interconnected

producer: an organism that uses an outside energy source, such as the Sun, and produces its own food

1. Review the terms and their definitions in the Mini Glossary. Write a sentence that tells the difference between a consumer and a producer.

_____

2. Write the words below at the correct levels in the energy pyramid. At each level, indicate whether the organisms are producers or consumers. Label the part of the pyramid that has more and less energy.

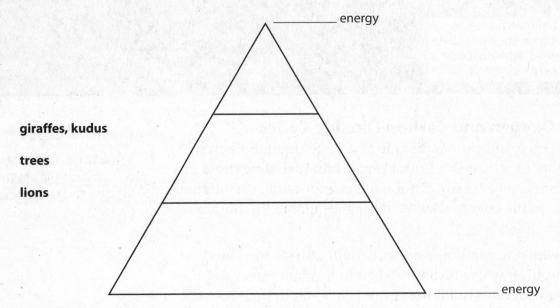

giraffes, kudus

trees

lions

3. Select one question that you wrote on a flash card and answer it.

_____

What do you think NOW?

Reread the statements at the beginning of the lesson. Fill in the After column with an A if you agree with the statement or a D if you disagree. Did you change your mind?

 **Connect ED**

Log on to ConnectED.mcgraw-hill.com and access your textbook to find this lesson's resources.

END OF LESSON

Interactions Within Ecosystems

Humans and Ecosystems

·············· **Before You Read** ··············

What do you think? Read the two statements below and decide whether you agree or disagree with them. Place an A in the Before column if you agree with the statement or a D if you disagree. After you've read this lesson, reread the statements to see if you have changed your mind.

Before	Statement	After
	5. Human actions can have unintended effects on the environment.	
	6. The only job of the U.S. Environmental Protection Agency is to enforce environmental laws.	

·············· **Read to Learn** ··············

Affecting the Environment

About 3.5 billion years ago, microscopic organisms called cyanobacteria (si an oh bak TIH ree uh) began to make a big change in Earth's environment. Cyanobacteria are unicellular organisms that probably were the first life-forms capable of photosynthesis.

Over a period of approximately one billion years, photosynthesis increased the amount of oxygen in the atmosphere. The oxygen level went from nearly zero to a level higher than today's. Without oxygen in the atmosphere, life on Earth might not have been possible.

All organisms change the environment in some way. For example, when beavers cut down trees and build dams, they reduce forest habitat. But at the same time, they create ponds—new habitats for dragonflies, frogs, ducks, and other species that live in or around water.

Humans also change the environment. Humans change ecosystems by replacing wildlife habitats with buildings, roads, farms, and mines. Our use of energy resources—such as coal and natural gas—can create pollutants that affect plant and animal life in the air, in the water, and on land.

Key Concepts

- In what ways do humans affect ecosystems?
- What can humans do to protect ecosystems and their resources?

Mark the Text

Ask Questions As you read, write questions you may have next to each paragraph. Read the lesson a second time and try to answer the questions. When you are done, ask your teacher any questions you still have.

✔ Reading Check

1. Explain How did cyanobacteria change Earth's environment?

_____

_____

_____

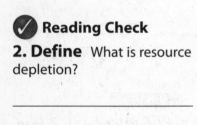
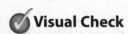 **Reading Check**

2. Define What is resource depletion?

_____

_____

_____

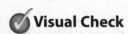 **Visual Check**

3. Locate The Ogallala Aquifer lies beneath most of the land area of which state?

_____

Using Natural Resources

Humans use many of the same resources as other species. They include renewable resources such as food, water, and oxygen. **Renewable resources** *are resources that can be replenished by natural processes at least as quickly as they are used.*

Unlike most species, humans also use nonrenewable resources, especially fossil fuels—coal, oil, and natural gas. **Nonrenewable resources** *are natural resources that are used up faster than they can be replaced by natural processes.* One example of a nonrenewable resource is fossil fuels. Earth's supply of fossil fuels is decreasing. People are looking for ways to replace fossil fuels with renewable sources.

Any resource becomes nonrenewable if it is used up faster than it can be replaced. **Resource depletion**—*the exhaustion of one or more resources in an area*—is happening in the United States and throughout the world. The Ogallala Aquifer is an example of resource depletion. The Ogallala Aquifer is a vast reservoir of underground water in the Great Plains region, as shown in the figure below. The aquifer is an important source of irrigation water for U.S. farmland. From the 1940s to the 1970s, people pumped water out of the aquifer much faster than natural processes could replenished it. Authorities predicted that by the year 2000 there would be little water left in the aquifer. In response, people began developing and using water-saving irrigation technologies. ✓

Ogallala Aquifer

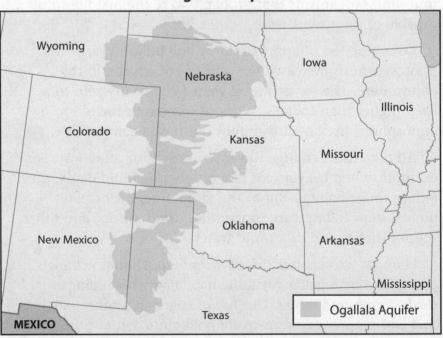

Understanding the Effects of Pollution

When people use resources, their actions can have unintended consequences. No one could have predicted that burning fossil fuels would lead to smog formation or acid rain. Over time people learned that the application of fertilizers to farmland and recreation areas could lead to harmful algal blooms in nearby lakes and streams. As people learn more about how their actions affect the environment, they can make better environmental choices. ✓

Unintended Consequences The invention of chlorofluorocarbons (klor oh flor oh KAR buhnz), commonly called CFCs, in the early 1900s is another example of unintended consequences. CFCs are chemical coolants used in refrigerators and air conditioners. In the 1970s, it was discovered that CFCs and related chemicals damage the ozone layer. The ozone layer is the layer of the atmosphere that shields Earth's surface from ultraviolet (UV) light from the Sun. High UV levels damage DNA, increase skin cancer rates, and disrupt photosynthesis.

An international treaty called the Montreal Protocol avoided potential further damage to the ozone layer. Signed by numerous countries in the 1980s, the treaty phased out the use of CFCs worldwide. Without the Montreal Protocol, most of Earth's ozone layer could have disappeared by the year 2028. 🔑

Global Climate Change Most scientists agree that fossil-fuel use also increases concentrations of carbon dioxide and other greenhouse gases in the atmosphere. The amount of greenhouse gases emitted by a person, an organization, an event, or a product is called its carbon footprint.

An increase in greenhouse gases is contributing to global warming—a rise in Earth's average surface temperature. Global climate change is a result of global warming and could lead to a variety of environmental consequences. Global climate change could change the types of crops that can be grown in various parts of the world. It could increase the number and severity of floods and droughts. Global climate change could even raise sea levels. ✓

✓ **Reading Check**
4. Evaluate Why is it important to learn about how human actions affect the environment?

_____

_____

_____

🔑 **Key Concept Check**
5. Summarize In what ways do humans affect ecosystems?

_____

_____

_____

✓ **Reading Check**
6. Identify What are some possible consequences of global climate change?

_____

_____

_____

Protecting the World

When people understand how human actions affect the environment, they can better avoid harming the environment. Scientists are working to develop renewable energy resources that can reduce pollution and people's dependence on fossil fuels. For example, using wind and solar power to generate electricity already is reducing our consumption of fossil fuels. Underwater turbines that capture the energy of tides and ocean currents also are being tested. Other strategies for protecting the environment include making and enforcing environmental laws and taking daily steps to reduce our impact on the planet. ✓

Enacting Environmental Laws

Who is responsible for preventing pollution, removing it when it happens, or repairing damage it causes to the environment? Protecting the environment can be expensive, and people don't always agree about who should pay the costs. The U.S. government passes laws to help protect the environment.

The U.S. Environmental Protection Agency (EPA) <u>enforces</u> many of these laws. Three of the most important environmental laws are the Endangered Species Act, the Clean Air Act, and the Clean Water Act.

The Endangered Species Act Signed into law in 1973, this act lists species threatened or endangered with extinction, designates habitat to protect them, and outlaws actions that would harm them.

The Clean Air Act Enacted in 1970, this law gives the EPA the power to create emissions standards for automobiles, industries, and power plants to reduce the amount of pollutants in the air. The law includes regulations designed to reduce carbon dioxide emissions.

The Clean Water Act First enacted in 1972 and expanded in 1977 and 1987, the Clean Water Act regulates the discharge of pollutants into waterways. This set of laws has helped make significant improvements to water quality.

The EPA also monitors environmental health, looks for ways to reduce human impacts, and develops plans for cleaning up polluted areas. The EPA supports environmental research at universities and national laboratories. ✓

Reading Check

7. Name some ways scientists are trying to reduce people's dependence on fossil fuels.

_____

_____

_____

ACADEMIC VOCABULARY

enforce
(verb) to carry out effectively

Reading Check

8. Recognize What is the EPA's role in keeping the environment healthy?

_____

_____

_____

_____

Making a Difference

Everyone can take action to help keep the environment healthy and to make sure that future generations of life on Earth have the resources they need to survive. Collectively, these actions are known as the 5Rs.

Restore Habitats and ecosystems that have been damaged can be restored—brought back to their original state. Examples include planting trees to restore forests or removing trash to clean up streams and beaches. ✓

Rethink Great ideas come from reconsidering, or rethinking, the way people carry out daily tasks. For example, a gym owner in Oregon invented a way for people exercising on spin bikes to generate electricity for the gym's video and sound systems. One train station in Tokyo uses tiles that convert people's footsteps into electricity. The tiles help power ticket gates, lights, and signs in the station.

Reduce People can reduce waste and pollution by using fewer resources whenever possible. Using fewer resources can mean turning off lights when they aren't being used, or wearing a sweater instead of increasing the setting on the thermostat. You could also walk or ride your bicycle instead of riding in a vehicle powered by fossil fuels.

Reuse There are many ways to reuse resources. One option is to have broken items repaired instead of replacing them. You could also buy used items instead of new items, or even invent new uses for objects instead of throwing them in the trash.

Recycle People can conserve resources by recycling—processing things so they can be used again for another purpose. Paper, plastic, glass, metal, yard waste, and used appliances and electronics can be recycled instead of dumped in landfills. ✓🔑

✓ Reading Check

9. Paraphrase What does it mean to *restore* a habitat?

_____

_____

_____

🔑 Key Concept Check

10. State How can people protect ecosystems and conserve resources?

_____

_____

_____

_____

Mini Glossary

nonrenewable resource: a natural resource that is used up faster than it can be replaced by natural processes

renewable resource: a resource that can be replenished by natural processes at least as quickly as it is used

resource depletion: the exhaustion of one or more resources in an area

1. Review the terms and their definitions in the Mini Glossary. Write a sentence explaining in your own words what a nonrenewable resource is and give at least one example.

_____

_____

2. The 5Rs are actions that everyone can take to conserve resources and keep the environment healthy. Identify each of the 5Rs described by the examples in the diagram.

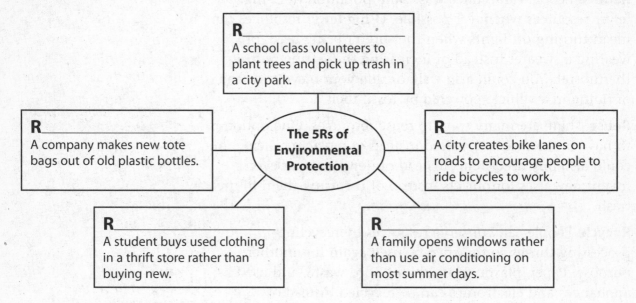

R_____
A school class volunteers to plant trees and pick up trash in a city park.

R_____
A company makes new tote bags out of old plastic bottles.

The 5Rs of Environmental Protection

R_____
A city creates bike lanes on roads to encourage people to ride bicycles to work.

R_____
A student buys used clothing in a thrift store rather than buying new.

R_____
A family opens windows rather than use air conditioning on most summer days.

3. People cut down a large section of forest to use the trees for building materials. What might be an unintended consequence of this action?

_____

What do you think NOW?

Reread the statements at the beginning of the lesson. Fill in the After column with an A if you agree with the statement or a D if you disagree. Did you change your mind?

 **Connect ED**

Log on to ConnectED.mcgraw-hill.com and access your textbook to find this lesson's resources.

END OF LESSON

Biomes and Ecosystems

Land Biomes

Copyright © Glencoe/McGraw-Hill, a division of The McGraw-Hill Companies, Inc.

················ **Before You Read** ···············

What do you think? Read the two statements below and decide whether you agree or disagree with them. Place an A in the Before column if you agree with the statement or a D if you disagree. After you've read this lesson, reread the statements and see if you have changed your mind.

Before	Statement	After
	1. Deserts can be cold.	
	2. There are no rain forests outside the tropics.	

··············· **Read to Learn** ···············

Land Ecosystems and Biomes

The living or once-living parts of an environment are the biotic parts. The biotic parts include people, trees, grass, birds, flowers, and insects. The nonliving parts of an environment are the abiotic parts. They include the air, sunlight, and water. The biotic parts of the environment need the abiotic parts to survive. The biotic and abiotic parts of an environment together make up an ecosystem.

Earth's continents have many different ecosystems. They range from deserts to rain forests. Scientists classify similar ecosystems in large geographic areas as biomes. *A* **biome** *is a geographic area on Earth that contains ecosystems with similar biotic and abiotic features.* You are already familiar with at least one of Earth's biomes—you live in one.

Biomes Earth has seven major land biomes. Areas classified as the same biome have similar climates and organisms. In this lesson you will learn about each of these land biomes.

Key Concepts

- How do Earth's land biomes differ?
- How do humans impact land biomes?

Mark the Text

Identify Main Ideas As you read this lesson, highlight the main ideas. Use the highlighted material to review the lesson.

✓ **Reading Check**
1. Define biome.

_____

_____

_____

_____

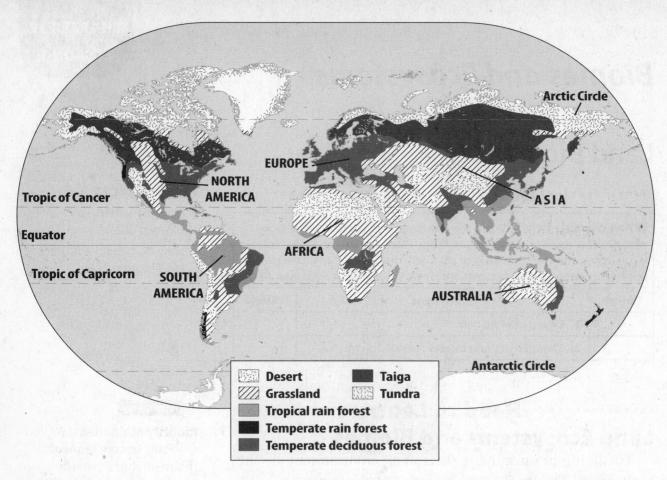

Copyright © Glencoe/McGraw-Hill, a division of The McGraw-Hill Companies, Inc.

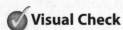 **Visual Check**

2. Locate Circle the areas where tropical rain forests are found.

Reading Check

3. State List four of the seven biomes found on Earth.

_____

_____

_____

_____

Earth's Biomes The seven types of land biomes—deserts, grasslands, tropical rain forests, temperate rain forests, temperate deciduous forests, taiga, and tundra—are shown above. Refer to this map as you read about each biome, its biodiversity, and the impact humans have on each of the biomes. ✓

Desert Biome

Deserts *are biomes that receive very little rain*. Deserts are on almost every continent. They are Earth's driest ecosystems.

- Most deserts are hot during the day and cold at night. Others, like those in Antarctica, are always cold.

- After a rain, rainwater drains away quickly because the soil is thin and porous. Large patches of ground in deserts are bare.

- Average monthly temperatures in deserts in the U.S. range from 6°C to 34°C. Average monthly precipitation ranges from 1 cm to 2.5 cm.

Biodiversity

- Animals of the desert biome include lizards, bats, owls, woodpeckers, and snakes. Most animals are not active during the hottest parts of the day.

- Plants include spiny cacti and thorny shrubs. Shallow roots enable a plant to absorb water quickly. Some plants have accordion-like stems that expand and store water. Small leaves or spines reduce the loss of water.

Human Impact

- Cities, farms, and recreational areas located in deserts use valuable water.

- Desert plants grow slowly. When they are damaged by people or livestock, it can take many years for them to recover.

Grassland Biome

Grassland *biomes are areas where grasses are the dominant plants.* Grasslands are also called prairies, savannas, and meadows. Grasslands are referred to as the world's "breadbaskets." This is because wheat, corn, oats, rye, barley, and other important cereal crops are grasses. They grow well in these areas.

- Grasslands have a wet and a dry season.

- Deep, fertile soil supports plant growth.

- Grass roots form a thick mass, called sod, which helps soil absorb water and hold it during periods of drought.

- Average monthly temperatures in U.S. grasslands range from −2°C to 38°C. Average monthly precipitation ranges from 2 cm to 6 cm. ✓

Biodiversity

- Trees grow along moist banks of streams and rivers in grassland biomes. Wildflowers bloom during the wet season.

- In North America, large herbivores, such as bison and elk, graze on grasses. Insects, birds, rabbits, prairie dogs, and snakes find shelter in the grasses.

- Predators in the grasslands of North America include hawks, ferrets, coyotes, and wolves.

- Giraffes, zebras, and lions are found in the African savannas. Australian grasslands are home to kangaroos, wallabies, and wild dogs. ✓

Copyright © Glencoe/McGraw-Hill, a division of The McGraw-Hill Companies, Inc.

💡 Think it Over

4. Describe the benefit of shallow roots for a desert plant.

_____

_____

_____

✓ Reading Check

5. Explain why grasslands are called "breadbaskets."

_____

_____

_____

✓ Reading Check

6. Name What are the grasslands in Africa called?

_____

_____

Human Impact

- People plow large areas of grassland to raise cereal crops. This reduces habitats for wild species.

- Because of hunting and the loss of habitats, large herbivores, such as bison, are now uncommon in many grasslands.

Tropical Rain Forest Biome

The forests that grow near the equator are called tropical rain forests. These forests receive large amounts of rain and have dense growths of tall, leafy trees.

- Weather in tropical rain forest biomes is warm and wet all year. ✓

- The soil is shallow and is easily washed away by rain.

- Less than 1 percent of the sunlight that reaches the top of forest trees reaches the forest floor.

- Half of Earth's species live in tropical rain forests. Most live in the canopy—the uppermost part of the forest.

- Average monthly temperatures range from 21°C to 32°C. Average monthly precipitation ranges from 14 cm to 26 cm. ✓

Biodiversity

- Few plants live on the dark forest floor.

- Vines climb the trunks of tall trees.

- Mosses, ferns, and orchids live on branches in the canopy.

- Insects make up the largest group of tropical animals. They include beetles, termites, ants, bees, and butterflies.

- Larger animals include parrots, toucans, snakes, frogs, flying squirrels, fruit bats, monkeys, jaguars, and ocelots. ✓

Human Impact

- People have cleared more than half of Earth's tropical rain forests for lumber, farming, and ranching. Poor soil does not support rapid growth of new trees in cleared areas.

- Some organizations are working to encourage people to use less wood harvested from rain forests.

✓ Reading Check

7. Describe the weather in a tropical rain forest.

_____

_____

_____

_____

✓ Reading Check

8. Identify Where do most organisms live in a tropical rain forest? (Circle the correct answer.)

a. on the forest floors

b. in the canopy

c. deep in the soil

✓ Reading Check

9. Name four kinds of animals that live in tropical rain forests.

_____

_____

_____

_____

Temperate Rain Forest Biome

Regions of Earth between the tropics and the polar circles are **temperate** *regions*. Temperate regions have relatively mild climates with distinct seasons.

Several biomes are in temperate regions, including rain forests. Temperate rain forests are moist ecosystems located mostly in coastal areas. Temperate rain forests are not as warm as tropical rain forests.

- Winters are mild and rainy.

- Summers are cool and foggy.

- Soil is rich and moist.

- Average monthly temperatures of U.S. temperate rain forests range from 0°C to 23°C. Average monthly precipitation ranges from 2 cm to 36 cm. ✅

Biodiversity

- Forests are made up mostly of spruce, hemlock, cedar, fir, and redwood trees. These trees can grow very large and tall.

- Fungi, ferns, mosses, vines, and small flowering plants grow on the moist forest floor.

- Animals include mosquitoes, butterflies, frogs, salamanders, woodpeckers, owls, eagles, chipmunks, raccoons, deer, elk, bears, foxes, and cougars.

Human Impact

- Temperate rain forest trees are a source of lumber. Logging can destroy the habitat of forest species.

- Rich soil enables cut forests to grow back. Tree farms help provide lumber without destroying the habitats. 🗝️

FOLDABLES

Make a horizontal two-tab book to record information about desert and temperate rain forest biomes.

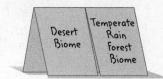

Desert Biome | Temperate Rain Forest Biome

✔️ **Reading Check**

10. Describe what the seasons are like in the temperate rain forest biome.

_____

_____

_____

🗝️ **Key Concept Check**

11. Describe In what ways do humans affect temperate rain forests?

_____

_____

_____

_____

Temperate Deciduous Forest Biome

Temperate deciduous forests grow in temperate regions where winter and summer climates have more variation than regions where temperate rain forests grow. Temperate deciduous forests are the most common forest ecosystems in the United States. They contain mostly deciduous trees. Deciduous trees lose their leaves in fall.

- Winter temperatures are often below freezing. Snow is common.

- Summers are hot and humid.

- Soil is rich in nutrients. A large variety of plants grow in great numbers in the soil.

- Average monthly temperatures in U.S. biomes range from −7°C to 29°C. Average monthly precipitation ranges from 4 cm to 10 cm.

Biodiversity

- Most plants, such as maples, oaks, birches, and other deciduous trees, stop growing during the winter and begin growing again in the spring.

- Animals include snakes, ants, butterflies, birds, raccoons, opossums, and foxes.

- Some animals, including chipmunks and bats, spend the winter in hibernation.

- Many birds and some butterflies, such as the monarch, migrate to warmer climates for the winter.

Human Impact

- Over the past several hundred years, humans have cleared thousands of acres of Earth's deciduous forests for farms and cities.

- Today, much of the clearing has stopped and some forests have regrown.

Think it Over

12. Describe how some animals in the temperate deciduous forest biome spend their winters.

_____

_____

Key Concept Check

13. Contrast How are temperate deciduous forests different from temperate rain forests?

_____

_____

_____

_____

Taiga Biome

A **taiga** (TI guh) *is a forest biome consisting mostly of cone-bearing evergreen trees*. The taiga biome exists only in the northern hemisphere. It occupies more space on Earth's continents than any other biome.

- Winters are long, cold, and snowy. Summers are short, warm, and moist.

- Soil is thin and acidic.

- Temperatures range from −16°C to 23°C. Precipitation ranges from 3 cm to 7 cm each month.

Biodiversity

- Evergreen trees, such as spruce, pine, and fir, are thin and shed snow easily.

- Animals include owls, mice, moose, bears, and other cold-adapted species.

- Abundant insects in summer attract many birds, which migrate south in winter. ✅

Human Impact

- Tree harvesting reduces taiga habitat.

Tundra Biome

A **tundra** (TUN druh) *biome is cold, dry, and treeless*. Most tundra is located south of the North Pole. It also exists in mountainous areas at high altitudes.

- Winters are dark and freezing, and summers are short and cool. The growing season is 50–60 days long.

- Permafrost—a layer of permanently frozen soil—prevents deep root growth.

- Temperatures in the Canadian tundra range from −28°C to 4°C. Precipitation ranges from 3 cm a month to 23 cm a month.

Biodiversity

- Plants include shallow-rooted mosses, lichens, and grasses.

- Many animals hibernate or migrate south during the winter. A few animals, including lemmings, live in tundras year-round.

Human Impact

- Drilling for oil and gas can interrupt migration patterns of animals. ✅

Copyright © Glencoe/McGraw-Hill, a division of The McGraw-Hill Companies, Inc.

✅ **Reading Check**

14. Name three kinds of trees found in the taiga biome.

_____

_____

_____

✅ **Reading Check**

15. Describe ways that humans can disturb the tundra biome.

_____

_____

_____

_____

Mini Glossary

biome: a geographic area on Earth that contains ecosystems with similar biotic and abiotic features

desert: a biome that receives very little rain

grassland: a biome where grasses are the dominant plants

taiga (TI guh): a forest biome consisting mostly of cone-bearing evergreen trees

temperate: a region between the tropics and the polar circles

tundra (TUN druh): a biome that is cold, dry, and treeless

1. Review the terms and their definitions in the Mini Glossary. Write a sentence that describes what a biome is and how biomes can differ.

_____

_____

2. Fill in the table below to identify the plant and animal life in the different biomes.

	Desert	Tropical Rain Forest	Temperate Rain Forest	Temperate Deciduous Forest	Taiga	Tundra	Grassland
Animals		Insects, parrots, toucans, snakes, monkeys, frogs, jaguars		Snakes, birds, raccoons, opossums, chipmunks, foxes		Lemmings, few others year-round	
Plants	Shrubs, cacti		Spruce, hemlock, cedar, fir, redwood, fungi, ferns, mosses, vines	Maples, oaks, birches	Cone-bearing evergreen trees		Trees along rivers, wildflowers, cereal crops, grasses

What do you think NOW?

Reread the statements at the beginning of the lesson. Fill in the After column with an A if you agree with the statement or a D if you disagree. Did you change your mind?

Log on to ConnectED.mcgraw-hill.com and access your textbook to find this lesson's resources.

END OF LESSON

Biomes and Ecosystems

Aquatic Ecosystems

···············Before You Read···············

What do you think? Read the two statements below and decide whether you agree or disagree with them. Place an A in the Before column if you agree with the statement or a D if you disagree. After you've read this lesson, reread the statements and see if you have changed your mind.

Before	Statement	After
	3. Estuaries do not protect coastal areas from erosion.	
	4. Animals form coral reefs.	

···············Read to Learn···············

Aquatic Ecosystems

Water is full of life. There are four main types of water, or aquatic, ecosystems. They are freshwater, wetland, estuary, and ocean. Each type of ecosystem contains a unique variety of organisms. Whales, dolphins, and corals live only in ocean ecosystems. Trout live only in freshwater ecosystems. Many other organisms, such as birds and seals, depend on aquatic ecosystems for food and shelter.

Temperature, sunlight, and oxygen gas that is dissolved in the water are important abiotic factors in aquatic ecosystems. Fish and other aquatic species have adaptations that enable them to use the oxygen from the water. For example, the gills of a fish separate oxygen from water and move the oxygen into the bloodstream of the fish. Mangrove plants take oxygen in through small pores in their leaves and roots.

Salinity (say LIH nuh tee) is another important abiotic factor in aquatic ecosystems. **Salinity** *is the amount of salt dissolved in water.* Water in saltwater ecosystems has high salinity compared to water in freshwater ecosystems. Freshwater contains little salt.

Key Concepts

- How do Earth's aquatic ecosystems differ?
- How do humans impact aquatic ecosystems?

◀ Study Coach

Create a Quiz Create a quiz about the types of aquatic ecosystems. Exchange quizzes. After taking the quizzes, discuss the answers.

Math Skills ×÷

Salinity is measured in parts per thousand (PPT). One PPT water contains 1 g of salt and 1,000 g of water. Use proportions to calculate salinity. What is the salinity of 100 g of water with 3.5 g of salt?

$$\frac{3.5 \text{ g salt}}{100 \text{ g seawater}} =$$

$$\frac{x \text{ g salt}}{1{,}000 \text{ g seawater}}$$

$$100\,x = 3500$$

$$x = \frac{3500}{100} = 35 \text{ PPT}$$

1. Use Proportions
A sample contains 0.1895 g of salt per 50 g of seawater. What is its salinity?

Freshwater: Streams and Rivers

Freshwater ecosystems include streams, rivers, ponds, and lakes. Streams and rivers contain flowing freshwater. Streams are usually narrow, shallow, and fast-flowing. Rivers are larger, deeper, and flow more slowly.

- Streams form from underground springs or from runoff from rain and melting snow.

- Stream water is often clear because soil particles are quickly washed downstream.

- Oxygen levels in streams are high because air mixes into the water as it splashes over rocks.

- Rivers form when streams flow together.

- Soil that washes into a river from streams or nearby land can make the river water muddy. Soil also introduces nutrients, such as nitrogen, into rivers.

- Slow-moving river water has higher levels of nutrients and lower levels of dissolved oxygen than fast-moving water.

Biodiversity

- Willows, cottonwoods, and other water-loving plants grow along streams and on riverbanks.

- Species adapted to fast-moving water include trout, salmon, crayfish, and many insects.

- Species adapted to slow-moving water include snails and catfish. ✓

Human Impact

- People take water from streams and rivers for drinking, doing laundry, bathing, irrigating crops, and industrial purposes.

- Hydroelectric plants use the energy in flowing water to generate electricity. Dams stop the natural flow of water.

- Runoff from cities, industries, and farms is a source of pollution. ✓

💡 **Think it Over**

2. Compare rivers and streams.

_____

_____

✓ **Reading Check**

3. Name three organisms found in a river ecosystem.

_____

_____

_____

✓ **Reading Check**

4. Identify possible sources of pollution to a river or a stream.

_____

_____

Freshwater: Ponds and Lakes

Ponds and lakes contain freshwater that is not flowing downhill. These bodies of water form in low areas on land.

- Ponds are warm and shallow.
- Sunlight reaches the bottoms of most ponds.
- Pond water is often high in nutrients.
- Lakes are larger and deeper than ponds.
- Sunlight penetrates into the top few feet of lake water. Deeper water is dark and cold.

Biodiversity

- Plants surround ponds and lake shores.
- Surface water in ponds and lakes contains plants, algae, and microscopic organisms that use sunlight for photosynthesis.
- Organisms living in shallow water along shorelines include cattails, reeds, insects, crayfish, frogs, fish, and turtles.
- Fewer organisms live in the deeper, colder water of lakes where there is little sunlight.
- Lake fish include perch, trout, bass, and walleye. ✓

Human Impact

- Humans fill in ponds and lakes with sediment to create land for houses and other structures.
- Runoff from farms, gardens, and roads washes pollutants into ponds and lakes. This disrupts the food webs that exist in these biomes.

Copyright © Glencoe/McGraw-Hill, a division of The McGraw-Hill Companies, Inc.

✓ Reading Check

5. Identify the statement that is true of ponds. (Circle the correct answer.)

a. Ponds have moving water.

b. Ponds have few nutrients.

c. Ponds are warm and shallow.

✓ Reading Check

6. Explain why few organisms live in the deep water of lakes.

_____

_____

🔑 Key Concept Check

7. Contrast How do ponds and lakes differ?

_____

_____

_____

Wetlands

Some types of aquatic ecosystems have mostly shallow water. **Wetlands** *are aquatic ecosystems that have a thin layer of water covering soil that is wet most of the time.* Wetlands contain freshwater, salt water, or both. They are among Earth's most fertile ecosystems.

- Freshwater wetlands form at the edges of lakes and ponds and in low areas on land. Saltwater wetlands form along ocean coasts.

- Nutrient levels and biodiversity are high.

- Wetlands trap sediments and purify water. Plants and microscopic organisms filter out pollution and waste materials. ✓

Biodiversity

- Plants that can live in wetlands include grasses and cattails. Few trees live in saltwater wetlands. Trees that can live in freshwater wetlands include cottonwoods, willows, and swamp oaks.

- Insects are numerous and include flies, mosquitoes, dragonflies, and butterflies.

- More than one-third of North American bird species use wetlands for nesting and feeding. Some of them are ducks, geese, herons, loons, warblers, and egrets.

- Other animals that depend on wetlands for food and breeding grounds include alligators, turtles, frogs, snakes, salamanders, muskrats, and beavers.

Human Impact

- In the past, people often thought wetlands were unimportant environments. Water was drained away to build homes and roads and to raise crops.

- Today, many wetlands are being preserved, and drained wetlands are being restored. ✓

Copyright © Glencoe/McGraw-Hill, a division of The McGraw-Hill Companies, Inc.

✓ Reading Check

8. Describe Where do wetlands form?

_____

_____

💡 **Think it Over**

9. Discuss Why is it important to protect wetlands?

_____

_____

🔑 Key Concept Check

10. Explain How do humans impact wetlands?

_____

_____

_____

Estuaries

Estuaries (ES chuh wer eez) *are regions along coastlines where streams or rivers flow into a body of salt water.* Most estuaries form along coastlines where freshwater in rivers mixes with salt water in oceans. The degree of salinity in estuary ecosystems varies.

- Salinity depends on rainfall, the amount of freshwater flowing from land, and the amount of salt water pushed in by tides.

- Estuaries help protect coastal land from flooding and erosion. Like wetlands, estuaries purify water and filter out pollution.

- Nutrient levels and biodiversity are high.

Biodiversity

- Plants that grow in salt water include mangroves, pickleweeds, and seagrasses.

- Animals include worms, snails, and many species that people use for food, including oysters, shrimp, crabs, and clams.

- Striped bass, salmon, flounder, and many other ocean fish lay their eggs in estuaries.

- Many species of birds depend on estuaries for breeding, nesting, and feeding.

Human Impact

- Large portions of estuaries have been filled with soil to make land for roads and buildings.

- Destruction of estuaries reduces habitats for estuary species. It also exposes the coastline to flooding and storm damage. ✓

Reading Check

11. Identify Where do estuaries form? (Circle the correct answer.)

a. in freshwater

b. in salt water

c. where freshwater and salt water mix

FOLDABLES®

Make a two-tab book to use to compare how biodiversity and human impact differ in wetlands and estuaries.

Reading Check

12. Name problems that can come from destroying estuaries.

_____

_____

_____

_____

Ocean: Open Oceans

Most of Earth's surface is covered by ocean water with high salinity. The oceans contain different types of ecosystems. If you took a boat trip several kilometers out to sea, you would be in the open ocean. This is one type of ecosystem. The open ocean extends from the steep edges of continental shelves to the deepest parts of the ocean. A cross section of the open ocean is shown in the figure. The amount of light in the water depends on depth.

- Photosynthesis takes place only in the uppermost, or sunlit, zone. Very little sunlight reaches the deeper twilight zone. None reaches the deepest water, known as the dark zone.

- Decaying matter and nutrients float down from the sunlit zone, through the twilight and dark zones, to the seafloor.

Biodiversity

- Microscopic algae and other producers (organisms that make their own food) in the sunlit zone form the base of most ocean food chains. Other organisms that live in the sunlit zone include jellyfish, tuna, mackerel, and dolphins.

- Many species of fish stay in the twilight zone during the day and go to the sunlit zone at night to feed.

- Sea cucumbers, brittle stars, and other bottom-dwelling organisms feed on decaying matter that drifts down from above.

- Many organisms in the dark zone live near cracks in the seafloor where lava erupts and new seafloor forms. ✓

Human Impact

- Overfishing threatens many ocean fish.

- Trash discarded from ocean vessels or washed into oceans from land is a source of pollution. Animals, such as seals, become tangled in plastic or mistake plastic for food.

✓ Visual Check

13. Draw arrows indicating the depth that sunlight reaches in the open ocean.

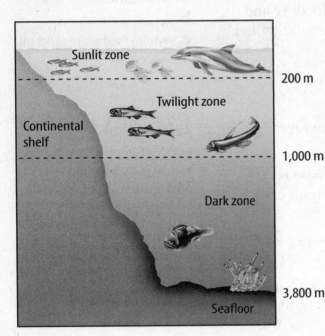

Sunlit zone

Twilight zone

Continental shelf

200 m

1,000 m

Dark zone

3,800 m

Seafloor

✓ Reading Check

14. Identify What is the base of most ocean food chains? (Circle the correct answer.)

a. microscopic algae

b. fish, such as tuna and mackerel

c. dolphins

Ocean: Coastal Oceans

Coastal oceans include several types of ecosystems, including continental shelves and the intertidal zones. *The intertidal zone is the ocean shore between the lowest low tide and the highest high tide.* ✓

- Sunlight reaches the bottom of shallow coastal ecosystems.

- Nutrients washed in from rivers and streams add to the high biodiversity.

Biodiversity

- The coastal ocean is home to mussels, fish, crabs, dolphins, and whales.

- Intertidal species have adaptations for surviving exposure to air during low tides and to heavy waves during high tides.

Human Impact

- Oil spills and other pollution harm coastal organisms.

Ocean: Coral Reefs

Another ocean ecosystem with high biodiversity is coral reefs. *A coral reef is an underwater structure made from outside skeletons of tiny, soft-bodied animals called coral.*

- Most coral reefs form in shallow tropical oceans.

- Coral reefs protect coastlines from storm damage and erosion. ✓

Biodiversity

- Coral reefs provide food and shelter for many animals, including parrotfish, groupers, angelfish, eels, shrimp, crabs, scallops, clams, worms, and snails.

Human Impact

- Pollution, overfishing, and harvesting of coral threaten coral reefs.

✓ **Reading Check**

15. Describe Where is the intertidal zone located?

_____

_____

_____

💡 **Think it Over**

16. Discuss why organisms in the intertidal zone must have special adaptations.

_____

_____

_____

_____

✓ **Reading Check**

17. Define coral reef.

_____

_____

_____

Mini Glossary

coral reef: an underwater structure made from skeletons of tiny, soft-bodied animals called coral

estuary (ES chuh wer ee): a region along a coastline where a stream or river flows into a body of salt water

intertidal zone: the ocean shore between the lowest low tide and the highest high tide

salinity: the amount of salt dissolved in water

wetland: an aquatic ecosystem with a thin layer of water covering soil that is wet most of the time

1. Review the terms and their definitions in the Mini Glossary. How is an estuary different from a wetland?

   _____

   _____

2. Fill in the chart below to review ocean ecosystems.

Ocean Ecosystem	Characteristics	Species That Live in the Ecosystem	One Human Impact
Open	Sunlit zone, twilight zone, dark zone	Algae, jellyfish, tuna, mackerel, dolphins, sea cucumbers, brittle stars, seals	
Coastal			Pollution
Coral Reef		Parrotfish, groupers, angelfish, eels, shrimp, crabs, scallops, clams, worms, snails	Harvesting of coral

3. You made your own quiz about aquatic ecosystems. How did writing quiz questions help you better understand aquatic ecosystems?

   _____

   _____

What do you think NOW?

Reread the statements at the beginning of the lesson. Fill in the After column with an A if you agree with the statement or a D if you disagree. Did you change your mind?

 Connect ED

Log on to ConnectED.mcgraw-hill.com and access your textbook to find this lesson's resources.

 END OF LESSON

Biomes and Ecosystems

How Ecosystems Change

·············· **Before You Read** ··············

What do you think? Read the two statements below and decide whether you agree or disagree with them. Place an A in the Before column if you agree with the statement or a D if you disagree. After you've read this lesson, reread the statements and see if you have changed your mind.

Before	Statement	After
	5. An ecosystem never changes.	
	6. Nothing grows in the area where a volcano has erupted.	

Key Concepts

- How do land ecosystems change over time?
- How do aquatic ecosystems change over time?

·············· **Read to Learn** ··············

How Land Ecosystems Change

Have you ever seen weeds growing up through cracks in a sidewalk? If they are not removed, the weeds will continue to grow. The crack will widen, making room for more weeds. Over time, the sidewalk will break apart. Shrubs and vines will sprout and grow. Their leaves and branches will grow large enough to cover the sidewalk. Eventually, trees could start to grow there.

This process is an example of **ecological succession**—*the process of one ecological community gradually changing into another.* Ecological succession occurs in a series of steps. These steps can usually be predicted. For example, small plants usually grow in an ecosystem before trees do.

The final stage of ecological succession in a land ecosystem is a **climax community**—*a stable community that no longer goes through major ecological changes.* Climax communities differ depending on the type of biome they are in. In a grassland biome, mature grassland is the climax community. Climax communities are usually stable over hundreds of years. As plants die, new plants of the same species grow as long as the climate stays the same. ✔️

Mark the Text

Identify Main Ideas
Highlight the main idea of each paragraph in the lesson. Reread the main ideas to review the lesson.

 **Key Concept Check**
1. Describe What is a climax community?

_____

_____

Copyright © Glencoe/McGraw-Hill, a division of The McGraw-Hill Companies, Inc.

Primary Succession

What do you think happens to a lava-filled landscape when a volcanic eruption is over? Volcanic lava eventually becomes new soil. This new soil supports plant growth. Ecological succession in new areas of land with little or no soil, such as on a lava flow, a sand dune, or exposed rock, is called primary succession. *The first species that colonize new or undisturbed land are* **pioneer species.** Lichens and mosses are <u>pioneer</u> species. The figure below shows what happens to an area during and after a volcanic eruption.

A. During a volcanic eruption, molten lava flows over the ground and into the water. After the eruption is over, the lava cools and hardens into bare rock.

B. Lichen spores carried on the wind settle on the rock. Lichens release acid that helps break down the rock and create soil. Lichens add nutrients to the soil as they die and decay.

C. Airborne spores from mosses and ferns settle on the thin soil and add to the soil when they die. The soil becomes thick enough to hold water. Insects and other small organisms begin living in the area.

D. After many years, the soil is deep and has enough nutrients for grasses, wildflowers, shrubs, and trees to grow. The new ecosystem provides habitats for animals. Eventually, a climax community develops.

SCIENCE USE V. COMMON USE

pioneer

Science Use the first species that colonize new or undisturbed land

Common Use the first human settlers in an area

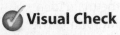

Visual Check

2. Specify Circle the stage below when a climax community develops.

A.

B.

C.

D.

Secondary Succession

Secondary succession occurs in areas where existing ecosystems have been disturbed or destroyed. One example is forestland in New England. Early colonists cleared land hundreds of years ago. Some of the cleared land was not planted with crops. This land gradually grew back to a climax forest community of beech and maple trees. The figure below shows what happens to an area of land as it is cleared and after it is cleared.

A. Settlers in North America cleared many acres of forests to plant their crops. On land where people stopped planting crops, the forests began to grow back.

B. Seeds of grasses, wildflowers, and other plants began to sprout and grow. Young shrubs and trees also started growing. These plants were habitats for insects and other small animals, such as mice.

C. White pines and poplars were the first trees in the area to grow tall. They provided shade and protection to trees that grow slower, such as beeches and maples.

D. Eventually, a climax community of beech and maple trees developed. As older trees die, new beech and maple trees grow and replace them.

Use a folded sheet of paper to describe and illustrate the before and after of secondary succession.

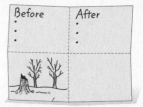

Reading Check

3. Identify Where does secondary succession occur?

_____

_____

Visual Check

4. Specify Circle the stage below when trees begin growing again.

A.

B.

C.

D.

How Freshwater Ecosystems Change

Like land ecosystems, freshwater ecosystems change over time in a natural, predictable process. This process is called aquatic succession.

Aquatic Succession

As shown in the figures below, aquatic succession begins with a body of water, such as a pond. Over time, sediments and decaying organisms carried by rainwater and streams build up and create soil on the bottoms of ponds, lakes, and wetlands.

Soil begins to fill the body of water. Eventually the body of water fills completely with soil. The water disappears and the area becomes land.

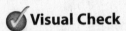
Key Concept Check

5. Describe What happens to a pond, a lake, or a wetland over time?

_____

_____

Visual Check

6. Identify Which image shows soil being created? (Circle the correct answer.)

a. A

b. B

c. C

A.

B.

C.

Eutrophication

Decaying organisms that fall to the bottom of a pond, a lake, or a wetland add nutrients to the water. **Eutrophication** (yoo troh fuh KAY shun) *is the process of a body of water becoming nutrient-rich*.

Eutrophication is a natural part of aquatic succession. However, humans also contribute to eutrophication. The fertilizers that farmers use on crops and waste from farm animals can be very high in nutrients. Other forms of pollution can also be very high in nutrients. When fertilizers and pollution run off into a pond or lake, concentrations of nutrients increase. Large populations of algae and microscopic organisms can grow using the nutrients. These organisms use up most of the dissolved oxygen in the water. Less oxygen is available for the fish and other organisms that live in the water. As a result, many of these organisms die. Their bodies decay and add to the buildup of the soil, speeding up succession.

WORD ORIGIN

eutrophication
from Greek *eutrophos*, means "nourishing"

Reading Check

7. Name two causes of eutrophication.

_____

_____

Mini Glossary

climax community: a stable community that no longer goes through major ecological changes

ecological succession: the process of one ecological community gradually changing into another

eutrophication (yoo troh fuh KAY shun): the process of a body of water becoming nutrient-rich

pioneer species: the first species that colonize new or undisturbed land

1. Review the terms and their definitions in the Mini Glossary. Write a sentence describing ecological succession. Use the term *climax community* in your sentence.

_____

_____

2. Fill in the table below to compare succession in aquatic ecosystems and land ecosystems.

	Aquatic Ecosystem	**Land Ecosystem**
How it changes		Grasses and other plants begin to grow. Young shrubs and trees then start growing.
What changes	Depth of water	
End result		Climax community

3. Summarize the characteristics of a climax community.

_____

_____

What do you think NOW?

Reread the statements at the beginning of the lesson. Fill in the After column with an A if you agree with the statement or a D if you disagree. Did you change your mind?

 Connect **ED**

Log on to ConnectED.mcgraw-hill.com and access your textbook to find this lesson's resources.

 END OF LESSON

Environmental Impacts

People and the Environment

What do you think? Read the two statements below and decide whether you agree or disagree with them. Place an A in the Before column if you agree with the statement or a D if you disagree. After you've read this lesson, reread the statements to see if you have changed your mind.

Before	Statement	After
	1. Earth can support an unlimited number of people.	
	2. Humans can have both positive and negative impacts on the environment.	

Population and Carrying Capacity

Have you ever seen a road sign that shows the population of a city you are entering? In this case, population means how many people live in the city. Scientists use the term *population*, too, but in a slightly different way. For scientists, *a* **population** *is all the members of a species living in a given area.* You are part of a population of humans. The other species in your area, such as birds or trees, each make up a separate population.

The Human Population

When the first American towns were settled, most had small populations. Today, some of those towns have become large cities with large populations. In a similar way, Earth was once home to relatively few humans. Today, about 6.7 billion people live on Earth. The greatest increase in human population occurred during the last few centuries.

Population Trends

Have you ever heard the phrase *population explosion?* Population explosion describes a sudden increase in a population. For most of history, the human population increased at a fairly steady rate. But in the 1800s, the population began to rise sharply.

Key Concepts

- What is the relationship between resource availability and human population growth?
- How do daily activities impact the environment?

> Study Coach

Make Flash Cards For each head in this lesson, write a question on one side of a flash card and the answer on the other side. Quiz yourself until you know all of the answers.

FOLDABLES®

Make a small vertical shutterfold, draw the arrows on each tab, and label as shown. Use the Foldable to discuss how human population growth relates to resources.

Visual Check

1. Compare the rate of human population growth from the years 200 to 1800 to the rate of growth from 1800 to 2000.

_____

_____

_____

_____

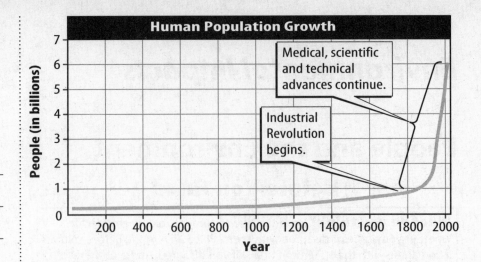

Human Population Growth

Medical, scientific and technical advances continue.

Industrial Revolution begins.

The graph above shows how the human population on Earth has changed. As you can see, the human population "exploded" in the last few hundred years.

What has caused this sharp increase in population? There are many reasons. Health care improved, cleaner water became available, and other improvements in technology led to people living longer and reproducing. In the next hour or so, about 15,000 babies will be born around the world. ✓

Population Limits

Every human being needs certain things to live, such as food, clean water, and shelter. People also need clothes, transportation, and other items.

Everything people use comes from resources found on Earth. Does Earth have enough resources to support an unlimited number of humans?

Earth has limited resources. Earth cannot support a population of any species in a particular environment beyond Earth's carrying capacity. **Carrying capacity** *is the largest number of individuals of a given species that Earth's resources can support and maintain for a long period of time*. If the human population continues to grow beyond Earth's carrying capacity, eventually Earth will not have enough resources to support humans.

Impact of Daily Actions

Each of the 6.7 billion people on Earth uses resources in some way. The use of these resources affects the environment. Think about the impact of just one activity that might be a part of your daily routine—a shower.

Reading Check

2. Identify What factors contributed to the increase in human population?

_____

_____

_____

_____

Key Concept Check

3. Assess What is the relationship between the availability of resources and human population growth?

_____

_____

_____

_____

Consuming Resources

Like many people, you might take a shower each day. The metal in the water pipes comes from <u>resources</u> mined from the ground. Mining can destroy habitats and pollute soil and water. Your towel might be made of cotton. Cotton is a resource obtained from plants. Fertilizers and other chemicals that are used on farms that produce cotton or other plant crops often run off into water. This runoff can affect the water's quality.

The water itself also is a resource. In some parts of the world, water is in short supply. The water for your shower is probably heated with fossil fuels. Remember that fossil fuels are nonrenewable resources. This means they are used up faster than they can be replaced by natural processes. Burning fossil fuels also releases pollution into the atmosphere.

Now, think about all the activities that you do in one day, such as going to school, eating meals, or playing computer games. All of these activities use resources. During your lifetime, your possible impact on the environment is great. Multiply this impact by 6.7 billion, and you can understand why it is important to use resources wisely. 🗝️➡️

Positive and Negative Impacts

Not all human activities have a negative impact on the environment. For example, cleaning up streams and picking up litter are ways people can have a positive impact on the environment. In the following lessons, you will learn how human activities affect soil, water, and air quality. You will also learn things you can do to help reduce the impact of your actions on the environment.

SCIENCE USE V. COMMON USE

resource

Science Use a natural source of supply or support

Common Use a source of information or expertise

🗝️➡️ **Key Concept Check**

4. Consider What are three things you did today that impacted the environment?

_____

_____

_____

💡 **Think it Over**

5. Describe one thing you will do tomorrow that will benefit the environment.

_____

_____

_____

Mini Glossary

carrying capacity: the largest number of individuals of a given species that Earth's resources can support and maintain for a long period of time

population: all the members of a species living in a given area

1. Review the terms and their definitions in the Mini Glossary. Write a sentence explaining the relationship between population and carrying capacity.

_____

_____

2. Use the graphic organizer below to identify three factors that are contributing to the human population explosion.

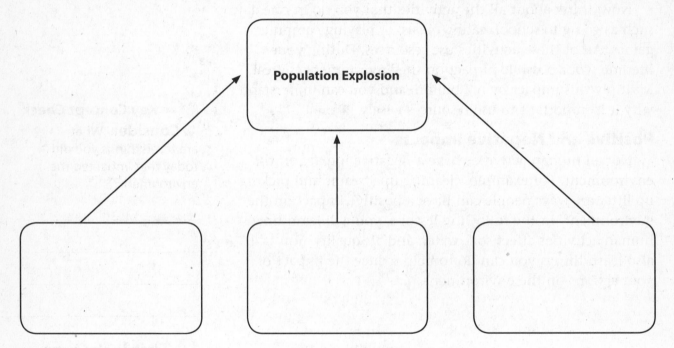

Population Explosion

3. Review the flash cards you created as you read the lesson. Select one and read the question. Then, without checking the lesson, write the answer below.

_____

_____

What do you think NOW?

Reread the statements at the beginning of the lesson. Fill in the After column with an A if you agree with the statement or a D if you disagree. Did you change your mind?

 **Connect** ED

Log on to ConnectED.mcgraw-hill.com and access your textbook to find this lesson's resources.

 END OF LESSON

Copyright © Glencoe/McGraw-Hill, a division of The McGraw-Hill Companies, Inc.

Environmental Impacts

Impacts on the Land

·············· **Before You Read** ··············

What do you think? Read the two statements below and decide whether you agree or disagree with them. Place an A in the Before column if you agree with the statement or a D if you disagree. After you've read this lesson, reread the statements to see if you have changed your mind.

Before	Statement	After
	3. Deforestation does not affect soil quality.	
	4. Most trash is recycled.	

·············· **Read to Learn** ··············

Using Land Resources

What do the metal in staples and the paper in your notebook have in common? Both come from resources found in or on land. People use land for timber production, agriculture or farming, and mining. All of these activities impact the environment.

Forest Resources

Humans cut down trees to make wood and paper products, such as your notebook. Trees are also cut for use as fuel and to clear land for growing crops, grazing, or building houses or highways. Sometimes entire forests are cut down for these purposes.

Deforestation *is the removal of large areas of forests for human purposes.* About 130,000 km^2 of tropical rain forests are cut down each year. This is an area equal in size to the state of Louisiana. Scientists estimate that 50 percent of all the species on Earth live in tropical rain forests. Deforestation destroys habitats, which can lead to species' extinction.

Deforestation also can affect soil quality. Plant roots hold soil in place. Without these natural anchors, soil erodes. In addition, deforestation affects air quality. Remember that when trees undergo photosynthesis, they remove carbon dioxide from the air. When there are fewer trees on Earth, more carbon dioxide remains in the atmosphere. ✓

Key Concepts
- What are the consequences of using land as a resource?
- How does proper waste management help prevent pollution?
- What actions help protect the land?

Mark the Text

Identify the Main Ideas To help you learn about impacts on the land, highlight each heading in one color. Then highlight the details that support and explain it in a different color. Refer to this highlighted text as you study the lesson.

✓ **Reading Check**
1. Identify three negative impacts of deforestation.

_____

_____

_____

Copyright © Glencoe/McGraw-Hill, a division of The McGraw-Hill Companies, Inc.

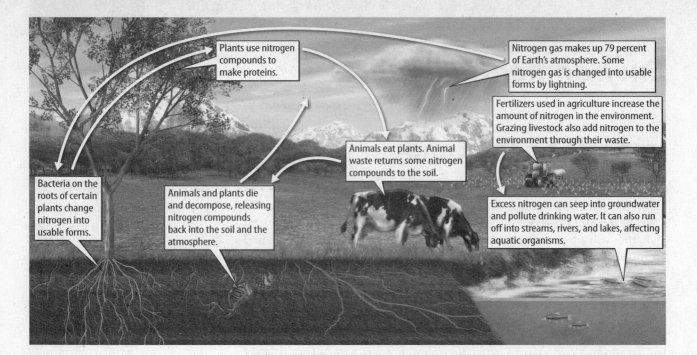

Plants use nitrogen compounds to make proteins.

Nitrogen gas makes up 79 percent of Earth's atmosphere. Some nitrogen gas is changed into usable forms by lightning.

Fertilizers used in agriculture increase the amount of nitrogen in the environment. Grazing livestock also add nitrogen to the environment through their waste.

Animals eat plants. Animal waste returns some nitrogen compounds to the soil.

Bacteria on the roots of certain plants change nitrogen into usable forms.

Animals and plants die and decompose, releasing nitrogen compounds back into the soil and the atmosphere.

Excess nitrogen can seep into groundwater and pollute drinking water. It can also run off into streams, rivers, and lakes, affecting aquatic organisms.

Visual Check

2. Describe How does the use of fertilizers affect the environment?

_____

_____

_____

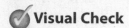

Make a horizontal two-tab concept map, and then label and draw arrows as shown to identify positive and negative factors that impact land.

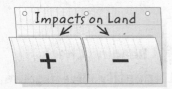

Agriculture and the Nitrogen Cycle

It takes a lot of food to feed 6.7 billion people. To grow enough food for the world's population, farmers often add fertilizers to soil. The fertilizers contain nitrogen and help increase crop yields.

As shown in the figure above, nitrogen is an element that naturally cycles through ecosystems. Living things use nitrogen to make proteins. And when these living things die and decompose or produce waste, they release nitrogen into the soil or the atmosphere. Nitrogen gas makes up about 79 percent of Earth's atmosphere. But most living things cannot use nitrogen in its gaseous form. Nitrogen must be converted into a form that organisms can use.

Bacteria that live on the roots of certain plants convert atmospheric nitrogen to a form that plants can use. Fertilizer that is added to soil to increase crop production contains a usable form of nitrogen.

Scientists estimate that human activities have doubled the amount of nitrogen cycling through ecosystems. Sources of much of this extra nitrogen are manufacturing facilities and fertilizers that have been applied to crops. The excess nitrogen can kill plants that have adapted to low nitrogen levels. It can also affect organisms that depend on those plants for food. Fertilizers can seep into groundwater supplies, which pollutes drinking water. Fertilizers can also run off into streams and rivers, affecting the organisms that live there.

Other Effects of Agriculture

Agricultural practices can have an effect on soil quality in other ways, too. Soil erosion can occur when land is overfarmed or overgrazed. High rates of soil erosion can lead to desertification. **Desertification** *is the development of desert-like conditions due to human activities and/or climate change.* When desertification occurs in a region of land, that land can no longer be used to produce food. ✅

Mining

Humans remove many useful rocks and minerals from the ground by mining. For example, copper is removed by digging a strip mine at the surface. Coal and other in-ground resources also can be removed by digging surface or underground mines.

Mines are necessary for obtaining much-needed resources. However, digging mines disturbs habitats and changes the landscape. Regulations have been established and must be followed to prevent water from being polluted by runoff that contains heavy metals from mines. 🗝️

Construction and Development

You have read about important resources that are found on or in land. But did you know that land itself is a resource? People use land for living space. Your home, your school, your favorite stores, and your neighborhood streets are built on land.

Urban Sprawl

In the 1950s, large areas of rural land in the United States were developed as suburbs. Suburbs are residential areas on the outside edges of a city. When the suburbs became crowded, people moved farther out into the country. More open land was cleared for still more development. *The development of land for houses and other buildings near a city is called* **urban sprawl.**

Urban sprawl can lead to habitat destruction and loss of farmland. People cut down forests to make room for housing developments. As large areas are paved for sidewalks and streets, less precipitation soaks into the ground. Runoff increases. An increase in runoff can reduce the water quality of streams, rivers, and groundwater, especially if the runoff contains sediments or chemical pollutants.

✅ **Reading Check**

3. Explain What causes desertification?

_____

_____

_____

_____

🗝️ **Key Concept Check**

4. Summarize What are some consequences of using land as a resource?

_____

_____

_____

💡 **Think it Over**

5. Infer What is the relationship between urban sprawl and population increase?

_____

_____

_____

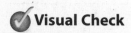
Roadways

Urban sprawl occurred at the same time as another trend in the United States—increased motor vehicle use. Before the 1940s, only a small percentage of Americans owned cars. But by 2005, there were 240 million vehicles for 295 million people. A greater number of cars increased the need for roadways.

In 1960, the United States had about 16,000 km of interstate highways. Today, the national highway system includes 256,000 km of paved roadways. Like urban sprawl, roadways increase runoff and disturb habitats.

Recreation

Not all of the land used by people is paved and developed. People also use land for recreation. They hike, bike, ski, and picnic, among other activities. In urban areas, some of these activities take place in public parks. As you will learn later in this lesson, parks and other green spaces help decrease runoff.

Waste Management

On a typical day, each person in the United States produces about 2.1 kg of trash. Altogether, people in the United States produce about 230 million metric tons of trash per year! Where does all that trash go?

Cross-Section of a Landfill

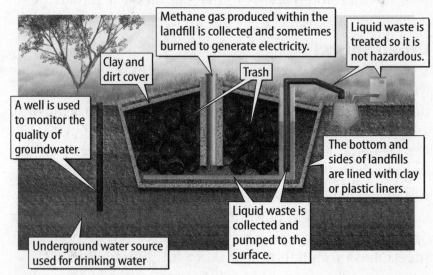

Methane gas produced within the landfill is collected and sometimes burned to generate electricity.

Liquid waste is treated so it is not hazardous.

Clay and dirt cover

Trash

A well is used to monitor the quality of groundwater.

The bottom and sides of landfills are lined with clay or plastic liners.

Liquid waste is collected and pumped to the surface.

Underground water source used for drinking water

Landfills

About 31 percent of the trash is recycled and composted. About 14 percent is burned. The remaining 55 percent is placed in landfills. Landfills are areas where trash is buried. Landfills are another way that people use land.

A landfill is carefully designed to meet government regulations. Trash is covered by soil to keep it from blowing away. Special liners help prevent pollutants from leaking into soil and groundwater supplies. The figure at the bottom of the previous page shows a cross-section of a solid-waste landfill. 🔑

Hazardous Waste

Some trash cannot be placed in landfills because it contains harmful substances that can affect soil, air, and water quality. This trash is called hazardous waste. The substances in hazardous waste also can affect the health of humans and other living organisms.

Industries and households generate hazardous waste. Hazardous waste from the medical industry includes used needles and bandages. Household hazardous waste includes used motor oil and batteries. The U.S. Environmental Protection Agency (EPA) works with state and local agencies to help people safely <u>dispose</u> of hazardous waste.

Positive Actions

Human actions can have negative effects on the environment, but they can have positive impacts as well. Governments, society, and individuals can work together to reduce the impact of human activities.

Protecting the Land

Yellowstone National Park was the first national park in the world. After it was created in 1872, other countries began setting aside land for preservation. State and local governments also followed this example. ✓

Protected forests and parks are important habitats for wildlife. They are enjoyed by millions of people. Mining and logging are allowed in some cases. However, the removal of resources must meet environmental regulations.

Reforestation and Reclamation

A forest is a complex ecosystem. But with careful planning, it can be managed as a renewable resource. Trees can be select-cut. That means that only some trees in an area are cut, rather than the entire forest. ✓

🔑 **Key Concept Check**

8. State What is done to prevent the trash in landfills from polluting air, soil, and water?

_____

_____

_____

_____

_____

ACADEMIC VOCABULARY
dispose
(verb) to throw away

✓ **Reading Check**

9. Name the first national park.

_____

_____

✓ **Reading Check**

10. Paraphrase Write your own definition of the term *select-cut*.

_____

_____

_____

_____

Reforestation People can practice reforestation. *Planting trees to replace trees that have been cut or burned down is called* **reforestation.** Reforestation can keep a forest healthy or help reestablish a deforested area.

Reclamation Mined land can be made environmentally healthy through reclamation. **Reclamation** *is the process of restoring land disturbed by mining.* In reclamation, mined areas are reshaped, covered with soil, and then replanted with trees and other vegetation. ✓

Green Spaces

Much of the land in urban areas is covered with parking lots, streets, buildings, and sidewalks. Many cities use green spaces to create natural environments. Green spaces are areas that are left undeveloped or lightly developed. They include parks within cities and forests around suburbs.

Green spaces provide recreational opportunities for people and shelter for wildlife. Green spaces also reduce runoff and improve air quality as plants remove excess carbon dioxide from the air.

How can you help?

Individuals can have a big impact on land-use issues by practicing the three Rs—reusing, reducing, and recycling. Composting is another way to lessen your impact on the land.

Reusing and Reducing Using an item for a new purpose is reusing. For example, you might have made a bird feeder from a used plastic milk jug. Reducing is using fewer resources. You can turn off the lights when you leave a room to reduce your use of electricity.

Recycling Making a new product from a used product is recycling. Plastic containers can be recycled into new plastic products. Recycled aluminum cans are used to make new aluminum cans. Using recycled paper to make new paper reduces deforestation. It also reduces the amount of water used during paper production.

Composting Another way people can lessen their environmental impact is composting. Compost is a mixture of decaying organic matter, such as leaves, food scraps, and grass clippings. It is used to improve soil quality by adding nutrients to soil. Composting and reusing, reducing, and recycling help reduce the amount of trash that ends up in landfills. 🔑

Copyright © Glencoe/McGraw-Hill, a division of The McGraw-Hill Companies, Inc.

✓ Reading Check

11. Explain How do reforestation and reclamation positively impact land?

_____

_____

_____

💡 Think it Over

12. Compare How might a green space in a city differ from a green space in a suburb?

_____

_____

_____

🔑 Key Concept Check

13. Consider What can you do to help lessen your impact on the land?

_____

_____

_____

Mini Glossary

deforestation: the removal of large areas of forests for human purposes

desertification: the development of desert-like conditions due to human activities and/or climate change

reclamation: the process of restoring land disturbed by mining

reforestation: planting trees to replace trees that have been cut or burned down

urban sprawl: the development of land for houses and other buildings near a city

1. Review the terms and their definitions in the Mini Glossary. Write a sentence comparing and contrasting deforestation and desertification.

_____

_____

2. Provide a definition for the term *runoff* in the box on the right. Then give three common sources of runoff in the boxes on the left.

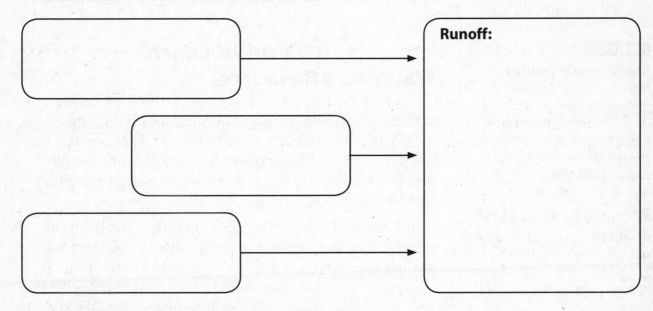

Runoff:

3. Why are green spaces important to urban areas?

_____

_____

What do you think NOW?

Reread the statements at the beginning of the lesson. Fill in the After column with an A if you agree with the statement or a D if you disagree. Did you change your mind?

Connect ED

Log on to ConnectED.mcgraw-hill.com and access your textbook to find this lesson's resources.

END OF LESSON

Environmental Impacts

Impacts on Water

Copyright © Glencoe/McGraw-Hill, a division of The McGraw-Hill Companies, Inc.

Key Concepts 🗝

- How do humans use water as a resource?
- How can pollution affect water quality?
- What actions help prevent water pollution?

Mark the Text

Identify the Main Ideas
Highlight two or three phrases in each paragraph that summarize the main ideas. After you have finished the lesson, review the highlighted text.

☑🗝 Key Concept Check
1. State How do humans use water as a resource?

_____

_____

_____

Interpreting Tables
2. Point Out What two activities account for most water use in the U.S.?

_____

_____

··············· **Before You Read** ···············

What do you think? Read the two statements below and decide whether you agree or disagree with them. Place an A in the Before column if you agree with the statement or a D if you disagree. After you've read this lesson, reread the statements to see if you have changed your mind.

Before	Statement	After
	5. Sources of water pollution are always easy to identify.	
	6. The proper method of disposal for used motor oil is to pour it down the drain.	

··············· **Read to Learn** ···············

Water as a Resource

Most of Earth's surface is covered with water. Living organisms on Earth are made mostly of water. All organisms, including humans, need water to survive. Humans also use water in ways that other organisms do not. For example, people wash cars, do laundry, and use water for recreation and transportation.

But household activities make up only a small part of human water use. As shown in the table below, most water in the United States is used by power plants. The water is used to generate electricity and to cool equipment. Like the land uses you learned about earlier, the use of water as a resource also impacts the environment. 🗝

Water Use in the United States	
Activity	**Percent of Use**
Electricity-generating power plants	48%
Irrigation of agricultural crops	34%
Public supply (includes houses)	11%
Industry	5%
Livestock, mining, aquaculture	less than 3%

Sources of Water Pollution

Water moves from Earth's surface to the atmosphere and back again in the water cycle. Thermal energy from the Sun causes water at Earth's surface to evaporate into the atmosphere. Water vapor in the air cools as it rises. Then it condenses and forms clouds.

Water returns to Earth's surface as precipitation. Runoff reenters oceans and rivers, or it can seep into the ground. Pollution from many sources can impact the quality of water as it moves through the water cycle.

Point-Source Pollution

Point-source pollution *is pollution from a single source that can be identified.* For example, a discharge pipe that releases industrial waste directly into a river is an example of point-source pollution.

Other examples of point-source pollution include oil spilling from a tanker and the runoff from a mining operation.

Nonpoint-Source Pollution

Pollution from several widespread sources that cannot be traced back to a single location is called **nonpoint-source pollution**. As precipitation runs over Earth's surface, the water picks up materials and substances from farms and urban developments. These different sources might be several kilometers apart. This makes it difficult to trace the pollution in the water back to one specific source.

Runoff from farms and urban developments is an example of nonpoint-source pollution.

Runoff from construction sites can contain excess amounts of sediment. This is another example of nonpoint-source pollution.

Most of the water pollution in the United States comes from nonpoint sources. This kind of pollution is harder to pinpoint and therefore harder to control. The figure at the top of the next two pages summarizes and illustrates some sources of point-source and nonpoint-source polution.

FOLDABLES®

Make a vertical three-tab book, and draw a Venn diagram on the front. Cut the folds to form three tabs, and then use it to compare and contrast sources of pollution.

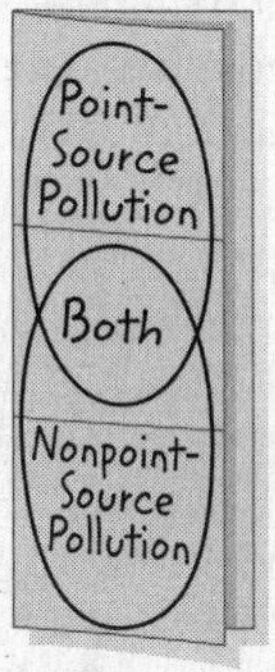

Key Concept Check

3. Explain How can pollution affect water quality?

____

____

____

Sources of Water Pollution

Runoff from mines can be acidic and can contain metals. It can contaminate surface water and groundwater drinking supplies and affect the growth and reproduction of aquatic organisms.

Runoff from construction sites might contain excess sediment, which makes water in streams and rivers cloudy. This reduces the amount of sunlight available for photosynthetic organisms.

Oil spills expose aquatic organisms to toxic chemicals. When oil covers the surface of the water, it can also reduce the amount of oxygen in the water.

Visual Check

4. Identify What are the point sources and nonpoint sources of pollution in this illustration?

_____

_____

_____

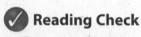

Reading Check

5. Summarize Why is it important to focus on preventing water pollution before it happens?

_____

_____

_____

Positive Actions

Once pollution enters water, it is difficult to remove. In fact, it can take decades to clean polluted groundwater! That is why most efforts to reduce water pollution focus on preventing it from entering the environment, rather than cleaning it up. ✓

International Cooperation

In the 1960s, Lake Erie, one of the Great Lakes, was heavily polluted by runoff from fertilized fields and industrial wastes. Rivers that flowed into the lake were polluted, too. Litter soaked with chemicals floated on the surface of one of these rivers—the Cuyahoga River. In 1969, the litter caught fire. The fire spurred Canada and the United States—the two countries that border the Great Lakes—to clean up the lakes.

The United States and Canada formed several agreements to clean up the Great Lakes. The goals of the countries are pollution prevention as well as cleanup and research.

The Great Lakes still face challenges from aquatic species that are not native to the lakes and from the impact of excess sediments. But pollution from toxic chemicals has decreased.

Agricultural runoff can contain fertilizers, which can upset the balance of nutrients in lakes, oceans, and other bodies of water.

Industrial waste can contain toxic chemicals that can harm aquatic organisms.

Urban runoff can contain pesticides and fertilizers from lawns, oil and gasoline from vehicles, and bacteria and viruses from waste, all of which can reduce the quality of surface water and groundwater.

National Initiatives

In addition to working with other governments, the United States has laws to help maintain water quality within its borders. The Clean Water Act, for example, regulates sources of water pollution, including sewage systems. The Safe Drinking Water Act protects supplies of drinking water throughout the country. ✔

How can you help?

Laws are effective ways to reduce water pollution. But simple actions taken by individuals can have positive impacts, too.

Reduce Use of Harmful Chemicals Many household products, such as paints and cleaners, contain harmful chemicals. Instead of using these, people can use alternative products that do not contain toxins. For example, baking soda and white vinegar are safe, inexpensive cleaning products. People can also use less artificial fertilizer on gardens and lawns. Instead, they can use compost to enrich soils without harming water quality.

✔ **Reading Check**

6. Paraphrase Use your own words to explain what the Clean Water Act does.

_____

_____

_____

_____

_____

Dispose of Waste Safely Sometimes using products that contain pollutants is necessary. For example, vehicles cannot run without motor oil. This motor oil has to be replaced regularly. People should never pour motor oil or other hazardous substances into drains, onto the ground, or directly into streams or lakes. These substances must be disposed of safely. Your local waste management agency has tips for safe disposal of hazardous waste.

Conserve Water Water pollution can be reduced simply by reducing water use. There are many easy ways to conserve water. For example, you can take shorter showers and turn off the water when you brush your teeth. You can keep your drinking water in the refrigerator instead of running water from a faucet until the water is cold. Sweeping leaves and branches from a deck instead of spraying them off with a hose is another way to reduce water use.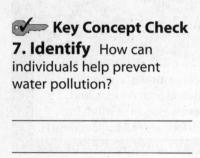

Key Concept Check

7. Identify How can individuals help prevent water pollution?

_____

_____

_____

_____

Mini Glossary

nonpoint-source pollution: pollution from several widespread sources that cannot be traced back to a single location

point-source pollution: pollution from a single source that can be identified

1. Review the terms and their definitions in the Mini Glossary. Write a sentence giving at least one example each of point-source pollution and nonpoint-source pollution.

_____

_____

2. In the graphic organizer, identify a specific action you can take to accomplish the goal.

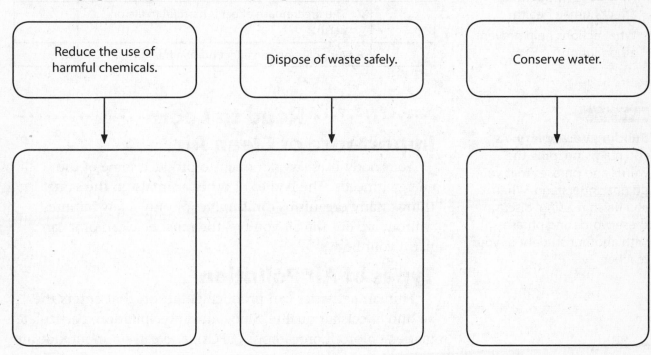

Reduce the use of harmful chemicals.	Dispose of waste safely.	Conserve water.

3. Which type of water pollution in the United States is the hardest to control? Why?

_____

_____

What do you think NOW?

Reread the statements at the beginning of the lesson. Fill in the After column with an A if you agree with the statement or a D if you disagree. Did you change your mind?

 **Connect ED**

Log on to ConnectED.mcgraw-hill.com and access your textbook to find this lesson's resources.

 END OF LESSON

CHAPTER 20
LESSON 4

Environmental Impacts

Impacts on the Atmosphere

Key Concepts

- What are some types of air pollution?
- How are global warming and the carbon cycle related?
- How does air pollution affect human health?
- What actions help prevent air pollution?

Mark the Text

Building Vocabulary As you read, underline the words and phrases that you do not understand. When you finish reading, discuss these words and phrases with another student or your teacher.

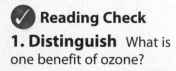 **Reading Check**

1. Distinguish What is one benefit of ozone?

_____

_____

_____

···········**Before You Read**···········

What do you think? Read the two statements below and decide whether you agree or disagree with them. Place an A in the Before column if you agree with the statement or a D if you disagree. After you've read this lesson, reread the statements to see if you have changed your mind.

Before	Statement	After
	7. The greenhouse effect is harmful to life on Earth.	
	8. Air pollution can affect human health.	

···········**Read to Learn**···········

Importance of Clean Air

Your body uses oxygen in air to produce some of the energy it needs. The bodies of other animals do the same thing. Many organisms can survive for only a few minutes without air. But the air you breathe must be clean or it can harm your body.

Types of Air Pollution

Human activities can produce pollution that enters the air and affects air quality. Smog, acid precipitation, particulate matter, chlorofluorocarbons (CFCs), and carbon monoxide are some types of air pollution.

Smog

The brownish haze that hangs in the sky over many cities is photochemical smog. **Photochemical smog** *is caused when nitrogen and carbon compounds in the air react in sunlight.* Nitrogen and carbon compounds are released when fossil fuels are burned to provide energy for vehicles and power plants.

When these compounds react in sunlight, they form other substances. One of these substances is ozone. Ozone high in the atmosphere helps protect living organisms from the Sun's ultraviolet radiation. However, ozone close to Earth's surface is a major component of smog. ✓

Acid Precipitation

Another form of pollution that occurs as a result of burning fossil fuels is acid precipitation. **Acid precipitation** *is rain or snow that has a lower pH than that of normal rainwater.* The pH of normal rainwater is about 5.6. Acid precipitation forms when gases containing nitrogen and sulfur react with water, oxygen, and other chemicals in the atmosphere.

Acid precipitation falls into lakes and ponds or onto the ground. It makes the water and the soil more acidic. Many living organisms cannot survive if the pH of water or soil becomes too low. Acid precipitation can kill trees and other plant life. ✓

Particulate Matter

The mix of both solid and liquid particles in the air is called **particulate matter.** Solid particles include smoke, dust, and dirt. These particles enter the air from natural processes, such as volcanic eruptions and forest fires. Human activities, such as burning fossil fuels at power plants and in vehicles, also release particulate matter. Inhaling particulate matter can cause coughing, difficulty breathing, and other respiratory problems.

CFCs

Ozone in the upper atmosphere absorbs harmful ultraviolet (UV) rays from the Sun. Using products that contain CFCs affects the ozone layer. Air conditioners and refrigerators made before 1996 are products that contain CFCs.

CFCs react with sunlight and destroy ozone molecules. As a result, the ozone layer becomes thinner. This allows more UV rays to reach Earth's surface. Studies have linked increased skin cancer rates to an increase in UV rays.

Carbon Monoxide

Carbon monoxide is a gas released from vehicles and industrial processes. Forest fires also release carbon monoxide into the air.

Wood-burning and gas stoves are sources of carbon monoxide indoors. Breathing carbon monoxide reduces the amount of oxygen that reaches the body's tissues and organs. ✓

Copyright © Glencoe/McGraw-Hill, a division of The McGraw-Hill Companies, Inc.

FOLDABLES

Make a two-tab book, label the tabs as shown, and use the Foldable to record factors that increase or decrease air pollution.

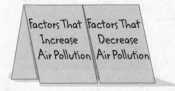

Factors That Increase Air Pollution | Factors That Decrease Air Pollution

✓ **Reading Check**

2. Identify What is the pH of normal rainwater?

_____

_____

⚷ **Key Concept Check**

3. Categorize What are some types of air pollution?

_____

_____

_____

The Carbon Cycle

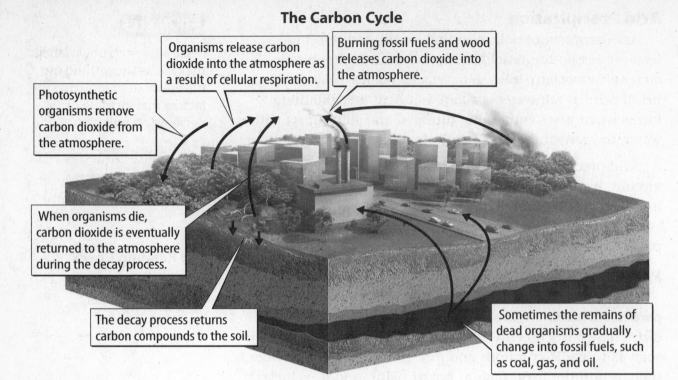

Organisms release carbon dioxide into the atmosphere as a result of cellular respiration.

Burning fossil fuels and wood releases carbon dioxide into the atmosphere.

Photosynthetic organisms remove carbon dioxide from the atmosphere.

When organisms die, carbon dioxide is eventually returned to the atmosphere during the decay process.

The decay process returns carbon compounds to the soil.

Sometimes the remains of dead organisms gradually change into fossil fuels, such as coal, gas, and oil.

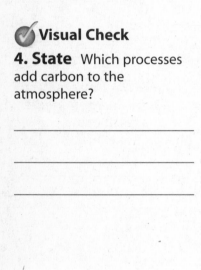

Visual Check

4. State Which processes add carbon to the atmosphere?

_____

_____

_____

Key Concept Check

5. Consider How are global warming and the carbon cycle related?

_____

_____

_____

_____

Global Warming and the Carbon Cycle

Air pollution affects natural cycles on Earth. For example, burning fossil fuels for electricity, heating, and transportation releases substances that cause acid precipitation. Burning fossil fuels also releases carbon dioxide into the atmosphere, as shown in the figure above. An increased concentration of carbon dioxide in the atmosphere can lead to global warming.

Global warming *is an increase in Earth's average surface temperature.* Earth's temperature has increased about 0.7°C over the past 100 years. Scientists hypothesize that the temperature will rise an additional 1.8 to 4.0°C over the next 100 years. Even a small increase in Earth's average surface temperature can cause widespread problems.

Effects of Global Warming

Warmer temperatures can cause ice to melt, making sea levels rise. Higher sea levels can cause flooding along coastal areas. Warmer temperatures can lead to warmer ocean waters. Warmer ocean waters might increase the strength and frequency of storms.

Global warming also can affect the kinds of living organisms found in ecosystems. For example, some hardwood trees do not grow well in warm environments. These trees will no longer be found in some areas if temperatures continue to rise.

The Greenhouse Effect

Why does too much carbon dioxide in the atmosphere increase Earth's temperature? *The* **greenhouse effect** *is the natural process that occurs when certain gases in the atmosphere absorb and reradiate thermal energy from the Sun.* This thermal energy warms Earth's surface. Without the greenhouse effect, Earth would be too cold for life as it exists now.

Carbon dioxide is a greenhouse gas. Other greenhouse gases include methane and water vapor. When the amount of greenhouse gases increases, more thermal energy is trapped and Earth's surface temperature rises. Global warming occurs. ✓

The figure below shows how greenhouse gases absorb and reradiate thermal energy from the Sun and warm Earth's surface.

The Greenhouse Effect

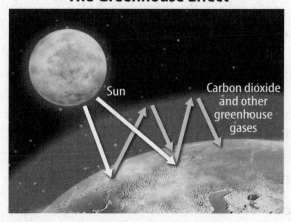

Sun

Carbon dioxide and other greenhouse gases

Health Disorders

Air pollution affects the environment. It also affects human health. Air pollution can cause respiratory problems, including triggering asthma attacks.

Asthma is a disorder of the respiratory system. During an asthma attack, breathing passageways narrow. This makes it hard for a person to breathe. ✓⌐

Measuring Air Quality

Some pollutants, such as smoke from forest fires, are easy to see. But other pollutants, such as carbon monoxide, are invisible. How can people know when levels of air pollution are high?

Copyright © Glencoe/McGraw-Hill, a division of The McGraw-Hill Companies, Inc.

✓ **Reading Check**

6. Explain How are the greenhouse effect and global warming related?

_____

_____

_____

✓ **Visual Check**

7. Identify What happens to thermal energy from the Sun as a result of the greenhouse effect?

_____

_____

⌐✓⌐ **Key Concept Check**

8. Assess How can air pollution affect human health?

_____

_____

_____

_____

Air Quality Index

Ozone Concentration (parts per million)	Air Quality Index Values	Air Quality Description	Preventative Actions
0.0 to 0.064	0 to 50	good	None needed.
0.065 to 0.084	51 to 100	moderate	Highly sensitive people should limit prolonged outdoor activity.
0.085 to 0.104	101 to 150	unhealthy for sensitive groups	Sensitive people should limit prolonged outdoor activity.
0.105 to 0.124	151 to 200	unhealthy	All should limit prolonged outdoor activity.
0.125 to 0.404	201 to 300	very unhealthy	Sensitive people should avoid outdoor activity. All should limit outdoor activity.

Interpreting Tables

9. Describe If the air quality index value is 25, what is the air quality description?

Visual Check

10. Label the source of the air pollution in the figure.

The EPA works with state and local agencies to measure and report air quality. *The* **Air Quality Index** *(AQI) is a scale that ranks levels of ozone and other air pollutants.* The AQI, shown above, ranks ozone levels on a scale of 0 to 300. Ozone in the upper atmosphere blocks harmful rays from the Sun. But ozone that is close to Earth's surface can cause health problems, including throat irritation, coughing, and chest pain. The figure below shows some health disorders caused by air pollution.

Health Effects of Air Pollution

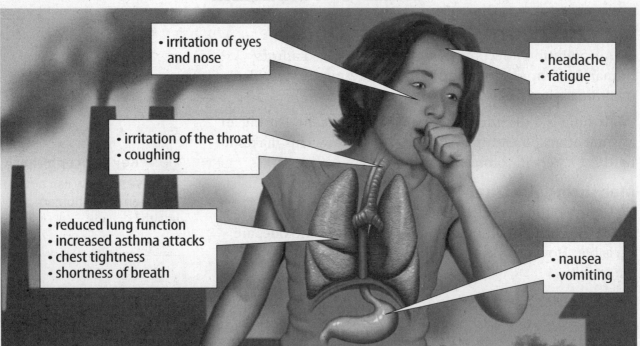

- irritation of eyes and nose
- headache
- fatigue
- irritation of the throat
- coughing
- reduced lung function
- increased asthma attacks
- chest tightness
- shortness of breath
- nausea
- vomiting

Positive Actions

Countries around the world are working together to reduce air pollution. For example, 190 countries, including the United States, have signed the Montreal Protocol to phase out the use of CFCs. Since it was signed, levels of CFCs have decreased. The Kyoto Protocol aims to reduce emissions of greenhouse gases. Currently, 184 countries have accepted the Kyoto Protocol.

National Initiatives

In the United States, the Clean Air Act sets limits on the amount of certain pollutants that can be released into the air. The Clean Air Act was passed in 1970. Since then, amounts of carbon monoxide, ozone near Earth's surface, and substances that produce acid precipitation have decreased by more than 50 percent. Toxins from industrial factories have gone down by 90 percent. ✅

Cleaner Energy

Using renewable energy resources such as solar power, wind power, and geothermal energy to heat homes helps reduce air pollution. Remember that renewable resources are resources that can be replaced by natural processes in a relatively short amount of time.

People also can invest in more energy-efficient appliances and vehicles. For example, some hybrid cars use both a battery and fossil fuels for power. They are more energy efficient and emit less pollution than vehicles that are powered by fossil fuels alone. Solar cars use only the Sun's energy for power.

How can you help?

Reducing energy use means that fewer pollutants are released into the air. You can turn the thermostat lower in the winter when you are heating your home and higher in the summer when the air conditioner is running to save energy. You can walk when it is safe to do so or use public transportation. Each small step you take to conserve energy helps improve air, water, and soil quality. ✅

Copyright © Glencoe/McGraw-Hill, a division of The McGraw-Hill Companies, Inc.

Think it Over

11. Predict What do you think will be the result of the Kyoto Protocol?

_____

_____

_____

✅ Reading Check

12. Evaluate What impact has the Clean Air Act had on air pollution?

_____

_____

_____

_____

_____

🔑 Key Concept Check

13. State How can people help prevent air pollution?

_____

_____

_____

Mini Glossary

acid precipitation: rain or snow that has a lower pH than that of normal rainwater

Air Quality Index (AQI): a scale that ranks levels of ozone and other air pollutants

global warming: an increase in Earth's average surface temperature

greenhouse effect: the natural process that occurs when certain gases in the atmosphere absorb and reradiate thermal energy from the Sun

particulate matter: the mix of both solid and liquid particles in the air

photochemical smog: caused when nitrogen and carbon compounds in the air react in sunlight

1. Review the terms and their definitions in the Mini Glossary. Write a sentence describing the effects of acid precipitation.

_____

_____

2. Use the graphic organizer to identify three effects of global warming.

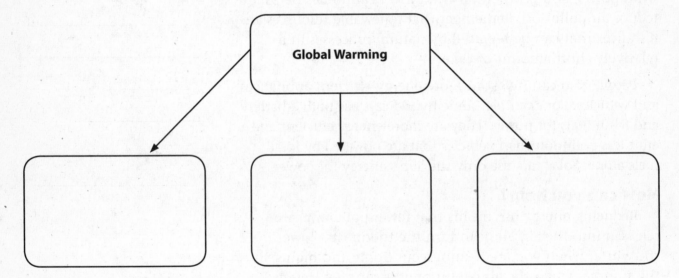

Global Warming

3. Describe the role that trees and other photosynthetic organisms play in the carbon cycle.

_____

_____

What do you think NOW?

Reread the statements at the beginning of the lesson. Fill in the After column with an A if you agree with the statement or a D if you disagree. Did you change your mind?

 **Connect ED**

Log on to ConnectED.mcgraw-hill.com and access your textbook to find this lesson's resources.

 END OF LESSON

Interactions of Human Body Systems

The Human Body

What do you think? Read the three statements below and decide whether you agree or disagree with them. Place an A in the Before column if you agree with the statement or a D if you disagree. After you've read this lesson, reread the statements to see if you have changed your mind.

Before	Statement	After
	1. Elements can be broken down into smaller parts.	
	2. Organic compounds are foods grown without pesticides.	
	3. Organ systems work together.	

Life and Chemistry

Have you ever modeled a volcanic eruption by mixing vinegar and baking soda? When baking soda—also called sodium bicarbonate ($NaHCO_3$)—combines with the acetic acid (CH_3COOH) in vinegar, a chemical reaction occurs. A chemical reaction is the process that occurs when compounds, called reactants, form one or more new substances, called products.

During a chemical reaction, bonds are broken and new bonds are formed. When sodium bicarbonate and acetic acid react, water (H_2O), carbon dioxide (CO_2), and sodium acetate ($NaCH_3COO$) form. You will see bubbles form when vinegar and baking soda mix. The bubbles are caused by the CO_2 gas and the water that are released as the products of this chemical reaction.

Chemical reactions are everywhere. Moldy bread and green pennies are the results of chemical reactions. Chemical reactions also occur in your body. These chemical reactions take place in your body's cells. They are essential for human life. ✓

Key Concepts

- What are the functions of inorganic substances in the human body?
- What are the functions of organic substances in the human body?
- How does the body's organization enable it to function?

◀ **Mark the Text**

Building Vocabulary As you read, circle all the words you do not understand. Highlight the part of the text that helps you understand these words. Review the marked words and their definitions after you finish reading the lesson.

✓ Reading Check

1. Name a common result of a chemical reaction.

_____

_____

Copyright © Glencoe/McGraw-Hill, a division of The McGraw-Hill Companies, Inc.

Elements and Compounds

Elements are the basic units that make up chemicals. Elements cannot be broken down or changed into another element during a chemical reaction. Elements have different physical properties. Of the almost 100 elements found in nature, six elements make up 99 percent of your body's mass. These elements are carbon, oxygen, hydrogen, nitrogen, calcium, and phosphorus. ✓

Compounds are substances made of two or more elements. Unlike elements, compounds can be broken down into simpler substances. Acetic acid is a compound. It is formed when the elements carbon, hydrogen, and oxygen combine.

Compounds form when elements bind together in one of two different ways. Compounds have either ionic bonds or covalent bonds. Ionic bonds are formed when electrons travel from one element to another. One element has a positive charge, and the other has a negative charge. The opposite charges attract. Table salt, or sodium chloride (NaCl), is an example of a compound with an ionic bond.

Covalent bonds are formed when the electrons in each element are shared. Many gases, such as oxygen (O_2) and nitrogen (N_2), form by covalent bonds.

Inorganic Substances

Inorganic compounds are everywhere on Earth. Inorganic compounds are substances that do not contain carbon-hydrogen bonds. Substances such as ammonia (NH_3) and NaCl are inorganic compounds. Many inorganic compounds, such as water and oxygen gas, are essential for human life.

Ionic Compounds As you have read, NaCl is an inorganic compound that forms by ionic bonding. Ionic bonds are formed when a positive ion is attracted to a negative ion. When a substance gives up or gains an electron, the substance is called an ion. In the compound NaCl, Na^+ is a positively charged ion, and Cl^- is a negatively charged ion. Many ions are important for human survival. For example, calcium (Ca^{2+}) helps nerves and muscle cells function. Calcium also makes up bone. These compounds rely on water to move through the body. ✓

Water You might know that water is called the universal solvent. A solvent is a substance that dissolves other substances. Ionic compounds dissolve well in water. That makes it possible for ions such as Na^+, Cl^-, and Ca^{2+} to travel through the body dissolved in water.

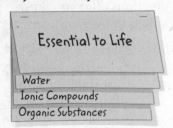

Reading Check

2. Name the elements that make up most of the body's mass.

_____

_____

_____

_____

_____

FOLDABLES

Make a layered book to record your notes on human body chemistry.

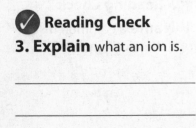

Essential to Life

Water
Ionic Compounds
Organic Substances

Reading Check

3. Explain what an ion is.

_____

_____

_____

Structure of a Water Molecule

Salt dissolved in water

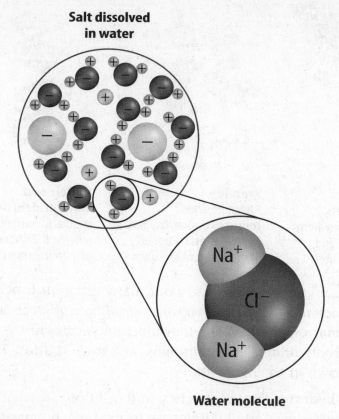

Water molecule

Water is able to dissolve many substances because of its polarity. Water molecules are formed by covalent bonds that link the hydrogen (H) to the oxygen (O).

As shown in the figure above, water is able to easily dissolve ionic substances. This is because the positive ions (Na^+) in the compound NaCl are attracted to the oxygen end of the water molecules and the negative ions (Cl^-) are attracted to the hydrogen end of the water molecules.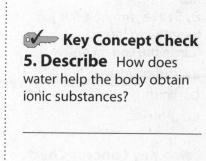

Look again at the water molecule above. The molecule has a bent shape with the large oxygen atom at one end of the molecule. This bent shape results in the oxygen end of the molecule having a negative charge. The hydrogen end has a positive charge.

Organic Substances

You may have heard the term *organic* used to describe certain fruits and vegetables. However, when the term *organic* is used in science, it describes certain compounds. Those compounds contain carbon and other elements, such as hydrogen, oxygen, phosphorus, nitrogen, or sulfur, held together by covalent bonds.

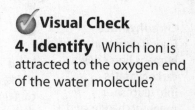

Visual Check
4. Identify Which ion is attracted to the oxygen end of the water molecule?

_____

_____

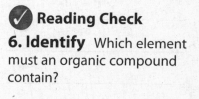

Key Concept Check
5. Describe How does water help the body obtain ionic substances?

_____

_____

_____

Reading Check
6. Identify Which element must an organic compound contain?

_____

Macromolecules

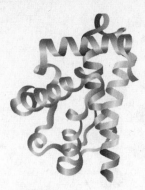

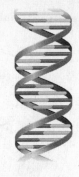

Carbohydrate
During digestion, humans break down carbohydrates into glucose, a monosaccharide, and store it as glycogen.

Lipid
Lipids, also called fats, contain fewer oxygen atoms than carbohydrates and do not dissolve in water.

Protein
All amino acids consist of carbon, oxygen, hydrogen, and nitrogen. Some also contain sulfur.

Nucleic acid
DNA is made of two strands of nucleotide polymers. RNA is made of a single strand.

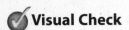 **Visual Check**

7. Identify Circle the macromolecule that is made of nucleotide polymers.

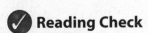 **Reading Check**

8. State Amino acids join together to form _____. (Circle the correct answer.)

a. carbohydrates

b. lipids

c. proteins

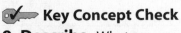 **Key Concept Check**

9. Describe What are the functions of organic compounds in the human body?

_____

_____

_____

_____

Organic compounds carry out many different functions. *Substances that form from joining many small molecules together are called* **macromolecules.** The four macromolecules in the body are carbohydrates, lipids, proteins, and nucleic acids. They are shown in the figure above.

Carbohydrates Sugars, starches, and cellulose are carbohydrates. Carbohydrates are formed when *simple sugars, called* **monosaccharides** (mah nuh SA kuh ridez), are joined together. Carbohydrates are the body's major source of energy.

Lipids Triglycerides and cholesterol are lipids (LIH pihdz). Like carbohydrates, they are made from carbon hydrogen, and oxygen. Lipids help insulate your body and are a major part of cell membranes.

Proteins The adult human body is made up of 10–20 percent protein. Proteins form when **amino** (uh MEE noh) **acids,** *the building blocks of protein,* join together. Some proteins give cells structure, some help cells communicate, and some are enzymes. ✓

Nucleic Acids Much as computer chips store information, nucleic acids are macromolecules that store information. This information is used by the body to perform different functions. Nucleic acids are formed when **nucleotides** (NEW klee uh tidez), *molecules made of a nitrogen base, a sugar, and a phosphate group,* join together. The body contains two types of nucleic acids, DNA and RNA. ⚷

The Body's Organization

To function and survive, macromolecules in the human body must be organized in different compartments. For example, most DNA is stored in the nuclei of cells. Cholesterol and other types of lipids are used to form cell membranes. Organizing macromolecules in specific locations helps cells carry out specific functions. ✔

Cells Recall that cells are the building blocks of all living organisms. Cells have different shapes depending on their function. Neurons are long and slender so they can carry information over long distances. Red blood cells are flexible disks that can move easily through blood vessels.

Tissues Cells that work together and perform a function make up tissues. Cardiac muscle cells form a tissue that helps the heart pump blood throughout the body.

Organs A group of tissues that work together and perform a function is an organ. The liver, spleen, and lungs are organs.

Organ Systems As shown in the figure to the right, an organ system is a group of organs that works together and performs a specific task. Organ systems work together and help the body communicate, defend itself, process energy, transport substances, and move. 🔑

Organ System

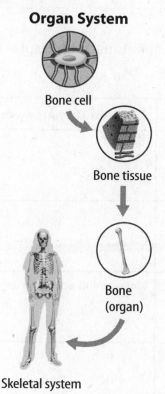

Bone cell

Bone tissue

Bone (organ)

Skeletal system

✔ **Reading Check**

10. Summarize Where is most DNA stored?

_____

_____

✔ **Visual Check**

11. Identify What is bone tissue made of?

_____

🔑 **Key Concept Check**

12. Explain How does the body's organization enable it to function?

_____

_____

_____

_____

_____

_____

Mini Glossary

amino (uh MEE noh) acid: a building block of protein

macromolecule: a substance that is formed from joining many small molecules together

monosaccharide (mah nuh SA kuh ride): a simple sugar

nucleotide (NEW klee uh tide): a molecule made of a nitrogen base, a sugar, and a phosphate group

1. Review the terms and their definitions in the Mini Glossary. Write a sentence that describes the importance of amino acids.

_____

_____

2. Use what you have learned about macromolecules to complete the table. The third item has been completed for you.

Macromolecules	How They Are Formed	Examples
Carbohydrates		
Lipids		
Proteins	when amino acids join together	enzymes
Nucleic acids		

3. Compare and contrast elements and compounds.

_____

_____

What do you think NOW?

Reread the statements at the beginning of the lesson. Fill in the After column with an A if you agree with the statement or a D if you disagree. Did you change your mind?

Connect ED

Log on to ConnectED.mcgraw-hill.com and access your textbook to find this lesson's resources.

END OF LESSON

Interactions of Human Body Systems

How Body Systems Interact

·············· **Before You Read** ··············

What do you think? Read the three statements below and decide whether you agree or disagree with them. Place an A in the Before column if you agree with the statement or a D if you disagree. After you've read this lesson, reread the statements to see if you have changed your mind.

Before	Statement	After
	4. Nutrients are processed by the skeletal system.	
	5. The nervous system moves oxygen through the body.	
	6. You do not control reflexes.	

·············· **Read to Learn** ··············

Homeostasis

Your body has a system to keep its internal temperature constant. That system works much like a thermostat that keeps a building's temperature constant. The endocrine system regulates body temperature. It sends messages through the nervous system.

For example, when temperatures in the body fall below 37°C, the nervous system signals the muscular system to cause the body to shiver. When you shiver, your muscles move. Tiny muscles attached to hairs on the skin contract and pull the hairs up straight, forming goose bumps. This movement generates thermal energy and helps raise body temperature. Keeping the body's temperature constant requires that the endocrine system, the nervous system, and the muscular system work together.

Your body's <u>organ</u> systems work together and maintain many types of homeostasis (hoh mee oh STAY sus). These include temperature, nutrient levels, oxygen, fluid levels, and pH. **Homeostasis** *is the ability to maintain constant internal conditions when outside conditions change.* In this lesson, you will read how organ systems work together and maintain homeostasis.

Key Concepts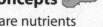
- How are nutrients processed in the body?
- How does the body transport and process oxygen and wastes?
- How does the body coordinate movement and respond to stimuli?
- How do feedback mechanisms help maintain homeostasis?

> **Study Coach**

Identify the Main Ideas
Work with a partner. Read a paragraph together. Then discuss what you learned in the paragraph. Continue until you and your partner understand the main ideas of the lesson.

SCIENCE USE v. COMMON USE ···
organ
Science Use a group of tissues performing a specific function

Common Use a keyboard instrument in which pipes are sounded by compressed air

Copyright © Glencoe/McGraw-Hill, a division of The McGraw-Hill Companies, Inc.

Processing Nutrients

Maintaining homeostasis keeps the internal environment in the body functioning properly. Many organ systems work together and maintain energy homeostasis.

The body gets most of its energy from carbohydrates. Lipids and proteins also provide energy. The food you eat is broken down by chemical and mechanical digestion.

Chemical digestion occurs when enzymes in saliva and acid in your stomach break down food. Mechanical digestion happens when you chew your food. The digestive system, the circulatory system, and the muscular system work together and process and obtain nutrients from food. The skeletal system, the endocrine system, and the lymphatic system also work with the digestive system and process those nutrients.

Muscles and Digestion

Food enters the body through the digestive system. There it is broken down into nutrients that can be absorbed into the body. However, the muscular system is needed to get food through the digestive system. Muscles that surround the stomach contract and move food to the small intestine. These contractions are called peristalsis (per uh STAHL sus).

Muscles help the jaw move when you chew. They help you swallow. Muscles also surround the esophagus, the stomach, the small intestine, and the large intestine. These muscles help move food through the digestive system.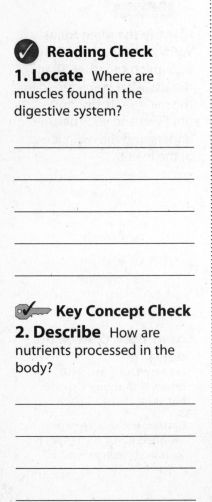

Circulation and Digestion

The small intestine has two important jobs. It breaks down food and absorbs nutrients. The muscular system and the circulatory system work with the small intestine.

The muscular system helps the small intestine break down food. The circulatory system works with the small intestine and gets nutrients to the rest of the body.

Nutrients are absorbed by small, fingerlike projections called villi (VIH li; singular, villus) in the small intestine. The villi have blood vessels inside them, which are part of the circulatory system. Nutrients enter these blood vessels and are then transported to the rest of the body. The muscular system also surrounds the blood vessels and helps blood and nutrients move through the body. The figure on the next page shows a close look at one villus and how the digestive system and the circulatory system work together.

Reading Check

1. Locate Where are muscles found in the digestive system?

_____

_____

_____

_____

_____

Key Concept Check

2. Describe How are nutrients processed in the body?

_____

_____

_____

_____

Copyright © Glencoe/McGraw-Hill, a division of The McGraw-Hill Companies, Inc.

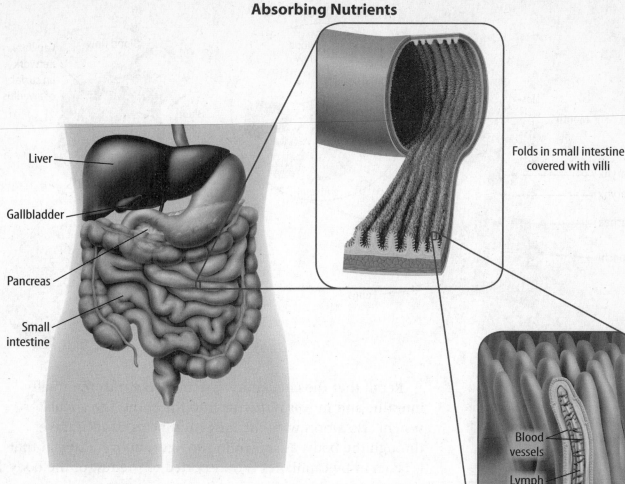

Liver

Gallbladder

Pancreas

Small
intestine

Folds in small intestine
covered with villi

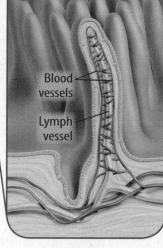

Blood
vessels

Lymph
vessel

One villus

Processing Oxygen and Wastes

The systems of the human body must work together for
the body to function properly. For example, humans require
oxygen to survive. Your lungs take in oxygen and release
carbon dioxide. The cells in your body use oxygen to help
process the energy in nutrients into energy that cells can use.
Oxygen helps the body obtain energy from nutrients by
performing cellular respiration. As discussed next, various
organ systems work together and help the body take in
oxygen and move it through the body.

Oxygen Transport

Oxygen enters the body through the respiratory system.
When you inhale, the respiratory system works with the
circulatory system and transports oxygen to all cells in the
body. The muscular system also helps the respiratory system
by expanding the chest so that cells in the lungs fill up with
oxygen.

☑ **Visual Check**
3. Identify What does
each villus contain?

_____

_____

Transporting Oxygen

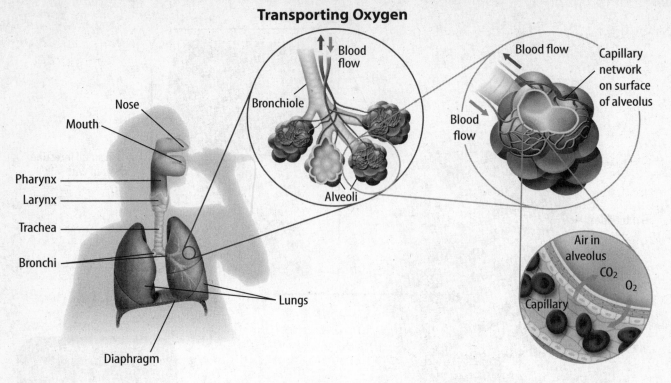

Nose
Mouth
Pharynx
Larynx
Trachea
Bronchi
Lungs
Diaphragm

Blood flow
Bronchiole
Alveoli

Blood flow
Blood flow
Capillary network on surface of alveolus

Air in alveolus
CO_2
O_2
Capillary

Visual Check

4. Name What do capillaries release?

_____

_____

Key Concept Check

5. Relate How does the body transport and process oxygen and wastes?

_____

_____

_____

_____

_____

_____

Recall that the circulatory system works with the small intestine and moves nutrients into the body. The circulatory system also works with the lungs and helps oxygen travel through the body, as shown in the figure above. Oxygen that is taken in by capillaries is transported to the rest of the body through larger blood vessels.

Eliminating Wastes

The excretory system works with several other organ systems and eliminates wastes. Recall that the body processes food, oxygen, and liquids.

Food and liquids are processed by the digestive system. After nutrients are absorbed during digestion, the excretory system removes solid waste products, or feces, through the rectum.

The excretory system also works with the respiratory and circulatory systems and removes carbon dioxide (CO_2) from the body. Oxygen is used in all organs of the body. The CO_2 produced by cells in the body enters capillaries and is transported to the lungs, where it is exhaled. These three systems work together and maintain oxygen homeostasis by making sure that CO_2 is removed.

The excretory system also maintains fluid homeostasis. Liquid waste travels through the circulatory system to the kidneys, as shown in the figure on the next page. The kidneys make urine. Liquid waste also travels to the skin where fluid is released during sweating.

Kidney Function

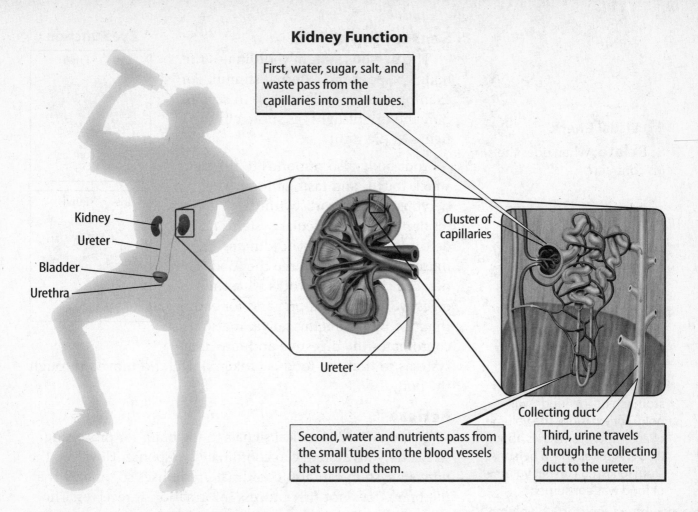

First, water, sugar, salt, and waste pass from the capillaries into small tubes.

Kidney

Ureter

Bladder

Urethra

Cluster of capillaries

Ureter

Collecting duct

Second, water and nutrients pass from the small tubes into the blood vessels that surround them.

Third, urine travels through the collecting duct to the ureter.

Control and Coordination

How does your heart beat without you thinking about it? The heart contains a group of specialized cells called pacemaker cells. Pacemaker cells control the rate at which the heart beats by responding to signals from the nervous system.

The nervous system speeds up your heartbeat when you are exercising. It slows the rate at which your heart beats when you are sleeping. The nervous system also works with other organ systems to control the body's functions.

The nervous system uses electrical signals to help organ systems of the body respond quickly to changes in internal and external environments. The body also uses the endocrine system to help it respond to changes in environments and maintain homeostasis.

The nervous system coordinates rapid changes. The endocrine system coordinates slower responses.

Visual Check

6. Interpret How are wastes transported to the kidneys?

_____

_____

_____

Key Concept Check

7. Explain How does the body coordinate movement and respond to stimuli?

_____

_____

_____

_____

Sensory Input

The nervous system coordinates the body's response to external stimuli. For example, your pupils change in size in dim and bright light, as shown in the figure to the right.

Your body also responds to the sight, smell, touch, and taste of food. The nervous system works with the respiratory and muscular systems to detect food smells. It coordinates muscles in the eyes to see the food. The nervous system also works with the digestive system and prepares for eating the food by producing saliva. It also coordinates the digestive and muscular systems so that the food is broken down and moved through the body.

Reflexes

Neurons send electrical signals to the brain for processing so the nervous system can coordinate a response. However, the nervous system can also coordinate responses so quickly that the brain does not first process information it receives. The response to touching a hot stove is so fast that you don't think before you remove your hand. This is because the nervous system has a rapid response system, called a reflex. A reflex reacts to stimuli without sending information to the brain for processing. Reflexes allow the nervous system to coordinate a rapid response and tell the muscular system and the skeletal system to move without thought.

Hormones

The endocrine system coordinates other organ systems by using chemical signals called hormones. Hormones are secreted from endocrine organs such as the thyroid gland, the adrenal gland, and the pancreas. ✓

These chemical signals travel through the circulatory system to organ systems such as the digestive and muscular systems. They also control processes that maintain homeostasis. In the beginning of this lesson, you read that temperature homeostasis is maintained by producing thermal energy. The endocrine, nervous, and muscular systems work together and maintain temperature homeostasis. Insulin, a hormone released from the pancreas, works with the digestive system and maintains energy homeostasis.

Iris Pupil

The iris contracts in dim light.

Iris Pupil

The iris expands in bright light.

Visual Check

8. Relate When does the iris contract?

_____

_____

Math Skills

Volume is a measure of the amount of matter that a hollow object, such as the stomach or the lungs, will hold. For example, the volume of an empty stomach is about 0.08 L. After eating, a person's stomach is 1.5 L. What volume of food was consumed?

Subtract the starting volume from the final volume.

1.5 L − 0.08 L = 1.42 L

9. Use Volume A certain person's bladder has a volume of 550 mL. The person has the urge to urinate when the bladder contains 200 mL of urine. What volume of the bladder remains empty?

_____

Reading Check

10. Define What are hormones?

_____

_____

_____

_____

Feedback Mechanisms

As stated earlier, homeostasis helps the body maintain a constant internal environment. The endocrine and nervous systems help detect changes in either the internal or the external environment and respond to those changes. Organ systems use feedback mechanisms to maintain homeostasis.

Negative Feedback

Negative feedback *is a control system that helps the body maintain homeostasis by sending a signal to stop a response.* Negative feedback is the type of control system in effect when you feel hungry and eat. The digestive system receives signals that it is time to eat. When you eat, the digestive and circulatory systems work together and increase the amount of nutrients in your body. As the nutrients are processed, your stomach sends signals to your brain to tell your body that you are full and to stop eating.

Positive Feedback

Positive feedback *is a control system that sends a signal to increase a response.* One example of a positive feedback system is blood clotting. When you are bleeding, the circulatory system maintains homeostasis by controlling blood loss. Blood cells called platelets move to the site of the wound. The platelets help control bleeding by forming a clot with a protein called fibrin. As the clot forms, more platelets travel to the clot. The figure below shows how the body uses positive feedback to clot blood.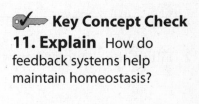

Childbirth is another example of positive feedback. The endocrine system signals the muscular system to contract. Signals from the muscular system tell the endocrine system to keep activating the muscular system. This continues until the baby is born.

Copyright © Glencoe/McGraw-Hill, a division of The McGraw-Hill Companies, Inc.

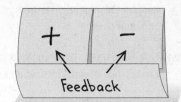

FOLDABLES

Make a two-tab book to compare types of feedback.

Key Concept Check

11. Explain How do feedback systems help maintain homeostasis?

_____

_____

_____

_____

Visual Check

12. Recognize What holds platelets in place when they form a blood clot?

_____

_____

Blood Clotting

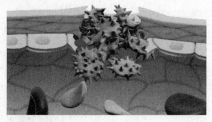

Step 1
Platelets rush to the tear and form a plug that stops the bleeding.

Step 2
A web of fibrin forms around the platelets and holds them in place.

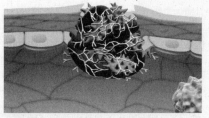

Step 3
The fibrin web catches more platelets and red blood cells, and these form a blood clot.

Mini Glossary

homeostasis (hoh mee oh STAY sus): the ability to maintain constant internal conditions when outside conditions change

negative feedback: a control system that helps the body maintain homeostasis by sending a signal to stop a response

positive feedback: a control system that sends a signal to increase a response

1. Review the terms and their definitions in the Mini Glossary. Write a sentence that contrasts negative and positive feedback.

_____

_____

2. Fill in the diagram below to show the systems that help transport oxygen.

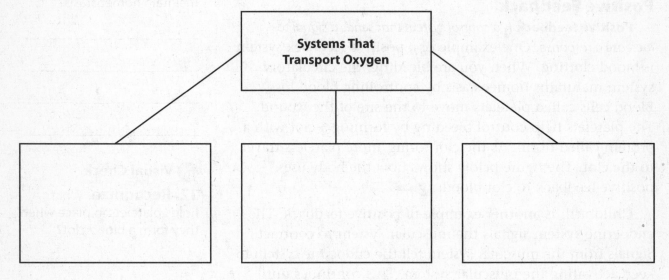

Systems That Transport Oxygen

3. In the space below, record one main idea you were able to better understand by working with your partner.

_____

_____

_____

What do you think NOW?

Reread the statements at the beginning of the lesson. Fill in the After column with an A if you agree with the statement or a D if you disagree. Did you change your mind?

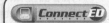

Log on to ConnectED.mcgraw-hill.com and access your textbook to find this lesson's resources.

END OF LESSON

Heredity and How Traits Change

How are traits inherited?

·············· **Before You Read** ··············

What do you think? Read the two statements below and decide whether you agree or disagree with them. Place an A in the Before column if you agree with the statement or a D if you disagree. After you've read this lesson, reread the statements to see if you have changed your mind.

Before	Statement	After
	1. Genes are on chromosomes.	
	2. Only dominant genes are passed on to offspring.	

············· **Read to Learn** ···············

From Parent to Offspring

You and your classmates are all the same species, *Homo sapiens*. But why do you all have different eye colors, hair colors, and heights? How do you inherit traits from your parents? **Heredity** (huh REH duh tee) *is the passing of traits from parents to offspring*. Scientists who study heredity have been answering questions like these for centuries.

What is genetics?

Genetics (juh NE tihks) *is the study of how traits pass from parents to offspring*. All organisms have genes. Genes determine an organism's shape and its life functions. Genes can even control how an organism behaves. For most organisms, genes are sections of DNA that contain information about a specific trait of that organism. This information can vary. A gene with different information for a trait is called an allele. For example, facial dimples result from alleles on a pair of chromosomes. The figure above shows the relationship between alleles on chromosomes and the traits they express. Each chromosome pair has genes for the same traits. Recall that chromosomes are coiled strands of DNA.

Chromosome Pair

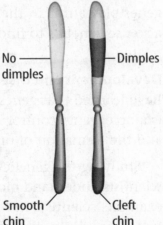

No dimples — — Dimples

Smooth chin — — Cleft chin

Key Concepts

- How are traits inherited?
- Why do scientists study genetics?
- What did Gregor Mendel investigate and discover about heredity?

◀ Study Coach

Make an outline as you read to summarize the information in the lesson. Use the main headings in the lesson as the main headings in your outline. Use your outline to review the lesson.

✓ Visual Check

1. Draw a circle around the alleles for dimples.

FOLDABLES

Make a horizontal two-tab book and use it to organize information about the study of genetics.

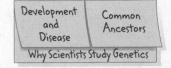

Development and Disease	Common Ancestors

Why Scientists Study Genetics

Copyright © Glencoe/McGraw-Hill, a division of The McGraw-Hill Companies, Inc.

REVIEW VOCABULARY

trait
a distinguishing characteristic
of an organism

How are traits inherited?

An organism passes its traits to its offspring in one of two ways: (1) through asexual reproduction or (2) through sexual reproduction. Unicellular organisms such as bacteria can reproduce asexually. Some multicellular organisms such as spider plants also reproduce asexually. In asexual reproduction, one organism makes a copy of its genes and itself. In sexual reproduction, offspring receive half of their genes from an egg cell and the other half from a sperm cell.

An individual organism expresses the traits in the genes it inherited. For example, if your eyes are blue, it is because you inherited the genes for blue eye color from your parents. In fact, the specific combination of the genes you inherited from your parents for all of your traits is unique. Unless you are an identical twin, triplet, or quadruplet, no other person has the same combination of all your genes! ✓

Inherited traits are different from traits that an individual acquires, or learns, during its lifetime. For example, a bird's size is mostly inherited from its parents. However, the song it sings is mostly learned. Young golden orioles will learn to sing by listening to their parents' songs. Obedience in tame animals is another example of an acquired trait. You can teach a dog to sit, but its puppies will not be born already knowing that they should sit when you tell them to. 🔑

Why do scientists study genetics?

Scientists began studying genetics to understand how traits are inherited. They soon learned that genes control how an organism develops. They also learned that sometimes genes play a role in the development of disease. Scientists also use genetics to find out more about how species are related.

Development and Disease By studying genetics, scientists have learned that genes control how organisms develop. For example, genes control limb development, body segmentation, and the formation of organs, such as eyes and ears.

Studying the genetics of development also can help scientists understand more about disease in humans. For example, scientists have learned how problems with genes in fish can result in diseases in those fish. Scientists can use what they have learned and apply it to the study of human genetics.

✓ **Reading Check**

2. Explain why each individual is unique.

_____

_____

_____

🔑 **Key Concept Check**

3. State How are traits inherited?

_____

_____

_____

Common Ancestors Studying genetics also can help scientists determine how organisms are related. The figure below shows how scientists placed a gene that controls eye development in mice into a fruit fly. The fly developed normal eyes! In another experiment, scientists placed the gene that controls eye development in fruit flies into a frog. Like the fly that received the mouse gene, the frog developed normal eyes. Because the genes from these organisms are similar enough to produce normal eyes when the genes are exchanged, scientists suspect that these species share a common, ancient ancestor. 🗝

☑🗝 **Key Concept Check**

4. Summarize Why do scientists study genetics?

_____

_____

_____

_____

_____

_____

Genetics Experiments

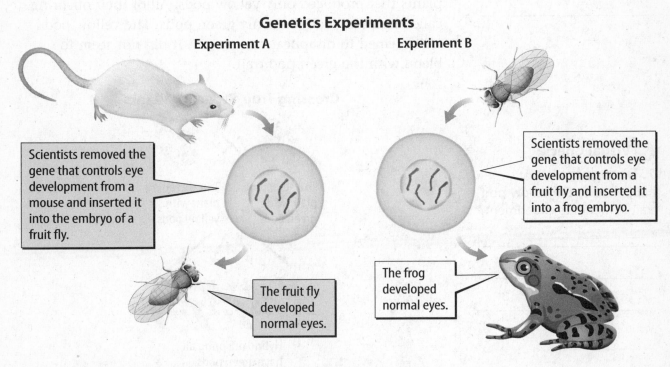

Experiment A

Experiment B

Scientists removed the gene that controls eye development from a mouse and inserted it into the embryo of a fruit fly.

Scientists removed the gene that controls eye development from a fruit fly and inserted it into a frog embryo.

The fruit fly developed normal eyes.

The frog developed normal eyes.

Heredity—The History and the Basics

For thousands of years, humans have been slowly improving crops and farm animals through selective breeding. **Selective breeding** *is the selection and breeding of organisms for desired traits.* Suppose you are a farmer who owns three hens. One produces 180 eggs per year, one produces 230, and one produces 280. If you wanted to breed hens that produced more eggs, which hen would you breed with the neighbor's rooster? Why?

Though people have successfully used selective breeding throughout history, they did not always get the results they expected. When an Austrian friar named Gregor Mendel began experimenting with pea plants, people finally learned more about how selective breeding works.

✓ **Visual Check**

5. Explain the difference between Experiment A and Experiment B.

_____

_____

_____

_____

_____

_____

Copyright © Glencoe/McGraw-Hill, a division of The McGraw-Hill Companies, Inc.

Mendel's Experiments

In 1856, Gregor Mendel began experimenting to answer the question of how traits are inherited. At the time, most scientists thought that traits blended from parents to offspring, similar to the way two colors of paint can be blended. But Mendel did not accept the blending hypothesis. ✔

Crossing True-Breeding Plants To test his ideas, Mendel carefully selected pea plants with specific traits. He then bred the plants. As shown in the figure below, Mendel chose plants that produced only green pods. These plants were called true-breeding plants. Mendel crossed the true-breeding plants that produced only green pods with true-breeding plants that produced only yellow pods. All of their offspring, called hybrids, produced only green pods. The yellow-pod trait seemed to disappear completely. It did not seem to blend with the green-pod trait.

Crossing True-Breeding Plants

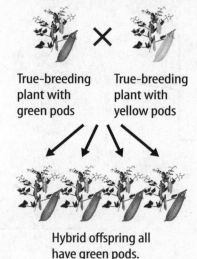

True-breeding plant with green pods × True-breeding plant with yellow pods

Hybrid offspring all have green pods.

Crossing Hybrids When Mendel crossed two hybrid plants with green pods, the cross resulted in some offspring with green pods and other offspring with yellow pods. The results of the cross are shown in the figure at the top of the next page. These offspring were in a ratio of about 3:1, green to yellow.

Mendel tested thousands of pea plants. He tracked traits such as seed shape and flower color. The crosses between hybrids for each trait produced a similar 3:1 ratio. Mendel proposed several ideas to explain his results. ✔

---

6. Relate Before Mendel, how did scientists believe traits were inherited?

_____

_____

_____

✓ **Visual Check**

7. Recognize What happened to the yellow-pod trait in Mendel's experiment?

_____

_____

_____

✓ **Reading Check**

8. Name three traits Mendel tracked in his hybrid experiments.

_____

_____

_____

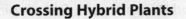

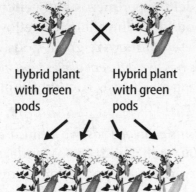

Hybrid plant with green pods × Hybrid plant with green pods

Offspring have green pods or yellow pods in a ratio of 3:1.

Dominant and Recessive Alleles

Mendel proposed that instead of blending, some traits of organisms are dominant, and others are recessive. *A **dominant trait** is a genetic factor that blocks another genetic factor. A **recessive trait** is a genetic factor that is blocked by the presence of a dominant factor.* When an individual has one dominant allele and one recessive allele for a trait, the dominant trait is expressed.

The presence of dominant and recessive traits explains why the offspring of the true-breeding green-pod plants and the true-breeding yellow-pod plants all produce green pods. Green pods are a dominant trait, and yellow pods are a recessive trait.

Mendel's Principles of Inheritance

Earlier in this lesson, you read that an allele is one form of a gene. Mendel did not know about genes or alleles. Mendel did suspect, however, that a physical factor was responsible for the traits of his pea plants. Mendel was right. The factors Mendel proposed now are called genes.

Genotype and Phenotype Which alleles are present on a pair of chromosomes determines whether an individual has the dominant or the recessive trait. *The alleles of all the genes on an organism's chromosomes make up the organism's* **genotype** (JEE nuh tipe). *How the traits appear, or are expressed, is the organism's* **phenotype** (FEE nuh tipe).

✔ Visual Check

9. State What was the ratio of offspring with green pods to offspring with yellow pods when two hybrids with green pods were crossed?

_____

✔ Reading Check

10. Differentiate What is the difference between a dominant and a recessive trait?

_____

_____

_____

_____

💡 Think it Over

11. Classify Which of the following is an example of a phenotype? (Circle the correct answer.)

a. an organism's chromosomes

b. eye color

c. alleles

Heterozygous and Homozygous Genotypes The hybrid pea plants from Mendel's experiments had genotypes of one allele for green pods and one allele for yellow pods. The phenotypes of these plants were green pods. *When an organism's genotype has two different alleles for a trait, it is called* **heterozygous** (he tuh roh ZI gus). The hybrid plants were heterozygous for pod color.

When an organism's genotype has two identical alleles for a trait, it is called **homozygous** (hoh muh ZI gus). Mendel's true-breeding plants were homozygous for pod color. Homozygous and heterozygous genotypes also affect the phenotypes of other organisms.

Next, you will read about the rediscovery of Mendel's work. Scientists have confirmed and built upon Mendel's ideas as they have learned more about genetics and heredity. 🗝

🗝 **Key Concept Check**

12. Explain What did Mendel investigate and discover about heredity?

_____

_____

_____

_____

Mini Glossary

dominant trait: a genetic factor that blocks another genetic factor

genetics (juh NE tihks): the study of how traits pass from parents to offspring

genotype (JEE nuh tipe): the alleles of all the genes on an organism's chromosomes

heredity (huh RE duh tee): the passing of traits from parents to offspring

heterozygous (he tuh roh ZI gus): an organism's genotype that has two different alleles for a trait

homozygous (hoh muh ZI gus): an organism's genotype that as two identical alleles for a trait

phenotype (FEE nuh tipe): how an organism's traits appear or are expressed

recessive trait: a genetic factor that is blocked by the presence of a dominant factor

selective breeding: the selection and breeding of organisms for desired traits

1. Review the terms and their definitions in the Mini Glossary. Write a sentence contrasting dominant and recessive traits.

_____

_____

2. Complete the graphic organizer below, which represents Mendel's pea-plant experiments.

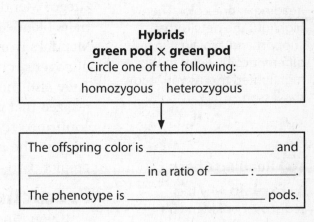

True-breeding **green pod × yellow pod** Circle one of the following: homozygous   heterozygous

The offspring color is _____ .

dominant color: _____

recessive color: _____

Hybrids **green pod × green pod** Circle one of the following: homozygous   heterozygous

The offspring color is _____ and _____ in a ratio of _____ : _____ .

The phenotype is _____ pods.

3. Select a word that appears in the main heading of the outline you created when you reviewed the lesson. In the space below, define that word.

_____

What do you think NOW?

Reread the statements at the beginning of the lesson. Fill in the After column with an A if you agree with the statement or a D if you disagree. Did you change your mind?

ConnectED

Log on to ConnectED.mcgraw-hill.com and access your textbook to find this lesson's resources.

END OF LESSON

Heredity and How Traits Change

Genetics After Mendel

Copyright © Glencoe/McGraw-Hill, a division of The McGraw-Hill Companies, Inc.

Key Concepts 🔑

- How can you use tools to predict genetic outcomes?
- What are the other patterns of inheritance?
- What role can mutations play in the inheritance of disease?

Mark the Text

Identify the Main Ideas
To help you learn about genetics, highlight each heading in one color. Then highlight the details that support and explain it in a different color. Refer to this highlighted text as you study the lesson.

✓ **Reading Check**

1. Explain why the rediscovery of Mendel's pea-plant experiments was important to genetics.

_____

_____

_____

_____

·············· **Before You Read** ··············

What do you think? Read the two statements below and decide whether you agree or disagree with them. Place an A in the Before column if you agree with the statement or a D if you disagree. After you've read this lesson, reread the statements to see if you have changed your mind.

Before	Statement	After
	3. Modern-day genetics disproved Gregor Mendel's ideas about inheritance.	
	4. Mutations can cause disease in an individual.	

·············· **Read to Learn** ··············

Rediscovering Mendel's Work

Soon after he published his results on pea-plant genetics, Gregor Mendel's work was mostly forgotten. The idea that traits blended from parents to offspring was still popular. Mendel's work was rediscovered in 1900. Scientists had discovered chromosomes and could see them inside cells. They also thought the cell nucleus contained genes but soon realized that genes were on chromosomes in the nucleus. They confirmed that genes were Mendel's dominant and recessive factors. The next step was to learn more about how to predict patterns of inheritance. ✓

Predicting Genetic Outcomes

If you flip a coin, the chance that it will land heads-up is one-half, or 50 percent. The chance that it will land tails-up is also 50 percent. The chance of it landing heads-up twice in a row is $\frac{1}{2} \times \frac{1}{2}$, which equals one-quarter, or 25 percent.

Probability

If you flip a coin ten times in a row, what is the chance that half of your flips will be heads? Using probability, you might predict a heads-to-tails ratio of 5:5. However, while the 5:5 ratio is the most probable outcome, any outcome is possible. Probabilities are predictions; they do not guarantee outcomes. Your coin flips could result in ten heads in a row.

Punnett Squares and Predicting Genetic Outcomes

With enough data, Mendel was able to predict the outcome of a monohybrid cross. *A cross between two individuals that are hybrids for one trait is a* **monohybrid cross.** Mendel predicted a 3:1 ratio of the dominant phenotype to the recessive phenotype. When Mendel crossed many sets of heterozygous plants that produced green pods, 428 of the offspring produced green pods. Only 152 produced yellow pods. The results were close to the 3:1 ratio he predicted. Mendel knew that the probability of getting green pods was three-quarters, or 75 percent.

Punnett Square

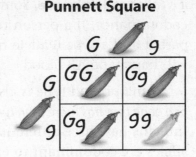

Phenotypes—3 green, 1 yellow
Genotypes—1 *GG*, 2 *Gg*, 1 *gg*

A **Punnett square** *shows the probability of all possible genotypes and phenotypes of offspring.* The figure above shows a Punnett square for a monohybrid cross between pea plants with green pods. The dominant allele *G* indicates green pods, and the recessive allele *g* indicates yellow pods. The darker pods are green; the lighter pod is yellow. The Punnett square predicts that 75 percent of the offspring will express the dominant phenotype of green pods.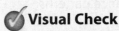

Other Patterns of Inheritance

Not all traits are dominant or recessive. Some traits are expressed by two alleles at the same time. This results in either incomplete dominance or codominance. Other traits are controlled by many alleles.

Incomplete Dominance

When an offspring's phenotype is a combination of its parents' phenotypes, it is called **incomplete dominance.** Neither allele is dominant. Instead, both alleles are expressed. This produces a phenotype that looks like a combination, or blend, of the parental traits.

One example is the oval radish. An oval radish can result from a cross between a radish plant that produces round radishes and one that produces oblong radishes. In this case, incomplete dominance produces offspring (oval radishes) that are a combination of the traits of the parents (the round radish plant and the oblong radish plant).

Visual Check
2. Calculate What percentage of the offspring will be heterozygous?

_____

Math Skills

Probability is a ratio that compares the number of ways a certain outcome occurs to the number of possible outcomes. If you have a regular six-sided die, what is the probability of rolling an even number?

$$\frac{\text{number of sides with even numbers}}{\text{total number of sides}} = \frac{3}{6}$$

Reduce to lowest terms.

$$\frac{3}{6} = \frac{1}{2}$$

3. Use Probability With the same six-sided die, what is the probability of rolling either a 3 or a 5?

_____

Key Concept Check
4. Describe How does a Punnett square help scientists predict genetic outcomes?

_____

_____

_____

Interpreting Tables

5. Identify the three different alleles for ABO blood type in humans.

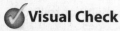

Visual Check

6. Identify What is the genotype of a white-eyed female fly?

_____

Codominance and Multiple Alleles

When both alleles can be independently observed in a phenotype, it is called **codominance.** Some human blood types show codominance. If a person receives a type A allele from one parent and a type B allele from the other parent, he or she will have type AB blood.

Human blood type is also an example of **multiple alleles,** *or a gene that has more than two alleles.* The ABO blood type in humans has three different alleles—I^A, I^B, and i. The I^A and I^B alleles are codominant to each other, but both are dominant to the i allele. Even though there are multiple alleles, a person can inherit only two of them—one from each parent, as shown in the following table.

Human ABO Blood Types	
Phenotype	**Possible Genotypes**
Type A	$I^A I^A$ or $I^A i$
Type B	$I^B I^B$ or $I^B i$
Type O	ii
Type AB	$I^A I^B$

Sex-Linked Traits

Sex chromosomes decide an organism's gender, or sex. Females have two X chromosomes. Males have an X and a Y chromosome. *When the allele for a trait is on an X or Y chromosome, it is called a* **sex-linked trait.** For example, normal fruit-fly eyes are red. If a white-eyed male breeds with normal females, all male and female offspring have red eyes. When these red-eyed females breed with normal males, all female offspring have red eyes, but half of the males have white eyes. Why?

Sex-Linked Traits

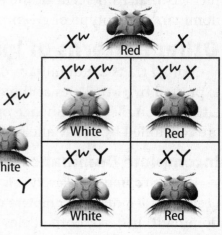

In fruit flies, the allele for eye color is on only the X chromosome, not on the Y chromosome, as shown in the figure above. In this Punnett square, the *w* on the X chromosome represents the sex-linked white-eye allele. Females with white eyes have two white-eye alleles. Males with white eyes have only one white-eye allele.

Polygenic Inheritance

Think about all your classmates and how their heights are different. Some traits, such as human height, are controlled by many genes. These genes express a range of outcomes.

Polygenic inheritance *occurs when multiple genes determine the phenotype of a trait.* Other examples of polygenic inheritance include the number of petals on a daisy and the length of flowers on tomato plants. 🔑

Inheritance of Disease

Understanding human diseases and how they are inherited is one area of genetic research. Scientists have long known that many diseases are more common in some families than in others.

Inherited diseases also tend to produce patterns across many generations. For example, a disease might skip every other generation or show up only in males.

Today, scientists use tools to understand disease patterns. A doctor might ask a patient if he or she has a family history of cancer. The doctor then might use a tool, called a pedigree, to help analyze the family history. *A* **pedigree** *shows genetic traits that were inherited by members of a family.* Pedigrees help determine if a trait has a genetic link. The figure below shows the pedigree for a family in which cancer occurred across four generations.

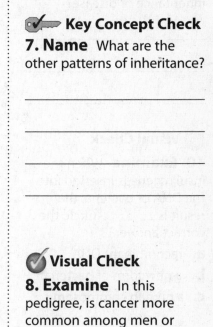

Key Concept Check

7. Name What are the other patterns of inheritance?

_____

_____

_____

_____

Visual Check

8. Examine In this pedigree, is cancer more common among men or women?

_____

Disease Inheritance

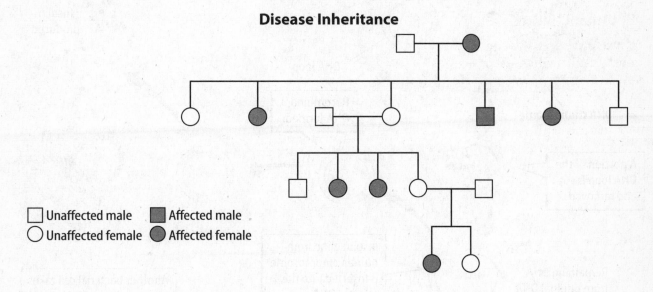

□ Unaffected male   ■ Affected male
○ Unaffected female ● Affected female

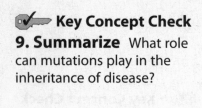

Mutations

Diseases are often the result of a mutation in one or more genes of an individual. *A* **mutation** *is any permanent change in the sequence of DNA in a gene or a chromosome of a cell*. Mutations often result in a change in the appearance or the function of an organism. But they do not always cause such a change.

Mutations can be dominant or recessive. If mutations occur in reproductive cells, they can be passed from parent to offspring. Cancer, diabetes, and birth defects result from mutations in genes. You will read more about the role of mutations in populations in Lesson 3.

Genetic Engineering

Scientists today are using what they have learned about genetics to help people. For example, some people with diabetes cannot produce the protein insulin. Insulin helps control sugar levels in blood. Scientists discovered the insulin gene in human cells. They inserted the gene into the DNA of bacteria. The bacteria now make human insulin, as shown in the figure below. This is an example of genetic engineering. *In* **genetic engineering,** *the genetic material of an organism is modified by inserting DNA from another organism.*

Genetically Engineered Human Insulin

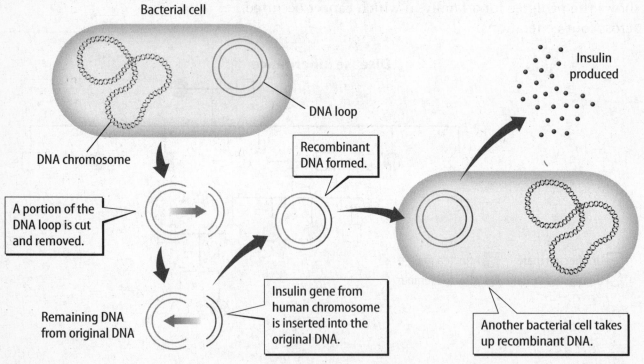

Bacterial cell

DNA loop

DNA chromosome

Recombinant DNA formed.

Insulin produced

A portion of the DNA loop is cut and removed.

Remaining DNA from original DNA

Insulin gene from human chromosome is inserted into the original DNA.

Another bacterial cell takes up recombinant DNA.

Using Genetic Engineering to Learn About Organisms

Genetic engineering today is used for many purposes. For example, in the 1960s, a unique gene was discovered in a species of jellyfish. The gene, called GFP, makes a protein that glows green.

Scientists can use GFP to learn how plants respond to changes in their environments. For example, a plant can be made to glow when it is under stress. Certain parts in animal embryos also can be made to glow when that part is beginning to grow.

Using Genetic Engineering to Fight Disease Scientists have been able to insert GFP into the genetic code of mosquito larvae. Because this experiment was successful, scientists hope to use genetic engineering to prevent mosquitoes from hosting the organism that causes malaria. Malaria kills millions of people each year and is transmitted by mosquito bites. This is one way scientists hope to use genetic engineering to fight disease. ✔

✔ **Reading Check**

11. State What are some different uses of genetic engineering?

_____

_____

_____

_____

Mini Glossary

codominance: occurs when both alleles can be independently observed in a phenotype

genetic engineering: when the genetic material of an organism is modified by inserting DNA from another organism

incomplete dominance: occurs when an offspring's phenotype is a combination of its parents' phenotypes

monohybrid cross: a cross between two individuals that are hybrids for one trait

multiple alleles: a gene that has more than two alleles

mutation: any permanent change in the sequence of DNA in a gene or a chromosome of a cell

pedigree: shows genetic traits that were inherited by members of a family

polygenic inheritance: occurs when multiple genes determine the phenotype of a trait

Punnett square: a table that shows the probability of all possible genotypes and phenotypes of offspring

sex-linked trait: occurs when the allele for a trait is on an X or Y chromosome

1. Review the terms and their definitions in the Mini Glossary. Write a sentence comparing and contrasting codominance and incomplete dominance.

_____

_____

2. Complete the Punnett square below, which shows a cross between two plants and their alleles for flower color. The dominant allele B indicates blue flowers; the recessive allele b indicates white flowers.

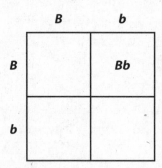

3. Identify the phenotypes and genotypes in the offspring from the cross in Question 2.

_____

What do you think NOW?

Reread the statements at the beginning of the lesson. Fill in the After column with an A if you agree with the statement or a D if you disagree. Did you change your mind?

Connect ED

Log on to ConnectED.mcgraw-hill.com and access your textbook to find this lesson's resources.

END OF LESSON

Reading Essentials

Heredity and How Traits Change

Adaptation and Evolution

·············· **Before You Read** ··············

What do you think? Read the two statements below and decide whether you agree or disagree with them. Place an A in the Before column if you agree with the statement or a D if you disagree. After you've read this lesson, reread the statements to see if you have changed your mind.

Before	Statement	After
	5. A population that lacks variation among its individuals might not be able to adapt to a changing environment.	
	6. Extinction occurs when the last individual of a species dies.	

·············· **Read to Learn** ··············

Mutations, Variation, and Natural Selection

Recall that mutations can lead to changes in traits. Therefore, mutations can produce differences among individuals. *Slight differences in inherited traits among individuals in a population are called* **variations.**

In 1976, scientists measured several traits in a population of medium ground finches on one of the Galápagos Islands. They discovered that the birds had variations in beak size. Most had smaller beaks, but all of the birds of this species preferred to eat small, soft seeds. The next year, it did not rain on the island. None of the plants reproduced, so no seeds formed. After all of the small, soft seeds had been eaten, many of the finches died. The few seeds left were fairly large and hard. The birds that survived were those that could crack and eat these larger seeds. These finches tended to be the birds with fairly large beaks.

In 1978, the scientists measured the beaks of the surviving birds' offspring. They compared the average beak size of birds hatched in 1978 to that of birds hatched in 1976. In just two years, the average beak size of birds in the population had increased. How did this happen?

Key Concepts 🗝

- How does natural selection occur?
- What is an adaptation?
- Why do traits change over time?

▸ **Mark the Text**

Building Vocabulary As you read, underline the words and phrases that you do not understand. When you finish reading, discuss these words and phrases with another student or your teacher.

FOLDABLES®

Make a vertical three-tab book and use it to organize notes on the different types of adaptations.

Sidebar: Copyright © Glencoe/McGraw-Hill, a division of The McGraw-Hill Companies, Inc.

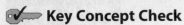

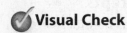 **Key Concept Check**

1. Explain How does natural selection occur?

_____

_____

_____

☑ **Visual Check**

2. Analyze What happened when individual sunflowers competed for limited resources?

_____

_____

Natural Selection

The process by which individuals with variations that help them survive in their environment live longer, compete better, and reproduce more than those individuals without these variations is called **natural selection.** For the finches, there was variation among individual birds for the trait of beak size. Some of the birds had small beaks, and some of them had larger beaks. When a change in the environment—a drought—occurred, the birds with larger beaks were better able to survive than birds with smaller beaks. The surviving individuals then passed on the favorable trait to their offspring. Over the two-year span of natural selection and reproduction, the average beak size of the birds in the population increased. The birds with larger beaks were naturally selected by environmental conditions and survived. Another example of natural selection in plants is shown in the figure below. ☞

Natural Selection: Sunflowers

❶ **Variation** Individuals in a population differ from one another. In this population, some sunflowers are taller than others.

❷ **Inheritance** Traits are inherited from parents. Tall sunflowers produce tall sunflowers. Short sunflowers produce short sunflowers.

❸ **Competition** Due to limited resources, not all offspring will survive. Individuals with a trait that better suits the environment are more likely to survive and reproduce. In this environment, short sunflowers are more successful.

❹ **Natural Selection** Over time, the average height of the sunflower population is short if the short sunflowers continue to reproduce successfully.

Adaptations

The traits of surviving individuals, such as larger beaks or shorter sunflowers, become more common as the survivors reproduce and pass the genes for their traits to their offspring. *An **adaptation** is an inherited trait that increases an organism's chance of surviving and reproducing in a particular environment.* Adaptations can be structural, functional, or behavioral. ✓

Structural Adaptations

A flying squirrel has a flap of skin between its forelegs and hind legs. This flap of skin enables the squirrel to glide distances of up to 45 m. The behavior possibly is used to escape from predators. This flap of skin is an example of a structural adaptation. Structural adaptations involve physical characteristics, such as color or shape.

Another example of a structural adaptation occurs in many desert plants. Some types of cactus plants have spines. Spines are leaves that are reduced in size. The adaptation of smaller leaves helps reduce water loss in a dry environment.

Functional Adaptations

Functional adaptations involve internal systems that affect an organism's physiology or biochemistry. For example, the alpine snowbell, a flower species, has adapted to survive in an environment with a short growing season.

In the high altitude of the mountains, where the alpine snowbell grows, light and temperature conditions for flowering are favorable for only a short period during the summer. The alpine snowbell produces flower buds at the end of the previous season. Over the winter, the buds remain dormant—alive, but not actively growing. In the spring, increased light triggers the plant to bloom even when it is still surrounded by snow. These adaptations enable the species to survive.

Behavioral Adaptations

Migration, and other behavioral adaptations, involve the ways an organism behaves or acts. For example, caribou migrate south for the winter. Other animals, such as birds, whales, and butterflies, also migrate. Animal species that migrate to find adequate food and suitable temperatures survive and reproduce more successfully. ✓

✓ Reading Check

3. Identify How can a species benefit from an adaptation?

_____

_____

_____

_____

💡 Think it Over

4. Hypothesize How might a monkey's long tail be an example of a structural adaptation?

_____

_____

_____

_____

🔑 Key Concept Check

5. Describe three types of adaptations.

_____

_____

_____

_____

_____

_____

Copyright © Glencoe/McGraw-Hill, a division of The McGraw-Hill Companies, Inc.

Evolution of Populations—Why Traits Change

Once an inherited trait has become more frequent in a population, the population has adapted and evolved. **Evolution** *is change over time*. Evolution by natural selection is a way that populations change over time. When populations evolve, species can look and behave differently than their ancestors. This happens because the frequency of genetic traits changes over time. As the environment changes, different inherited traits might enable survival, and the population can evolve again.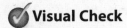

A Modern Example of Change Over Time

Bacteria can cause infections in your body, such as strep throat or pneumonia. Sometimes a doctor might prescribe an antibiotic to help you fight an infection. Antibiotics are drugs that kill bacteria. Although antibiotics often kill bacteria effectively, variation exists within a population of bacteria. As shown in the figure below, some bacteria in a population might have a mutation that lets them <u>survive</u> when exposed to an antibiotic. When the surviving bacteria reproduce, that trait passes to their offspring. Soon, most individuals in the population survive when exposed to the antibiotic. Bacteria that survive when exposed to an antibiotic are called antibiotic-resistant. Antibiotic-resistant bacteria have caused deadly infections and are of great concern to scientists.

🔑 Key Concept Check

6. State Why do traits change over time?

_____

_____

_____

ACADEMIC VOCABULARY
survive
(verb) to remain alive

✓ Visual Check

7. Draw a circle around the dish with the largest number of resistant bacteria.

Antibiotic Resistance in Bacteria

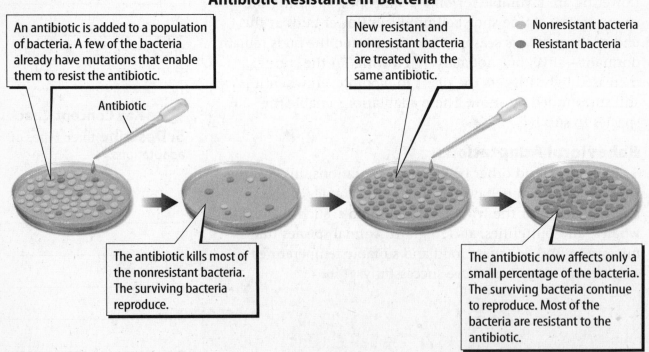

An antibiotic is added to a population of bacteria. A few of the bacteria already have mutations that enable them to resist the antibiotic.

Antibiotic

New resistant and nonresistant bacteria are treated with the same antibiotic.

● Nonresistant bacteria
● Resistant bacteria

The antibiotic kills most of the nonresistant bacteria. The surviving bacteria reproduce.

The antibiotic now affects only a small percentage of the bacteria. The surviving bacteria continue to reproduce. Most of the bacteria are resistant to the antibiotic.

Extinction and Conservation Biology

You already have read that for evolution by natural section to occur, variation within a population is needed. A population with diversity can survive changes in its environment and persist through time. But what happens when a population lacks variation among its individuals and the environment changes? The population might lose its ability to reproduce successfully and fail to survive. *When the last individual of a species dies, the species has undergone* **extinction.**

Today, many species are threatened with extinction. A species' habitat might have been altered or destroyed. Some species have been hunted to extinction. For others, new species <u>introduced</u> into many habitats make it difficult for some native species to survive and reproduce.

A relatively new field of science focuses on saving species. **Conservation biology** *is a branch of biology that studies why many species are in trouble and what can be done to save them.* Sometimes scientists' knowledge of genetics helps species that are in danger of extinction. For example, by 1995, the population of Florida panthers was between 20 and 30 individuals. The population had lost much of its natural variation and was struggling to survive.

Fortunately, scientists' understanding of genetics and heredity saved the population from extinction. Scientists introduced into the Florida population several female panthers from a population in Texas. This was done to increase genetic diversity in the Florida population. By 2003, the Florida panther population had increased to 80 individuals, and the effort was considered a success. ✓

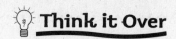

Think it Over

8. Name one animal species that is in danger of becoming extinct.

_____

SCIENCE USE V. COMMON USE

introduce

Science Use to bring a substance or organism into a habitat or a population

Common Use to make someone known to others

✓ **Reading Check**

9. Explain Why did scientists introduce panthers from Texas into the Florida panther population?

_____

_____

_____

Mini Glossary

adaptation: an inherited trait that increases an organism's chance of surviving and reproducing in a particular environment

conservation biology: a branch of biology that studies why many species are in trouble and what can be done to save them

evolution: change over time

extinction: occurs when the last individual of a species dies

natural selection: the process by which individuals with variations that help them survive in their environment live longer, compete better, and reproduce more than those individuals without these variations

variation: a slight difference in inherited traits among individuals in a population

1. Review the terms and their definitions in the Mini Glossary. Write a sentence that explains the difference between an adaptation and a variation.

_____

_____

2. Use the graphic organizer below to identify three factors that contribute to extinction.

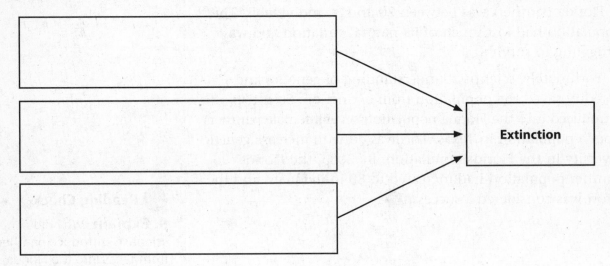

3. An environment supports a species of beetle. Some beetles are green, some brown. Because of a drought, birds see green beetles and eat them more often than brown beetles. Over time, which color would you expect to become more common? Why?

_____

What do you think NOW?

Reread the statements at the beginning of the lesson. Fill in the After column with an A if you agree with the statement or a D if you disagree. Did you change your mind?

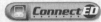 **Connect** ED

Log on to ConnectED.mcgraw-hill.com and access your textbook to find this lesson's resources.

END OF LESSON

PERIODIC TABLE OF THE ELEMENTS

Element — Hydrogen
Atomic number — 1
Symbol — **H**
Atomic mass — 1.01

State of matter

- Gas
- Liquid
- Solid
- Synthetic

A column in the periodic table is called a **group**.

A row in the periodic table is called a **period**.

	1	2	3	4	5	6	7	8	9
1	Hydrogen 1 **H** 1.01								
2	Lithium 3 **Li** 6.94	Beryllium 4 **Be** 9.01							
3	Sodium 11 **Na** 22.99	Magnesium 12 **Mg** 24.31							
4	Potassium 19 **K** 39.10	Calcium 20 **Ca** 40.08	Scandium 21 **Sc** 44.96	Titanium 22 **Ti** 47.87	Vanadium 23 **V** 50.94	Chromium 24 **Cr** 52.00	Manganese 25 **Mn** 54.94	Iron 26 **Fe** 55.85	Cobalt 27 **Co** 58.93
5	Rubidium 37 **Rb** 85.47	Strontium 38 **Sr** 87.62	Yttrium 39 **Y** 88.91	Zirconium 40 **Zr** 91.22	Niobium 41 **Nb** 92.91	Molybdenum 42 **Mo** 95.96	Technetium 43 **Tc** (98)	Ruthenium 44 **Ru** 101.07	Rhodium 45 **Rh** 102.91
6	Cesium 55 **Cs** 132.91	Barium 56 **Ba** 137.33	Lanthanum 57 **La** 138.91	Hafnium 72 **Hf** 178.49	Tantalum 73 **Ta** 180.95	Tungsten 74 **W** 183.84	Rhenium 75 **Re** 186.21	Osmium 76 **Os** 190.23	Iridium 77 **Ir** 192.22
7	Francium 87 **Fr** (223)	Radium 88 **Ra** (226)	Actinium 89 **Ac** (227)	Rutherfordium 104 **Rf** (267)	Dubnium 105 **Db** (268)	Seaborgium 106 **Sg** (271)	Bohrium 107 **Bh** (272)	Hassium 108 **Hs** (270)	Meitnerium 109 **Mt** (276)

The number in parentheses is the mass number of the longest lived isotope for that element.

Lanthanide series	Cerium 58 **Ce** 140.12	Praseodymium 59 **Pr** 140.91	Neodymium 60 **Nd** 144.24	Promethium 61 **Pm** (145)	Samarium 62 **Sm** 150.36	Europium 63 **Eu** 151.96
Actinide series	Thorium 90 **Th** 232.04	Protactinium 91 **Pa** 231.04	Uranium 92 **U** 238.03	Neptunium 93 **Np** (237)	Plutonium 94 **Pu** (244)	Americium 95 **Am** (243)